Civil Engineering
Hydraulics

CIVIL ENGINEERING HYDRAULICS

Essential Theory with Worked Examples

Fourth Edition

C. Nalluri
R.E. Featherstone

b

**Blackwell
Science**

© R.E. Featherstone & C. Nalluri 1982, 1988, 1995, 2001

Blackwell Science Ltd
Editorial Offices:
Osney Mead, Oxford OX2 0EL
25 John Street, London WC1N 2BS
23 Ainslie Place, Edinburgh EH3 6AJ
350 Main Street, Malden
  MA 02148 5018, USA
54 University Street, Carlton
  Victoria 3053, Australia
10, rue Casimir Delavigne
  75006 Paris, France

Other Editorial Offices:

Blackwell Wissenschafts-Verlag GmbH
Kurfürstendamm 57
10707 Berlin, Germany

Blackwell Science KK
MG Kodenmacho Building
7–10 Kodenmacho Nihombashi
Chuo-ku, Tokyo 104, Japan

Iowa State University Press
A Blackwell Science Company
2121 S. State Avenue
Ames, Iowa 50014-8300, USA

First edition published by
Granada Publishing 1982
Reprinted with minor amendments 1983
Reprinted by Collins Professional
and Technical Books 1984
Second edition published by
BSP Professional Books 1988
Reprinted 1991, 1994
Third edition published by
Blackwell Science 1995
Reprinted 1995, 1997, 1998
Fourth edition 2001

Printed and bound in Great Britain at the
Alden Press Ltd, Oxford and Northampton

The Blackwell Science logo is a trade mark of Blackwell Science Ltd, registered at the United Kingdom Trade Marks Registry

DISTRIBUTORS

Marston Book Services Ltd
PO Box 269
Abingdon
Oxon OX14 4YN
(*Orders:* Tel: 01235 465500
         Fax: 01235 465555)

USA
Blackwell Science, Inc.
Commerce Place
350 Main Street
Malden, MA 02148 5018
(*Orders:* Tel: 800 759 6102
              781 388 8250
         Fax: 781 388 8255)

Canada
Login Brothers Book Company
324 Saulteaux Crescent
Winnipeg, Manitoba R3J 3T2
(*Orders:* Tel: 204 837-2987
         Fax: 204 837-3116)

Australia
Blackwell Science Pty Ltd
54 University Street
Carlton, Victoria 3053
(*Orders:* Tel: 03 9347 0300
         Fax: 03 9347 5001)

A catalogue record for this title is available from the British Library

ISBN 0-632-05514-6

Library of Congress
Cataloging-in-Publication Data
Nalluri, C.
   Civil engineering hydraulics: essential theory with worked examples/C. Nalluri, R.E. Featherstone.—4th ed.
       p.   cm.
   Featherstone's name appears first on the earlier ed.
   Includes bibliographical references and index.
   ISBN 0-632-05514-6 (alk. paper)
   1. Hydraulic engineering.   2. Hydraulics.
I. Featherstone, R.E.   II. Featherstone, R.E.
Civil engineering hydraulics.   III. Title.

TC145 .N35   2001
627—dc21
                                    00-068876

For further information on Blackwell Science, visit our website: www.blackwell-science.com

Contents

Preface to fourth edition   xi

Chapter 1   Properties of Fluids   **1**
   1.1   Introduction   1
   1.2   Engineering units   1
   1.3   Mass density and specific weight   1
   1.4   Relative density   2
   1.5   Viscosity of fluids   2
   1.6   Compressibility and elasticity of fluids   2
   1.7   Vapour pressure of liquids   3
   1.8   Surface tension and capillarity   3
   Worked examples   3
   Recommended reading   5
   Problems   5

Chapter 2   Fluid Statics   **7**
   2.1   Introduction   7
   2.2   Pascal's law   7
   2.3   Pressure variation with depth in a static incompressible fluid   7
   2.4   Pressure measurement   9
   2.5   Hydrostatic thrust on plane surfaces   11
   2.6   Pressure diagrams   14
   2.7   Hydrostatic thrust on curved surfaces   15
   2.8   Hydrostatic buoyant thrust   17
   2.9   Stability of floating bodies   18
   2.10 Determination of metacentre   19
   2.11 Periodic time of rolling (or oscillation) of a floating body   21
   2.12 Liquid ballast and the effective metacentric height   21
   2.13 Relative equilibrium   22
   Worked examples   24
   Recommended reading   42
   Problems   42

Chapter 3   Fluid Flow Concepts and Measurements   47
  3.1   Kinematics of fluids   47
  3.2   Steady and unsteady flows   48
  3.3   Uniform and non-uniform flows   48
  3.4   Rotational and irrotational flows   49
  3.5   One, two and three dimensional flows   49
  3.6   Streamtube and continuity equation   49
  3.7   Accelerations of fluid particles   50
  3.8   Two kinds of fluid flow   52
  3.9   Dynamics of fluid flow   53
  3.10 Energy equation for an ideal fluid flow   53
  3.11 Modified energy equation for real fluid flows   55
  3.12 Separation and cavitation in fluid flow   57
  3.13 Impulse-momentum equation   58
  3.14 Energy losses in sudden transitions   58
  3.15 Flow measurement through pipes   59
  3.16 Flow measurement through orifices and mouthpieces   62
  3.17 Flow measurement in channels   66
  Worked examples   70
  Recommended reading   86
  Problems   86

Chapter 4   Flow of Incompressible Fluids in Pipelines   91
  4.1   Resistance in circular pipelines flowing full   91
  4.2   Resistance to flow in non-circular sections   95
  4.3   Local losses   96
  Worked examples   97
  Recommended reading   118
  Problems   118

Chapter 5   Pipe Network Analysis   121
  5.1   Introduction   121
  5.2   The head balance method ('Loop' method)   122
  5.3   The quantity balance method ('nodal' method)   123
  5.4   Newton Raphson method   125
  5.5   The linear theory method   125
  5.6   The gradient method   126
  Worked examples   128
  Recommended reading   147
  Problems   148

Chapter 6   Pump-pipeline System Analysis and Design   154
  6.1   Introduction   154
  6.2   Hydraulic gradient in pump-pipeline systems   155
  6.3   Multiple pump systems   156

6.4   Variable speed pump operation   157
6.5   Suction lift limitations   158
Worked examples   160
Recommended reading   173
Problems   173

Chapter 7   Boundary Layers on Flat Plates and in Ducts   177
7.1   Introduction   177
7.2   The laminar boundary layer   177
7.3   The turbulent boundary layer   178
7.4   Combined drag due to both laminar and turbulent boundary
      layers   179
7.5   The displacement thickness   179
7.6   Boundary layers in turbulent pipe flow   180
7.7   The laminar sub-layer   182
Worked examples   184
Recommended reading   192
Problems   192

Chapter 8   Steady Flow in Open Channels   194
8.1   Introduction   194
8.2   Uniform flow resistance   195
8.3   Channels of composite roughness   196
8.4   Channels of compound section   197
8.5   Channel design   198
8.6   Uniform flow in part-full circular pipes   200
8.7   Steady, rapidly varied channel flow-energy principles   203
8.8   The momentum equation and the hydraulic jump   204
8.9   Steady gradually varied open channel flow   206
8.10 Computations of gradually varied flow   207
8.11 Graphical and numerical integration methods   208
8.12 The direct step method   208
8.13 The standard step method   208
8.14 Canal delivery problems   210
8.15 Culvert flow   212
8.16 Spatially varied flow in open channels   213
Worked examples   215
Recommended reading   252
Problems   253

Chapter 9   Dimensional Analysis, Similitude and Hydraulic Models 258
9.1   Introduction   258
9.2   Dimensional analysis   259
9.3   Physical significance of non-dimensional groups   259
9.4   The Buckingham π theorem   260

9.5	Similitude and model studies	260
	Worked examples	261
	Recommended reading	275
	Problems	275

Chapter 10 Ideal Fluid Flow and Curvilinear Flow **277**
10.1	Ideal fluid flow	277
10.2	Streamlines, the stream function	277
10.3	Relationship between discharge and stream function	279
10.4	Circulation and the velocity potential function	279
10.5	Stream functions for basic flow patterns	280
10.6	Combinations of basic flow patterns	281
10.7	Pressure at points in the flow field	282
10.8	The use of flow nets and numerical methods	282
10.9	Curvilinear flow of real fluids	286
10.10	Free and forced vortices	287
	Worked examples	287
	Recommended reading	299
	Problems	299

Chapter 11 Gradually Varied Unsteady Flow from Reservoirs **302**
11.1	Discharge between reservoirs under varying head	302
11.2	Unsteady flow over a spillway	304
11.3	Flow establishment	305
	Worked examples	306
	Problems	316

Chapter 12 Mass Oscillations and Pressure Transients in Pipelines **318**
12.1	Mass oscillation in pipe systems – surge chamber operation	318
12.2	Solution, neglecting tunnel friction and throttle losses for sudden discharge stoppage	320
12.3	Solution, including tunnel and surge chamber losses for sudden discharge stoppage	321
12.4	Finite difference methods in the solution of the surge chamber equations	321
12.5	Pressure transients in pipelines (waterhammer)	323
12.6	The basic differential equations of waterhammer	324
12.7	Solutions of the waterhammer equations	326
12.8	The Allievi equations	327
12.9	Alternative formulation	330
12.10	The Schnyder–Bergeron graphical method	331
12.11	Pipeline friction and other losses	333
12.12	Pressure transients at interior points	334
12.13	Pressure transients caused by pump stoppage	335
	Worked examples	337

Recommended reading   348
Problems   348

Chapter 13   Unsteady Flow in Channels   **350**
13.1   Introduction   350
13.2   Gradually varied unsteady flow   350
13.3   Surges in open channels   351
13.4   The upstream positive surge   352
13.5   The downstream positive surge   354
13.6   Negative surge waves   354
13.7   The dam break   356
Worked examples   357
Recommended reading   360
Problems   360

Chapter 14   Uniform Flow in Loose-boundary Channels   **362**
14.1   Introduction   362
14.2   Flow regimes   362
14.3   Incipient (threshold) motion   362
14.4   Resistance to flow in alluvial (loose bed) channels   364
14.5   Velocity distributions in loose-boundary channels   366
14.6   Sediment transport   366
14.7   Bed load transport   368
14.8   Suspended load transport   370
14.9   Total load transport   372
14.10 Regime channel design   373
14.11 Rigid bed channels with sediment transport   378
Worked examples   380
Recommended reading   392
Problems   394

Chapter 15   Hydraulic Structures   **397**
15.1   Introduction   397
15.2   Spillways   397
15.3   Energy dissipators and downstream scour protection 403
Worked examples   405
Recommended reading   417
Problems   418

Answers   420
Index   425
Conversion Table (Metric to Imperial Units)   431

Preface to fourth edition

This book is designed to meet the special, but not exclusive, needs of civil engineering students reading for degrees and diplomas and for Engineering Council Parts I and II examinations at universities and colleges of higher education.

Having as its main feature a substantial number of worked examples, its purpose is to augment lecture courses and standard textbooks on fluid mechanics and hydraulics by illustrating the application of the underlying theory to a wide range of practical situations. The inclusion of exercise problems, with answers, will enable students to assess their understanding of the theory and methods of analysis and design. Aspects of relevant theory, in concise form and accompanied by explanation, are included at the beginning of each chapter. The book should prove useful not only to students but also to practising engineers as a concise working reference.

A Solutions Manual to the problems (to be solved) listed under each chapter in the text is also available for those lecturing in this subject. To find out more log on to www.blackwell-science.com/nalluri/.

The contents are concentrated on the types of problem commonly encountered by civil engineers in the field of hydraulic engineering as compared with the wider fields covered by fluid mechanics texts. Due to space limitations, however, the text does not extend into the specialist fields of mathematical simulation models required, for example, for esturial flow computations and detailed waterhammer analysis.

SI units are used throughout and standard symbols for physical properties employed. The numerical procedures illustrated are appropriate to the use of electronic calculators, but readers may find it instructive to write programs for execution on microcomputers for problems requiring repetitive solutions or numerical methods.

The revised fourth edition has been enlarged to include the gradient method in solving pipe networks (Chapter 5) and a new chapter (Chapter 15) on hydraulic structures. These new additions have been written by Professor J. Saldarriaga, Visiting Professor at Newcastle University from the University of Los Andes, Colombia. Also, additional worked examples, reading publications and problems to be solved are provided where appropriate.

The authors are indebted to Professor P. Novak, Emeritus Professor of Civil and Hydraulic Engineering, University of Newcastle-upon-Tyne, for his helpful criticisms and advice throughout the preparation of this book.

C. Nalluri
2001

Chapter 1
Properties of Fluids

C. Nalluri

1.1 Introduction

A fluid is a substance which deforms continuously, or flows, when subjected to shear stresses. The term fluid embraces both gases and liquids; a given mass of liquid will occupy a definite volume whereas a gas will fill its container. Gases are readily compressible; the low compressibility, or elastic volumetric deformation, of liquids is generally neglected in computations except those relating to large depths in the oceans and in pressure transients in pipelines.

This text however deals exclusively with liquids and more particularly with Newtonian liquids, i.e. those having a linear relationship between shear stress and rate of deformation.

1.2 Engineering units

MKS (Metre-Kilogramme-Second) system is the internationally agreed version of metric system (SI) of units. All physical quantities can be described by a set of three primary units, mass (kg), length (m) and time (s) designated by M, L and T respectively.

The unit of force is called newton (N) and 1 N is the force which accelerates a mass of 1 kg at a rate of 1 m/s^2. 1 N = 1 kg m/s^2 ($:MLT^{-2}$).

The unit of work is called the joule (J) and it is the energy needed to move a force of 1 N over a distance of 1 m, i.e. 1 Nm($:ML^2T^{-2}$). Power is the energy or work done per unit time and its unit is the watt (W). 1 W = 1 Nm/s = 1 J/s($:ML^2T^{-3}$).

1.3 Mass density and specific weight

Mass density (ρ) or density of a substance is defined as the mass of the substance per unit volume (kg/m^3: ML^{-3}) and is different from specific weight (γ), which is the force exerted by the earth's gravity (g) upon a unit volume of the substance ($\gamma = \rho g$: N/m^3: $ML^{-2}T^{-2}$). In a satellite where there is no gravity, an object has no specific weight but possesses the same density that it has on the earth.

1.4 Relative density

Relative density (σ) of a substance is the ratio of its mass density to that of water at a standard temperature (4°C) and pressure (atmospheric) and is dimensionless ($M^0L^0T^0$).

For water: $\rho = 10^3$ kg/m^3, $\gamma = 10^3 \times 9 \cdot 81 \simeq 10^4$ N/m^3 and $\sigma = 1$.

1.5 Viscosity of fluids

Viscosity is that property of a fluid which, by virtue of cohesion and inter-action between fluid molecules offers resistance to shear deformation. Different fluids deform at different rates under the action of the same shear stress. Fluids with high viscosity such as syrup deform relatively more slowly than low viscosity fluids such as water.

All fluids are viscous and 'Newtonian fluids' obey the linear relationship

$$\tau = \mu \frac{du}{dy} \text{ (Newton's law of viscosity)} \tag{1.1}$$

where τ is the shear stress (N/m^2:ML^{-1}T^{-2}), du/dy the velocity gradient, or rate of deformation (radians/s:T^{-1}), and μ the coefficient of dynamic (or absolute) viscosity (Ns/m^2 or kg/ms: ML^{-1}T^{-1}).

A smaller unit of viscosity, called the poise, is 1 gm/cm s. 1 kg/ms = 10 poises.

Kinematic viscosity (ν) is the ratio of dynamic viscosity to mass density expressed in m^2/s(L^{2}T^{-1}).

A smaller unit of kinematic viscosity is the stoke (1 cm^2/s). 1 m^2/s = 10^4 stokes.

Water is a Newtonian fluid having a dynamic viscosity of 10^{-3}Ns/m^2 or kinematic viscosity of 10^{-6} m^2/s at 20°C.

1.6 Compressibility and elasticity of fluids

All fluids are compressible under the application of an external force and when the force is removed they expand back to their original volume exhibiting the property that stress is proportional to volumetric strain.

The bulk modulus of elasticity, K = pressure change/volumetric strain

$$= dp/(dV/V):(N/m^2:ML^{-1}T^{-2}) \tag{1.2}$$

Water with a bulk modulus of $2 \cdot 1 \times 10^9$ N/m^2 at 20°C is 100 times more compressible than steel, but it is ordinarily considered incompressible.

1.7 Vapour pressure of liquids

A liquid in a closed container is subjected to partial vapour pressure due to the escaping molecules from the surface; it reaches a stage of equilibrium when this pressure reaches saturated vapour pressure. Since this depends upon molecular activity, which is a function of temperature, the vapour pressure of a fluid also depends upon its temperature and increases with it. If the pressure above a liquid reaches the vapour pressure of the liquid, boiling occurs; for example if the pressure is reduced sufficiently boiling may occur at room temperature.

The saturated vapour pressure for water at 20°C is $2 \cdot 45 \times 10^3 \text{N/m}^2$.

1.8 Surface tension and capillarity

Liquids possess the properties of cohesion and adhesion due to molecular attraction. Due to the property of cohesion, liquids can resist small tensile forces at the interface between the liquid and air, known as surface tension (σ:N/m:MT^{-2}). If the liquid molecules have greater adhesion than cohesion, then the liquid sticks to the surface of the container with which it is in contact resulting in a capillary rise of the liquid surface; a predominating cohesion on the other hand causes capillary depression. The surface tension for water is 73×10^{-3}N/m at 20°C.

The capillary rise or depression h of a liquid in a tube of diameter d can be written as

$$h = 4 \, \sigma \cos \theta / \rho g d \tag{1.3}$$

where θ is the angle of contact between liquid and solid.

Surface tension increases the pressure within a droplet of liquid. The internal pressure, p, balancing the surface tensional force of a small spherical droplet of radius, r, is given by

$$p = \frac{2\sigma}{r} \tag{1.4}$$

Worked examples

Example 1.1
The density of an oil at 20°C is 850 kg/m^3. Find its relative density and kinematic viscosity if the dynamic viscosity is 5×10^{-3}kg/ms.

Solution:

$$\text{Relative density, } \sigma \quad = \rho \text{ of oil}/\rho \text{ of water}$$

$$= 850/10^3$$

$$= 0 \cdot 85$$

Kinematic viscosity, $v = \mu/\rho$

$$= 5 \times 10^{-3}/850$$

$$= 5\cdot88 \times 10^{-6} \mathrm{m}^2/\mathrm{s}.$$

Example 1.2

If the velocity distribution of a viscous liquid ($\mu = 0\cdot9$ Ns/m^2) over a fixed boundary is given by $u = 0\cdot68y - y^2$ in which u is the velocity in m/s at a distance y metres above the boundary surface, determine the shear stress at the surface and at $y = 0\cdot34$ m.

Solution:

$$u = 0\cdot68y - y^2$$

$\therefore$ du/dy $= 0\cdot68 - 2y$; hence $(\mathrm{du/dy})_{y\,=\,0} = 0\cdot68\ \mathrm{s}^{-1}$

and $(\mathrm{du/dy})_{y\,=\,0\cdot34\ \mathrm{m}} = 0$

Dynamic viscosity of the fluid, $\mu = 0\cdot9$ Ns/m^2

From equation 1.1

$\tau = \mu\mathrm{du/dy}$, shear stress $(\tau)_{y\,=\,0} = 0\cdot9 \times 0\cdot68$

$$= 0\cdot612\ \mathrm{N/m}^2$$

and at $y = 0\cdot34$ m, $\tau = 0$.

Example 1.3

At a depth of 8·5 km in the ocean the pressure is 90 MN/m^2.

The specific weight of the sea water at the surface is 10·2 kN/m^3 and its average bulk modulus is $2\cdot4 \times 10^6$kN/m^2. Determine (a) the change in specific volume, (b) the specific volume, and (c) the specific weight of sea water at 8·5 km depth.

Solution:

Change in pressure dp at a depth of 8·5 km $= 90$ MN/m^2

$$= 9 \times 10^4\ \mathrm{kN/m}^2$$

Bulk modulus, $K = 2\cdot4 \times 10^6$ kN/m^2

From $K = \mathrm{dp}/(\mathrm{dV/V})$, $\mathrm{dV/V} = 9 \times 10^4/2\cdot4 \times 10^6$

$$= 3\cdot75 \times 10^{-2}$$

Defining specific volume as $1/\gamma$ (m^3/kN), the specific volume of sea water at the surface $= 1/10\cdot2 = 9\cdot8 \times 10^{-2}$m^3/kN.

$\therefore$ Change in specific volume between that at the surface and at $8.5\,\text{km}$ depth, $dV = 3.75 \times 10^{-2} \times 9.8 \times 10^{-2}$

$$= 36.75 \times 10^{-4}\ \text{m}^3/\text{kN}$$

The specific volume of sea water at $8.5\,\text{km}$ depth

$$= 9.8 \times 10^{-2} - 36.75 \times 10^{-4}$$

$$= 9.44 \times 10^{-2}\,\text{m}^3/\text{kN}$$

$\therefore$ The specific weight of sea water at $8.5\,\text{km}$ depth

$$= 1/\text{specific volume}$$

$$= 1/9.44 \times 10^{-2}$$

$$= 10.6\ \text{kN/m}^3.$$

Recommended reading

1. Brown, R.C. (1950) *Mechanics and properties of matter*. London: Longman.
2. Massey, B.S. (B. Stanford) (1998) *Mechanics of fluids*, 7th edn. London: Van Nostrand-Reinhold.

Problems

1. (a) Explain why the viscosity of a liquid decreases while that of a gas increases with an increase of temperature.
    (b) The following data refer to a liquid under shearing action at a constant temperature. Determine its dynamic viscosity.

du/dy (rad/s):	0	0.2	0.4	0.6	0.8
τ (N/m^2):	0	0	1.9	3.1	4

2. A 300 mm wide shaft sleeve moves along a 100 mm diameter shaft at a speed of 0.5 m/s under the application of a force of 250 N in the direction of its motion. If 1000 N of force is applied what speed will the sleeve attain? Assume the temperature of the sleeve to be constant and determine the viscosity of the Newtonian fluid in the clearance between the shaft and its sleeve if the radial clearance is estimated to be 0.075 mm.

3. A shaft of 100 mm diameter rotates at 120 rad/s in a bearing 150 mm long. If the radial clearance is 0.2 mm and the absolute viscosity of the lubricant is 0.20 kg/ms find the power loss in the bearing.

4. A block of dimensions 300 mm $\times$ 300 mm $\times$ 300 mm and mass 30 kg slides down a plane inclined at $30°$ to the horizontal, on which there is a thin

film of oil of viscosity $2·3 \times 10^{-3}$ Ns/m^2. Determine the speed of the block if the film thickness is estimated to be 0·03 mm.

5.   Calculate the capillary effect in mm in a glass tube of 6 mm diameter when immersed in (i) water, and (ii) mercury, both liquids being at 20°C. Assume σ to be 73×10^{-3} N/m for water and 0·5 N/m for mercury. The contact angles for water and mercury are zero and 130° respectively.

6.   Calculate the internal pressure of a 25 mm diameter soap bubble if the tension in the soap film is 0·5 N/m.

Chapter 2
Fluid Statics

C. Nalluri

2.1 Introduction

Fluid statics is the study of pressures throughout a fluid at rest and the pressure forces on finite surfaces. Since the fluid is at rest there are no shear stresses in it. Hence the pressure, p, at a point on a plane surface (inside the fluid or on the boundaries of its container), defined as the limiting value of the ratio of normal force to surface area as the area approaches zero size, always acts normal to the surface and is measured in N/m^2 (pascals, Pa) or in bars (1 bar = 10^5 N/m^2 or 10^5 Pa).

2.2 Pascal's law

Pascal's law states that the pressure at a point in a fluid at rest is the same in all directions. This means it is independent of the orientation of the surface around the point.

Consider a small triangular prism of unit length surrounding the point in a fluid at rest (fig. 2.1).

Since the body is in static equilibrium, we can write:

$$p_1 (AB \times 1) - p_3 (BC \times 1) \cos \theta = 0 \qquad \text{(i)}$$

$$\text{and } p_2 (AC \times 1) - p_3 (BC \times 1) \sin \theta - W = 0 \qquad \text{(ii)}$$

From (i) $p_1 = p_3$, since $\cos \theta = AB/BC$
and (ii) gives $p_2 = p_3$, since $\sin \theta = AC/BC$ and $W = 0$ as the prism shrinks to a point.

$$\therefore p_1 = p_2 = p_3$$

2.3 Pressure variation with depth in a static incompressible fluid

Consider an elementary cylindrical volume of fluid (of length, L, and cross-sectional area, dA) within the static fluid mass (fig. 2.2), p being the pressure at an elevation of y and dp being the pressure variation corresponding to an elevation variation of dy.

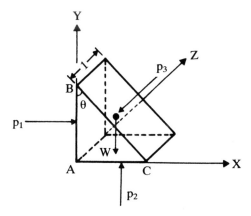

Figure 2.1   Pressure at a point

For equilibrium of the elementary volume

$$p \, dA - \rho g \, dA \, L \sin \theta - (p + dp) \, dA = 0$$

$$\text{or } dp = - \rho g \, dy \text{ (since } \sin \theta = dy/L) \tag{2.1}$$

ρ being constant for incompressible fluids, we can write

$$\int dp = - \rho g \int dy$$

which gives $p = - \rho g y + C$ (i)

When $y = y_o$, $p = p_a$, the atmospheric pressure (fig. 2.3).

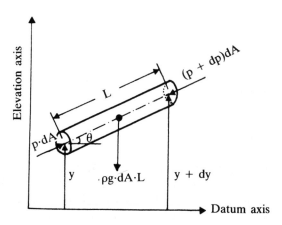

Figure 2.2   Pressure variation with elevation

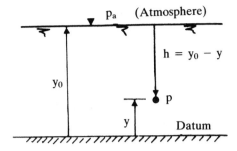

Figure 2.3 Pressure and pressure head at a point

$\therefore$ From (i) $p - p_a = \rho g(y_o - y)$

$$= \rho g h$$

or the pressure at a depth h, $p = p_a + \rho g h$

$$= \rho g h \text{ above atmospheric pressure}$$

$$(2.2)$$

Note (a) If $p = \rho g h$, $h = p/\rho g$ and is known as the pressure head in metres of fluid of density, ρ.

(b) Equation (i) can be written as $p/\rho g + y = $ constant which shows that any increase in elevation is compensated by corresponding decrease in pressure head.

$(p/\rho g + y)$ is known as piezometric head and such a variation is known as hydrostatic pressure distribution.

If the static fluid is a compressible liquid ρ is no longer constant and it is dependent on the degree of its compressibility. Equations 2.1 and 1.2 yield the relationship

$$\frac{1}{\rho} = \frac{1}{\rho_o} - \frac{gh}{K} \tag{2.3}$$

where ρ is the density at a depth, h, below the free surface at which its density is ρ_o.

2.4 Pressure measurement

The pressure at the earth's surface depends upon the air column above it. At sea level this atmospheric pressure is about 101 kN/m² equivalent to 10·3 m of water or 760 mm of mercury columns. A perfect vacuum is an empty space where the pressure is zero. Gauge pressure is the pressure measured above or below atmospheric pressure. The pressure below atmosphere is also called negative or partial vacuum pressure. Absolute pressure is the pressure measured above a perfect vacuum, the absolute zero.

(a) A simple vertical tube fixed to a system, whose pressure is to be measured, is called a piezometer (fig. 2.4a). The liquid rises to such a level that the liquid column's height balances the pressure inside.
(b) A bent tube in the form of U, known as a U-tube manometer is much more convenient than a simple piezometer. Heavy immiscible mano-meter liquids are used to measure large pressures and small pressures are measured by using lighter liquids (fig. 2.4b).
(c) An inclined tube or U-tube (fig. 2.4c) is used as a pressure measuring device when the pressures are very small. The accuracy of measurement is improved by providing suitable inclination.
(d) A differential manometer (fig. 2.4d) is essentially a U-tube manometer containing a single liquid capable of measuring large pressure differences between two systems. If the pressure difference is very small the mano-meter may be modified by providing enlarged ends and two different liquids in the two limbs and is called differential micromanometer.

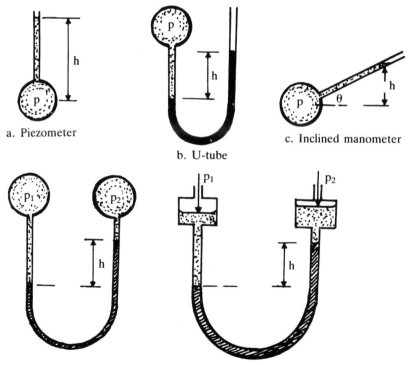

a. Piezometer

b. U-tube

c. Inclined manometer

d. Differential manometers

Figure 2.4 Pressure measurement devices

If the density of water is ρ, a water column of height, h, produces a pressure $p = \rho g h$ and this can be expressed in terms of any other liquid column h_1 as $\rho_1 \, g h_1$, ρ_1 being its density.

$$\therefore \text{ h in water column} = (\rho_1/\rho)h_1 = \sigma h_1 \qquad (2.4)$$

where σ is the relative density of the liquid.

For each one of the above pressure measurement devices an equation can be written using the principle of hydrostatic pressure distribution, expressing the pressures in metres of water column (equation 2.4) for convenience.

2.5 Hydrostatic thrust on plane surfaces

Let the plane surface be inclined at an angle of θ to the free surface of water as shown in fig. 2.5.

If the plane area, A, is assumed to consist of elemental areas, dA, the elemental forces, dF, always normal to the surface area, are parallel. Therefore the system is equivalent to one resultant force, F, known as the hydrostatic thrust. Its point of application, C, which would produce the same moment effects as the distributed thrust, is called the centre of pressure.

$$\text{We can write, } F = \int_A dF = \int_A \rho g h \; dA = \rho g \sin \theta \int_A dA \; x$$

$$= \rho g \sin \theta \, A \, \bar{x}$$

$$= \rho g \bar{h} \, A \qquad (2.5)$$

where $\bar{h}$ is the vertical depth of the centroid, G.

Taking moments of these forces about $0 - 0$

$$F \, x_0 = \rho g \sin \theta \int_A dA \; x^2$$

$\therefore$ The distance to the centre of pressure, C

$$x_0 = \int_A dA \; x^2 \bigg/ \int_A dA \; x$$

$$= \frac{\text{second moment of the area about } 0 - 0}{\text{first moment of the area about } 0 - 0} \qquad (2.6)$$

$$= I_0/A\bar{x}$$

But $I_0 = I_g + A\bar{x}^2$ (parallel axis rule) where I_g is the second moment of area of the surface about an axis through its centroid and parallel to axis $0 - 0$.

$$\therefore x_0 = \bar{x} + I_g/A\bar{x} \qquad (2.7)$$

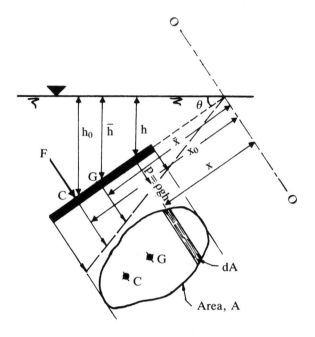

Figure 2.5 Hydrostatic thrust on a plane surface

which shows that the centre of pressure is always below the centroid of the area.

Depth of centre of pressure below free surface, $h_o = x_o \sin \theta$

$$\therefore h_o = \bar{h} + I_g \sin^2 \theta / A\bar{h} \tag{2.8}$$

For a vertical surface $\theta = 90°$

$$\therefore h_o = \bar{h} + I_g / A\bar{h} \tag{2.9}$$

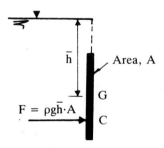

Figure 2.6 Vertical plane surface

Table 2.1   Second moments of plane areas

Shape	Size	Second moment of area I_g, about an axis GG through centroid
Rectangle	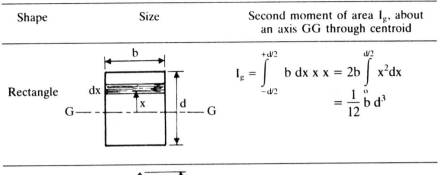	$I_g = \int_{-d/2}^{+d/2} b\ dx \times x = 2b \int_0^{d/2} x^2 dx$ $= \dfrac{1}{12} b\ d^3$
Triangle	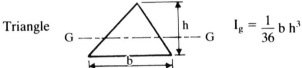	$I_g = \dfrac{1}{36} b\ h^3$
Circle		$I_g = \pi d^4/64$

The distance between centroid and centre of pressure,

$$GC = I_g / A\bar{h} \text{ (fig. 2.6)} \tag{2.10}$$

∴ The moment of F about the centroid,

$$F \times GC = \rho g \bar{h}\ A \times I_g / A\bar{h}$$

$$= \rho g\ I_g$$

which is independent of depth of submergence.

Note (a) Radius of gyration of the area about G,

$$k_g = \sqrt{I_g / A} \tag{2.11}$$

$$\text{giving } h_o = \bar{h} + k_g^2 / \bar{h} \tag{2.12}$$

  (b) When the surface area is symmetrical about its vertical centroidal axis, the centre of pressure always lies on this symmetrical axis but below the centroid of the area.

If the area is not symmetrical, an additional co-ordinate, y_o, must be fixed to locate the centre of pressure completely.

By moments (fig. 2.7),

$$y_o \int_A dF = \int_A dF \, y$$

$$\text{or } y_o \, \rho g \bar{x} \sin \theta \, A = \int_A \rho g x \sin \theta \, dA \, y$$

$$\therefore y_o = \frac{1}{A \bar{x}} \int_A xy \, dA$$

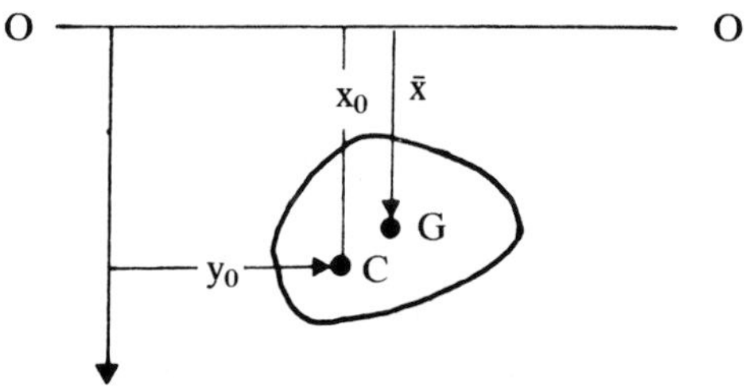

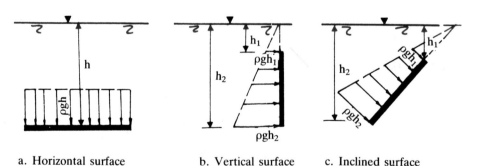

Figure 2.7   Centre of pressure of an asymmetrical plane surface

2.6 Pressure diagrams

Another approach to determine hydrostatic thrust and its location is by the concept of pressure distribution over the surface (fig. 2.8).

a. Horizontal surface   b. Vertical surface   c. Inclined surface

Figure 2.8   Pressure diagrams

Total thrust on a rectangular vertical surface subjected to water pressure on one side (fig. 2.9) by pressure diagram:

Average pressure on the surface $= \rho g H / 2$

$\therefore$ Total thrust, $F =$ average pressure $\times$ area of surface

$$= (\rho gH/2)\ H \times B$$

$$= \frac{1}{2}\ \rho gH^2 \times B$$

$$= \text{volume of the pressure prism} \qquad (2.14)$$

or total thrust/unit width $= \dfrac{1}{2}\ \rho gH^2$

$$= \text{area of the pressure diagram} \qquad (2.15)$$

and the centre of pressure is the centroid of the pressure prism.

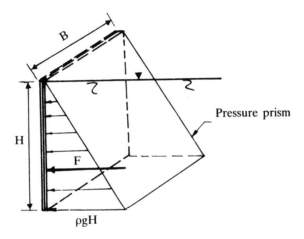

Figure 2.9 Pressure prism

2.7 Hydrostatic thrust on curved surfaces

Consider a curved gate surface subjected to water pressure as in fig. 2.10:

The pressure at any point h, below the free water surface is ρgh and is normal to the gate surface and the nature of its distribution over the entire surface makes the analytical integration difficult.

However, the total thrust acting normally on the surface can be split into two components and the problem of determining the thrust approached indirectly by combining these two components.

Considering an elementary area of the surface, dA (fig. 2.11), at an angle θ to the vertical, pressure intensity on this elementary area $= \rho gh$

$\quad\therefore$ Total thrust on this area, $dF \quad = \rho gh\ dA$

$\quad$ Horizontal component of dF, $dF_x = \rho gh\ dA \cos\theta$

$\quad$ Vertical component of dF, $dF_y \quad = \rho gh\ dA \sin\theta$

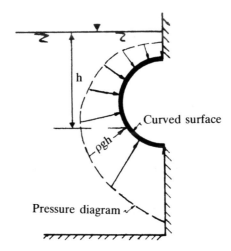

Figure 2.10 Hydrostatic thrust on curved surface

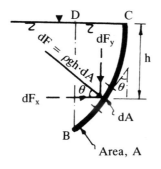

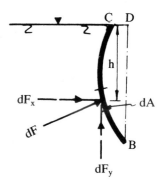

a. Surface containing liquid   b. Surface displacing liquid

Figure 2.11 Thrust components on curved surfaces

∴ Horizontal component of the total thrust on the curved area A,

$$F_x = \int_A \rho gh \, dA \cos \theta = \rho g\bar{h} \, A_v$$

where A_v is the vertically projected area of the curved surface;

or F_x = pressure intensity at the centroid of a vertically projected area
(BD) × vertically projected area   (2.16)

and vertical component, $F_y = \int_A \rho gh \, dA \sin \theta$

$$= \rho g \int_A dV, \text{ dV being the volume of the water}$$

prism (real or virtual) over the area dA.

$\therefore F_y = \rho g \, V$

= the weight of water (real or virtual) above the curved surface BC bounded by the vertical BD and the free water surface CD   (2.17)

$\therefore$ The resultant thrust, $F = \sqrt{F_x^2 + F_y^2}$ (2.18)

acting normally to the surface at an angle,

$\alpha = \tan^{-1} (F_y/F_x)$ to the horizontal.   (2.19)

2.8 Hydrostatic buoyant thrust

When a body is submerged or floating in a static fluid various parts of the surface of the body are exposed to pressures dependent on the depths of submergence.

Consider two elemental cylindrical volumes (one vertical and one horizontal) of the body shown (fig. 2.12) submerged in a fluid, the cross-sectional area of each cylinder being dA.

Vertical upthrust on the cylinder BC = $(p_c - p_b) \, dA$

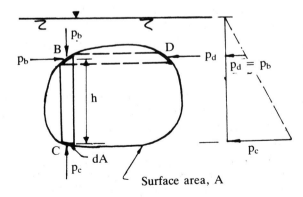

Figure 2.12 Submerged body and buoyant thrust

$$\therefore \text{ Total upthrust on the body} = \int_A (p_c - p_b) \, dA$$

$$= \int_A \rho g h \, dA$$

$$= \int_A \rho g \, dV$$

$$= \rho g \, V = \text{weight of fluid}$$
$$\text{displaced} \qquad (2.20)$$

where V is the volume of the submerged body displacing the fluid.

Horizontal thrust on the cylinder $BD = (p_b - p_d) \, dA$

$$\therefore \text{ Total horizontal thrust on the body} = \int_A (p_b - p_d) \, dA$$

$$= 0 \text{ (since } p_b = p_d)$$

Hence it can be concluded that the only force acting on the body is the vertical upthrust known as the buoyant thrust or force which is equal to the weight of fluid displaced by the body (Archimedes' principle). This buoyant thrust acts through the centroid of the displaced fluid volume.

2.9 Stability of floating bodies

The buoyant thrust on a body of weight, W, and centroid, G, acts through the centroid of the displaced fluid volume and this point of application of the buoyant force is called the centre of buoyancy, B, of the body. For the body to be in equilibrium, the weight, W, must equal the buoyant thrust, F_b both acting along the same vertical line (fig. 2.13).

For small angles of heel, the intersection point of the vertical through the new centre of buoyancy, B', and the line, BG, produced is known as the metacentre, M, and the body thus disturbed tends to oscillate about, M. The distance between G and M is the metacentric height.

Conditions of equilibrium:

(a) Stable equilibrium (fig. 2.14a): If M lies above G, i.e. positive metacentric height, the couple so produced sets in a restoring moment equal to W GM sin θ opposing the disturbing moment and thereby bringing the body back to its original position and the body is said to be in stable equilibrium; this is achieved when BM> BG.
(b) Unstable equilibrium (fig. 2.14b): If M is below G, i.e. negative metacentric height, the moment of the couple further disturbs the displacement and the body is then in unstable equilibrium. This condition therefore exists when BM < BG.

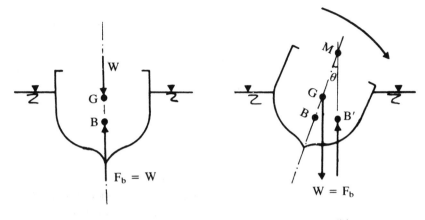

a. Equilibrium condition   b. Disturbed condition

Figure 2.13   Centre of buoyancy and metacentre

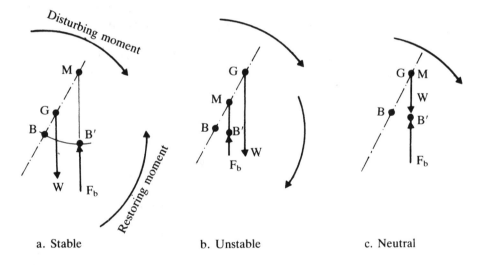

a. Stable   b. Unstable   c. Neutral

Figure 2.14   Conditions of equilibrium

(c) Neutral equilibrium (fig. 2.14c): If G and M coincide, i.e. zero meta-
    centric height, the body floats stably in its displaced position. This
    condition of neutral equilibrium exists when BM = BG.

2.10 Determination of metacentre

In fig. 2.15, AA is the water-line and when the body is given a small tilt θ°
two wedge forces, due to the submergence and the emergence of the wedge

areas AOA' on either side of the axis of rolling, are imposed on the body forming a couple which tends to restore the body to its undisturbed condition. The effect of this couple is the same as the moment caused by the shift of the total buoyant force F_b from B to B', the new centre of buoyancy.

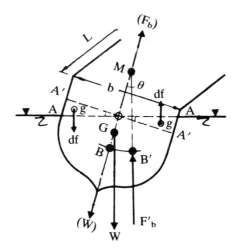

Figure 2.15   Determination of metacentre

The buoyant force acting through B',

$$F'_b = W + df - df = W = F_b$$

By moments about B, $F'_b \times BB' = df \times \overline{gg}$

$$\therefore BB' = df \times \overline{gg}/F'_b = df \times \overline{gg}/W$$
$$= df \times \overline{gg}/\rho g\ V \qquad (i)$$

where V is the volume of the displaced fluid.
The wedge force, $df = \rho g \times \frac{1}{2}\ AA' \times \frac{1}{2}b \times L$ (for small angles) where L is the length of the body.

$$AA' = \frac{1}{2}b\ \theta \text{ and } \overline{gg} = \frac{2}{3}\left(\frac{1}{2}b\right) + \frac{2}{3}\left(\frac{1}{2}b\right) = \frac{2}{3}b$$

$$\therefore BB' = BM\ \theta = \frac{\rho g \times \dfrac{1}{4}b\ \theta \times \dfrac{1}{2}b \times L \times \dfrac{2}{3}b}{\rho g\ V} \quad \text{from (i)}$$

$$\text{or } BM = \frac{1}{12}\ Lb^3/V = I/V \qquad (2.21)$$

where I is the second moment of the plan area of the body at water level about its longitudinal axis.

Hence the metacentric height, GM = BM − BG

$$= I/V - BG \qquad\qquad (2.22)$$

2.11 Periodic time of rolling (or oscillation) of a floating body

For a small displacement θ, restoring moment, T_r = W GM θ = W m θ, where W is the weight of the body

$\therefore$ Angular acceleration due to T_r,

$\alpha = T_r/I$, where I is the mass moment of inertia given by M k^2, M being the mass of body (W/g) and k its radius of gyration about the centroid.

$\therefore \alpha$ = W m $\theta/(W/g)k^2$ = m g θ/k^2 or α is proportional to θ.

Hence the motion is simple harmonic and its periodic time,

$$T = 2\pi \sqrt{\text{displacement/acceleration}}$$

$$= 2\pi \sqrt{\frac{\theta}{m\ g\ \theta/k^2}} = 2\pi \sqrt{k^2/g\ m} \qquad\qquad (2.23)$$

For larger values of m, the floating body will no doubt be stable (BM > BG) but the period of oscillation decreases, thereby increasing the frequency of rolling which may be uncomfortable to passengers and also the body may be subjected to damage. Hence the metacentric height must be fixed, by experience, according to the type of vessel.

2.12 Liquid ballast and the effective metacentric height

For a tilt angle θ, the fluid in the tank (fig. 2.16a) is displaced thereby shifting its centroid from G to G'. This is analogous to the case of a floating vessel, the centre of buoyancy of which shifts from B to B' through a small heel angle θ.

Hence we can write:

GG' = GM θ = θ I/V (since in the case of floating vessel

                         BB' = BM θ = θ I/V)

When the vessel heels, the centroids of the volumes V_1 and V_2 in the compartments (fig. 2.16b) will move by θ I_1/V_1 and θ I_2/V_2 thus shifting the centroid of the vessel from G to G'.

$\therefore$ By taking moments:

W GG' = W_1 θ I_1/V_1 + W_2 θ I_2/V_2

or ρg V GG' = $\rho_1 g$ V_1 θ I_1/V_1 + $\rho_1 g$ V_2 θ I_2/V_2

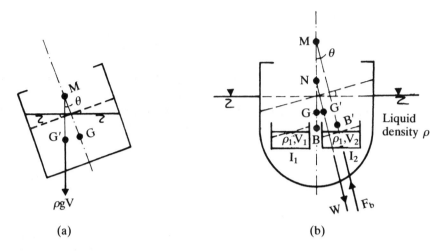

Figure 2.16   Floating vessel with liquid ballast

Hence $GG' = \rho_1 \theta (I_1 + I_2)/\rho V$,

V being the volume of displaced fluid by the vessel whose second moment of area at floating level about its longitudinal axis is I. B' is the new centroid of the displaced liquid through which the buoyant force F_b $(= \rho g\, V)$ acts, thereby setting a restoring moment $= \rho g\, V\, NM\, \theta$, NM being the effective metacentric height.

$$NM = BM - GN - BG$$

$$BM = I/V; \quad GN = GG'/\theta = (\rho_1/\rho)(I_1 + I_2)/V$$

$$\therefore NM = \frac{I - (\rho_1/\rho)(I_1 + I_2)}{V} - BG \tag{2.24}$$

and if $\rho_1 = \rho$, $NM = \dfrac{I - (I_1 + I_2)}{V} - BG$

2.13 Relative equilibrium

If a body of fluid is subjected to motion such that no layer moves relative to an adjacent layer, shear stresses do not exist within the fluid. In other words, in a moving fluid mass if the fluid particles do not move relative to each other, they are said to be in static condition and a relative or dynamic equilibrium exists between them under the action of accelerating force and fluid pressures are everywhere normal to the surfaces on which they act.

Uniform linear acceleration
A liquid in an open vessel subjected to a uniform acceleration adjusts to the acceleration after some time so that it moves as a solid and the whole mass of liquid will be in relative equilibrium.

A horizontal acceleration (fig. 2.17a) a_x causes the free liquid surface to slope upward in a direction opposite to a_x and the entire mass of liquid is then under the action of gravity force, hydrostatic forces and the accelerating or inertial force ma_x, m being the liquid mass.

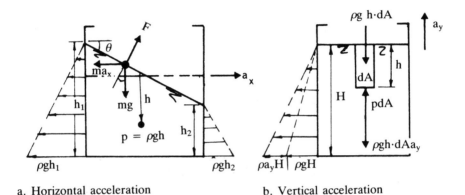

a. Horizontal acceleration   b. Vertical acceleration

Figure 2.17   Fluid subjected to linear accelerations

For equilibrium of a particle of mass m, say on the free surface:

$F \sin \theta = ma_x$ and $F \cos \theta - mg = 0$ or $F \cos \theta = mg$

∴ Slope of free surface, $\tan \theta = ma_x/mg = a_x/g$ (2.25)

and the lines of constant pressure will be parallel to the free liquid surface.

A vertical acceleration (fig. 2.17b) (positive upwards) a_y causes no disturbance to the free surface and the fluid mass is in equilibrium under gravity, hydrostatic forces and the inertial force ma_y.

For equilibrium of a small column of liquid of area dA

$p \, dA = \rho h \, dA \, g + \rho h \, dA \, a_y$

∴ p, the pressure intensity at a depth h below free surface
   $= \rho gh \, (1 + a_y/g)$ (2.26)

Radial acceleration
Fluid particles moving in a curved path experience radial acceleration. When a cylindrical container partly filled with a liquid is rotated at a constant angular velocity ω about a vertical axis the rotational motion is transmitted to different parts of the liquid and after some time the whole fluid mass

assumes the same angular velocity as a solid and the fluid particles experience no relative motion.

A particle of mass, m, on the free surface (fig. 2.18) is in equilibrium under the action of gravity, hydrostatic force and the centrifugal accelerating force $m\omega^2 r$, $\omega^2 r$ being the centrifugal acceleration due to rotation.

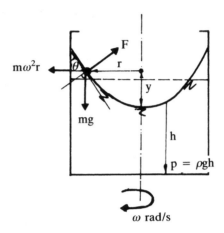

Figure 2.18 Fluid subjected to radial acceleration

The gradient of the free surface,

$$\tan\theta = dy/dr = m\omega^2 r/mg = \omega^2 r/g$$

$$\therefore \quad y = \omega^2 r^2/2g + \text{Constant, C}$$

When $r = 0$, $y = 0$ and hence $C = 0$

$$\therefore y = \omega^2 r^2/2g \tag{2.27}$$

which shows that the free liquid surface is a paraboloid of revolution and this principle is used in a hydrostatic tachometer.

Worked examples

Example 2.1
A hydraulic jack having a ram 150 mm in diameter lifts a weight of 20 kN under the action of a 30 mm diameter plunger. The stroke length of the plunger is 250 mm and if it makes 100 strokes per minute, find by how much the load is lifted per minute and what power is required to drive the plunger.

Solution:
Since the pressure is the same in all directions and is transmitted through the fluid in the hydraulic jack (fig. 2.19),

pressure intensity, $p = F/a = W/A$

$\therefore$ Force on the plunger, $F = W(a/A)$

$$= 20 \times 10^3 (30^2/150^2)$$

$$= 800 \text{ N}$$

Distance moved per minute by the plunger

$$= 100 \times 0.25$$

$$= 25 \text{ m}$$

$\therefore$ Distance through which the weight is lifted per minute

$$= 25 (30^2/150^2)$$

$$= 1 \text{ m}$$

$\therefore$ Power required $= 20 \times 10^3 \times 1/60 \text{ Nm/s}$

$$= 333.3 \text{ W}.$$

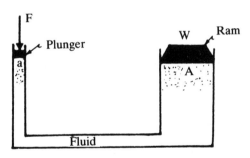

Figure 2.19 Hydraulic jack

Example 2.2
A U-tube containing mercury (relative density 13.6) has its right-hand limb open to atmosphere and the left-hand limb connected to a pipe conveying water under pressure, the difference in levels of mercury in the two limbs being 200 mm. If the mercury level in the left limb is 400 mm below the centre line of the pipe, find the absolute pressure in the pipeline in kPa. Also find the new difference in levels of the mercury in the U-tube, if the pressure in the pipe falls by 2 kN/m².

Solution:
Starting from the left-hand side end (fig. 2.20a)

$p/\rho g + 0.40 - 13.6 \times 0.20 = 0$ (atmosphere)

$\therefore p/\rho g = 2.32$ m of water

or $p = 10^3 \times 9.81 \times 2.32 = 22.76$ kN/m²

The corresponding absolute pressure $= 101 + 22.76$

$$= 123.76 \text{ kN/m}^2$$

$$= 123.76 \text{ kPa.}$$

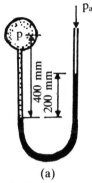
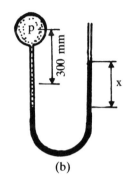

(a)   (b)

Figure 2.20 U-tube manometer

When the manometer is not connected to the system the mercury levels in both the limbs equalise and are 300 mm below centre line of pipe and writing the manometer equation for new conditions (fig. 2.20b)

$$20.76 \times 10^3/10^3 \times 9.81 + 0.30 + x/2 - 13.6 \, x = 0$$

$\therefore$ x, the new difference in mercury levels $= 0.184$ m or 184 mm.

Example 2.3
A double column enlarged ends manometer is used to measure a small pressure difference between two points of a system conveying air under pressure, the diameter of U-tube being 1/10 of the diameter of the enlarged ends. The heavy liquid used is water and the lighter liquid in both limbs is oil of relative density 0·82. Assuming the surfaces of the lighter liquid to remain in the enlarged ends, determine the difference in pressure in millimetres of water for a manometer displacement of 50 mm.

What would be the manometer reading if carbon tetrachloride (relative density 1.6) were used in place of water, the pressure conditions remaining the same?

Solution:
Referring to fig. 2.21, the manometer equation can be written as:

$$p_1/\rho g + 0.82h - 0.05 - (h - 0.05 + 2 \, dx) \, 0.82 = p_2/\rho g$$

and by volumes displaced

$$dx\ A = (0.05/2)a$$

$$\text{or } 2\ dx = 0.05(a/A) = 0.05(1/10)^2$$

$$\therefore (p_1 - p_2)/\rho g = 9.41 \times 10^{-3} \text{ m or } 9.41 \text{ mm of water.}$$

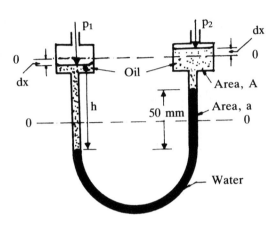

Figure 2.21 Differential micromanometer

For the same pressure conditions if y is the manometer reading using carbon tetrachloride, the manometer equation is:

$$p_1/\rho g + 0.82h - 1.6\ y - 0.82\ (h - y + 2\ dx) = p_2/\rho g$$

$$\text{and } 2\ dx = y/10^2$$

$$\therefore (p_1 - p_2)/\rho g = 9.41 \times 10^{-3} = 0.788y$$

Hence y, the manometer displacement

$$= 9.41 \times 10^{-3}/0.788$$

$$= 11.94 \times 10^{-3} \text{ m or } 11.94 \text{ mm of carbon tetrachloride.}$$

Example 2.4
One end of an inclined U-tube manometer is connected to a system carrying air under a very small pressure. If the other end is open to atmosphere and the angle of inclination is 3° to the horizontal and the tube contains oil of relative density 0.8, calculate (i) the air pressure in the system for a manometer reading of 500 mm along the slope, and (ii) the equivalent vertical water column height.

Solution:
The manometer equation gives (fig. 2.22):

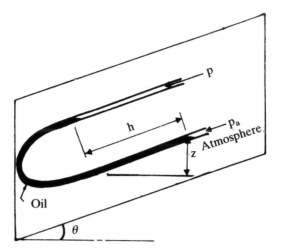

Figure 2.22   Inclined U-tube manometer

$p/\rho_{o}g - z = 0$ and $z = h \sin \theta$

$p = \rho_{o}gh \sin \theta$, ρ_{o} being the density of oil

$\quad = 0{\cdot}8 \times 1000 \times 9{\cdot}81 \times 0{\cdot}5 \times \sin 3°$

$\quad = 205{\cdot}36 \ \text{N/m}^2$

If h' is the equivalent water column height and ρ the density of water, we can write:

$p \ = \rho gh' = \rho_{o}gh \sin \theta$

$\therefore h' = (\rho_{o}/\rho) \ h \sin \theta$

$\quad = \sigma \ h \sin \theta$

$\quad = 0{\cdot}8 \times 0{\cdot}5 \times \sin 3°$

$\quad = 2{\cdot}09 \times 10^{-2} \ \text{m}.$

Example 2.5
(a) An open steel tank of base 4 m square has its sides sloping outwards such that its top is 7 m square. If the tank is 2 m high and is filled with water, determine the total thrust and its location (i) on the base, and (ii) on one of the sloping sides.
(b) If the four sides of the tank slope inwards so that its top is 1 m square, find the thrust and its location on the base when it is filled with water.

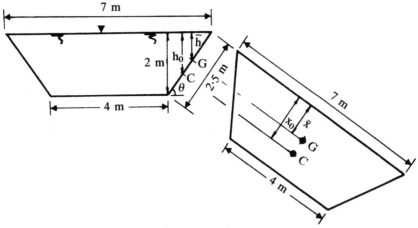

a. Sides sloping outwards

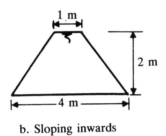

b. Sloping inwards

Figure 2.23 Open tank with sloping sides

Solution:

   Pressure intensity on the base, $p = \rho g \times 2$ N/m^2

   Hence total thrust on the base, $P = p \times A$

   Referring to fig. 2.23a and b, thrust $P = \rho g \times 2 \times 4 \times 4 = 314$ kN
for both cases (Pascal's or hydrostatic paradox), and by symmetry this acts
through the centroid of the base.
Total thrust on a side (fig. 2.23a):

   Length of sloping side $= \sqrt{1.5^2 + 2^2} = 2.5$ m

   By moments,

$$\frac{(7+4)}{2} \times 2.5 \times \bar{x} = 4 \times 2.5 \times \frac{2.5}{2} + 2 \times \frac{1}{2} \times 1.5 \times 2.5 \times \frac{2.5}{3}$$

$$13.75\,\bar{x} = 12.5 + 3.125$$

$$\therefore \bar{x} = 1.136 \text{ m}$$

∴ Depth of immersion,

$$\bar{h} = \bar{x} \sin \theta$$

$$= 1 \cdot 136 \times \frac{2}{2 \cdot 5}$$

$$= 0 \cdot 91 \text{ m}$$

Hence total thrust, $F = \rho g \bar{h} A$

$$= 10^3 \times 9 \cdot 81 \times 0 \cdot 91 \times \frac{(7 + 4)}{2} \times 2 \cdot 5$$

$$= 122 \cdot 75 \text{ kN}$$

Centre of pressure: $h_o = \bar{h} + I_g \sin^2 \theta / A \bar{h}$

$$I_g = (1/12)4 \times 2 \cdot 5^3 + 2 \times (1/36)1 \cdot 5 \times 2 \cdot 5^3$$

$$= 5 \cdot 208 + 1 \cdot 302$$

$$= 6 \cdot 51 \text{ m}^4$$

$$\therefore h_o = 0 \cdot 91 + \frac{6 \cdot 51 \ (2/2 \cdot 5)^2}{\dfrac{(7 + 4)}{2} \ 2 \cdot 5 \times 0 \cdot 91}$$

$$= 0 \cdot 91 + 0 \cdot 333$$

$$= 1 \cdot 243 \text{ m}$$

Example 2.6
A 2 m × 2 m tank with vertical sides contains oil of density 900 kg/m³ to a depth of 0·8 m floating on 1·2 m depth of water. Calculate the total thrust and its location on one side of the tank. (See fig. 2.24.)

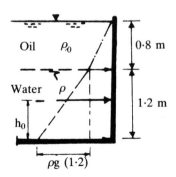

Figure 2.24   Oil and water thrusts on a side of a tank

Solution:

Total thrust on one vertical side, F = volume of the pressure prism

$$= \left[\frac{1}{2} \rho_o g \ (0 \cdot 8)^2 + \rho_o g \times 0 \cdot 8 \times 1 \cdot 2 + \frac{1}{2} \rho g \ (1 \cdot 2)^2 \right] 2$$

$$= 5 \cdot 65 + 16 \cdot 95 + 14 \cdot 13$$

$$= 36 \cdot 73 \ \text{kN}$$

Centre of pressure, h_o: by moments,

$$36 \cdot 73 \times h_o = 5 \cdot 65 \ (1 \cdot 2 + 0 \cdot 8/3) + 16 \cdot 95 \times 1 \cdot 2/2 + 14 \cdot 13 \times 1 \cdot 2/3$$

$$= 8 \cdot 29 + 10 \cdot 17 + 5 \cdot 65$$

$$= 24 \cdot 11$$

∴ $h_o = 24 \cdot 11/36 \cdot 73$

$$= 0 \cdot 656 \ \text{m above the base.}$$

Example 2.7

(a) A circular butterfly gate pivoted about a horizontal axis passing through its centroid is subjected to hydrostatic thrust on one side and counterbalanced by a force, F, applied at the bottom as shown in fig. 2.25. If the diameter of the gate is 4 m and the water depth is 1 m above the gate determine the force F, required to keep the gate in position.

(b) If the gate is to retain water to its top level on the other side also, determine the net hydrostatic thrust on the gate and suggest the new conditions for the gate to be in equilibrium. (See fig. 2.26.)

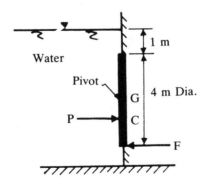

Figure 2.25 Circular gate

Solution:

(a) Water on one side only:

     Hydrostatic thrust on the gate,

$$P = \rho g\bar{h} \ A = 10^3 \times 9\cdot81 \times 3 \times \frac{1}{4}\pi 4^2$$

$$= 369.83 \ kN$$

and the distance, $CG = I_g/A\bar{h}$

$$= \frac{\pi}{64} \ 4^4/\frac{1}{4}\pi 4^2 \times 3$$

$$= 0\cdot333 \ m.$$

Taking moments about G,

$$369\cdot83 \times 0\cdot333 = F \times 2$$

$$F = 61\cdot64 \ kN.$$

(b) If the gate is retaining water on the other side also, the net hydrostatic thrust is due to the resultant pressure diagram with a uniform pressure distribution of intensity equal to ρgh (fig. 2.26).

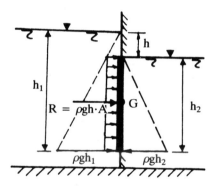

Figure 2.26 Gate retaining water on both sides

∴ Net hydrostatic thrust $R = \rho gh \ A$

$$= 10^3 \times 9\cdot81 \times 1 \times \frac{1}{4}\pi 4^2$$

$$= 123\cdot28 \ kN.$$

This acts through centroid of the gate, G, and since its moment about G is zero, $F = 0$ for the gate to be in equilibrium, for any depth, h, of water above the gate on the other side.

Example 2.8
An open tank 3 m × 1 m in cross-section (fig. 2.27a) holds water to a depth of 3 m. Determine the magnitude, direction and line of action of the forces exerted upon the plane surfaces AB and CD and the curved surface BC of the tank.

Solution:

Force on face AB/m length = area of pressure diagram (fig. 2.27b)

$$= \frac{1}{2} \rho g \times 2^2 = 19 \cdot 62 \text{ kN/m}$$

acting normal to the face AB at a depth of $(2/3) \times 2 = 1 \cdot 33$ m from water surface.

Force on curved surface BC/m length:

Horizontal component, F_x (from pressure diagram)

$$= \rho g \times 2 \times 1 + \frac{1}{2} \rho g \times 1^2 = 24.52 \text{ kN/m}$$

Vertical component, F_y = weight of water above the surface

$$= \rho g \times 2 \times 1 \times 1 + \rho g \times \frac{\pi \times 1^2}{4} \times 1$$

$$= 27 \cdot 32 \text{ kN/m}$$

∴ Resultant thrust, F $= \sqrt{24 \cdot 52^2 + 27 \cdot 32^2}$

$$= 36 \cdot 71 \text{ kN/m}$$

acting at an angle, $\alpha = \tan^{-1} (27 \cdot 32/24 \cdot 52) = 48°5'$, to the horizontal and passing through the centre of curvature of the surface BC.

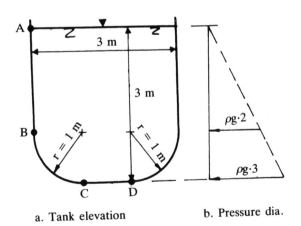

a. Tank elevation   b. Pressure dia.

Figure 2.27 Open tank with plane and curved surfaces

Force on surface CD/m length:

Uniform pressure intensity on CD = $\rho g \times 3 \text{ N/m}^2$

∴ Total thrust on CD = uniform pressure × area

$$= \rho g \times 3 \times 1 \times 1$$

$$= 29{\cdot}43 \text{ kN/m}$$

acting vertically downwards (normal to CD) through the mid-point of the surface CD.

Example 2.9
A 3 m diameter roller gate retains water on both sides of a spillway crest as shown in fig. 2.28. Determine (i) the magnitude, direction and location of the resultant hydrostatic thrust acting on the gate per unit length, and (ii) the horizontal water thrust on the spillway per unit length.

Solution:
Thrust on the gate:

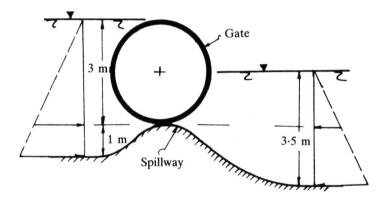

Figure 2.28   Roller gate on a spillway

Left side: horizontal component $= \dfrac{1}{2} \rho g \times 3^2$

$$= 44{\cdot}14 \text{ kN/m}$$

vertical component $= \rho g \times \dfrac{1}{2} \dfrac{\pi}{4} 3^2 \times 1$

$$= 34{\cdot}67 \text{ kN/m}$$

Right side: horizontal component $= \dfrac{1}{2} \rho g \, (1{\cdot}5)^2$

$$= 11{\cdot}03 \text{ kN/m}$$

vertical component $= \rho g \times \dfrac{1}{4} \dfrac{\pi}{4} 3^2 \times 1$

$$= 17{\cdot}34 \text{ kN/m}$$

∴ Net horizontal component on the gate (left to right)

$$= 44 \cdot 14 - 11 \cdot 03$$

$$= 33 \cdot 11 \text{ kN/m}$$

and net vertical component (upwards) $= 34 \cdot 67 + 17 \cdot 34$
$$= 50 \cdot 01 \text{ kN/m}$$

∴ Resultant hydrostatic thrust on the gate

$$= \sqrt{(33.11)^2 + (50.01)^2}$$

$$= 60 \text{ kN/m}$$

acting at an angle, $\alpha = \tan^{-1} (33 \cdot 11/50 \cdot 01) = 33°30'$ to the vertical and passes through the centre of the gate (normal to the surface).

∴ Depth of centre of pressure $= r + r \cos \alpha$

$$= 1 \cdot 5 (1 + \cos 33°30')$$

$$= 2 \cdot 75 \text{ m below the free surface of left}$$
side.

Horizontal thrust on the spillway:
From pressure diagrams (see fig. 2.28), thrust from left-hand side

$$= \frac{1}{2} (\rho g \times 3 + \rho g \times 4) \times 1$$

$$= 34 \cdot 33 \text{ kN/m}$$

and from right-hand side $= \frac{1}{2} (\rho g \times 1 \cdot 5 + \rho g \times 3 \cdot 5) \times 2$

$$= 49 \cdot 05 \text{ kN/m}$$

∴ Resultant thrust (horizontal) on the spillway

$$= 49 \cdot 05 - 34 \cdot 33$$

$$= 14 \cdot 72 \text{ kN/m towards left.}$$

Example 2.10
The gates of a lock (fig. 2.29) are 5 m high and when closed include an angle of 120°. The width of the lock is 6 m. Each gate is carried on two hinges placed on the top and bottom of the gate. If the water levels are 4·5 m and 3 m on the upstream and downstream sides respectively, determine the magnitudes of the forces on the hinges due to water pressure.

Solution:
Forces on any one gate (say AB) are: F, the resultant water thrust, T, the thrust of gate, BC, normal to contact surface and, R, the resultant of hinge

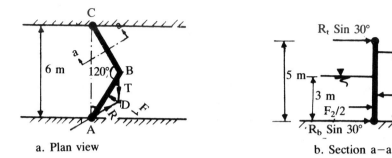

a. Plan view   b. Section a−a

Figure 2.29 Lock gates

forces. Since these three forces keep the gate in equilibrium they should meet at a point, D (fig. 2.29a).

Resolution of forces along AB and normal to AB gives (A $\hat{B}$ D = B $\hat{A}$ D = 30°):

$$T \cos 30° = R \cos 30° \text{ or } T = R \tag{i}$$

and $F = R \sin 30° + T \sin 30°$ or $F = R$ $\qquad$ (ii)

Length of gate = 3/sin 60° = 3.464 m.

The resultant of water thrusts on either side of the gate, $F = F_1 - F_2$

$$F_1 = \frac{1}{2} \rho g \ (4 \cdot 5)^2 \times 3 \cdot 464$$

$\qquad$ = 344 kN acting at 4·5/3 = 1·5 m from the base

and $F_2 = \frac{1}{2}\rho g \times 3^2 \times 3 \cdot 464$

$\qquad$ = 153 kN acting at 3/3 = 1 m from the base

∴ Resultant water thrust, F = 344 − 153

$\qquad\qquad\qquad\qquad$ = 191 kN = R from (ii)

Total hinge reaction, $R = R_t + R_b$ (sum of top and bottom hinge forces) $\qquad$ (iii)

from (ii) F/2 = R sin 30°

or $(F_1 - F_2)/2 = R_t \sin 30° + R_b \sin 30°$ $\qquad$ (iv)

Taking moments about the bottom hinge (fig. 2.29b):

$$\frac{344}{2} \times 1\cdot5 - \frac{153}{2} \times 1 = R_t \sin 30° \times 5$$

$$\therefore \quad R_t = 72\cdot6 \text{ kN}$$

and hence from (iii) $R_b = 191 - 72 \cdot 6$
$$= 118 \cdot 4 \text{ kN}$$

Example 2.11
A rectangular block of wood floats in water with 50 mm projecting above the water surface. When placed in glycerine of relative density 1·35, the block projects 75 mm above the surface of glycerine. Determine the relative density of the wood.

Solution:
Weight of wooden block, W = upthrust in water = upthrust in glycerine
$$= \text{weight of fluid displaced}$$

$$W = \rho_w g \, A \, h = \rho g \, A \, (h - 50 \times 10^{-3}) = \rho_G g \, A(h - 75 \times 10^{-3})$$

ρ, ρ_w and ρ_G being the densities of water, wood and glycerine respectively and A, the cross-sectional area of the block and h, its height.

$\therefore$ The relative density of glycerine, $\rho_G/\rho = \dfrac{h - 50 \times 10^{-3}}{h - 75 \times 10^{-3}} = 1 \cdot 35.$

$\therefore \; h = 146 \cdot 43 \times 10^{-3}$ m or 146·43 mm.

Hence the relative density of wood, $\rho_w/\rho = \dfrac{146 \cdot 43 - 50}{146 \cdot 43} = 0 \cdot 658.$

Example 2.12
(a) A ship of 50 MN displacement has a weight of 100 kN moved 10 m across the deck causing a heel angle of 5°. Find the metacentric height of the ship.
(b) A homogeneous circular cylinder of radius, R, and height, H, is to float stably in a liquid. Show that R must not be less than $\sqrt{2r(1-r)}$ H in order to float with its axis vertical, where r is the ratio of relative densities of the cylinder and the liquid. Hence establish the condition for R/H to be minimum.

Solution:
Referring to fig. 2.30a,

$$\text{Moment heeling the ship} = 100 \times 10 = 1000 \text{ kN m}$$

$$= \text{moment due to the shifting of W from G to G}'$$

$$= W \times GG'$$

$$\therefore \; GG' = GM \sin \theta = \frac{1000}{50 \times 10^3} = \frac{1}{50} \text{ m}$$

$$\text{Hence GM, the metacentric height} = \frac{1}{50 \times \sin 5°} = 0 \cdot 23 \text{ m}$$

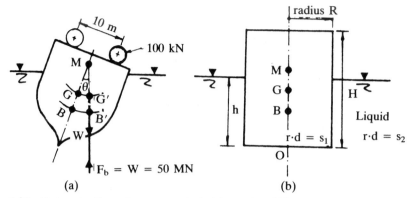

(a)   (b)

Figure 2.30   Determination of metacentric height and stability conditions

Referring to fig. 2.30b,

      Weight of cylinder = weight of liquid displaced.

If the depth of submergence is h, we can write

$$\rho g s_1 \pi R^2\, H = \rho g s_2 \pi R^2\, h, \qquad\qquad \rho \text{ being the density of water}$$

and hence, $s_1/s_2 = r = h/H$ (i)

$$OG = H/2 \qquad\qquad\qquad\qquad\qquad (ii)$$

$$\text{and } OB = h/2 = r\,H/2 \qquad\qquad\qquad (iii)$$

$\therefore$ BG = OG − OB = H/2 − r H/2

$$= (H/2)\,(1 - r)$$

and BM = I/V = $\dfrac{\pi R^4}{4}\bigg/ \pi R^2\, rH$

$$= R^2/4rH$$

For stable condition, BM > BG

$$R^2/4rH > (H/2)\,(1 - r)$$

or $R^2/H^2 > 2r\,(1 - r)$

$\therefore$ R/H > $\sqrt{2r\,(1 - r)}$

and hence, R > $\sqrt{2r\,(1 - r)}$ H

For limiting value of R/H, r (1 − r) is to be minimum

or d[r (1 − r)] = 0

$\therefore$ 1 − 2r = 0 and hence r = $\dfrac{1}{2}$

Example 2.13
An oil tanker 3 m wide, 2 m deep and 10 m long contains oil of density 800 kg/m³ to a depth of 1 m. Determine the maximum horizontal acceleration that can be given to the tanker such that the oil just reaches its top end.

If this tanker is closed and completely filled with the oil and accelerated horizontally at 3 m/s² determine the total liquid thrust (i) on the front end, (ii) on the rear end, and (iii) on one of its longitudinal vertical sides. (See fig. 2.31.)

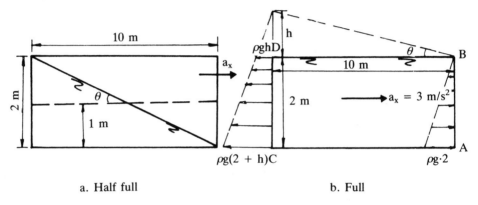

a. Half full   b. Full

Figure 2.31   Oil tanker subjected to accelerations

Solution:
From fig. 2.31a, maximum possible surface slope = $1/5 = a_x/g$

$\therefore$ a_x, the maximum horizontal acceleration = $(1/5) \times 9.81$

$$= 1 \cdot 962 \text{ m/s}^2.$$

When the tanker is completely filled and closed, there will be pressure built up at the rear end equivalent to the virtual oil column (h) that would assume a slope of a_x/g (fig. 2.31b).

(i)   total thrust on front end AB = $\frac{1}{2} \rho g \times 2^2 \times 3 = 58 \cdot 86$ kN

(ii) total thrust on rear end CD:
      Virtual rise of oil level at rear end,

$$h = 10 \times \tan \theta = 10 \times a_x/g = 10 \times 3/9 \cdot 81 = 3 \cdot 06 \text{ m}$$

$\therefore$ Total thrust on CD = $\dfrac{\rho g \ (3 \cdot 06) + \rho g \ (2 + 3 \cdot 06)}{2} \times 2 \times 3$

$$= 239 \text{ kN}$$

(iii) total thrust on side ABCD = volume of the pressure prism

$$= \frac{1}{2} \rho g \times 2^2 \times 10 + \frac{1}{2} \rho g \times 3 \cdot 06 \times 10 \times 2$$

$$= \frac{1}{2} \rho g \, (2 + 3 \cdot 06) \times 2 \times 10$$

$$= 496 \text{ kN.}$$

Example 2.14
A vertical hoist carries a square tank of 2 m × 2 m containing water to the top of a construction scaffold with a varying speed of 2 m/s per second. If the water depth is 2 m, calculate the total hydrostatic thrust on the bottom of the tank.

If this tank of water is lowered with an acceleration equal to that of gravity, what are the thrusts on the floor and sides of the tank?

Solution:

Vertical upward acceleration, a_y = 2 m/s^2

Pressure intensity at a depth   h = $\rho gh \, (1 + a_y/g)$

$$= \rho gh \, (1 + 2/9 \cdot 81)$$

$$= 1 \cdot 204 \times gh \text{ kN/m}^2$$

∴ Total hydrostatic thrust on the floor

= intensity × area

= $1 \cdot 204 \times 9 \cdot 81 \times 2 \times 2 \times 2$

= 94·5 kN

Downward acceleration = $- 9 \cdot 81$ m/s^2

Pressure intensity at a depth h = $\rho gh \, (1 - 9 \cdot 81/9 \cdot 81)$

$$= 0$$

∴ There exists no hydrostatic thrust on the floor nor on the sides.

Example 2.15
A 375 mm high open cylinder, 150 mm in diameter, is filled with water and rotated about its vertical axis at an angular speed of 33·5 rad/s. Determine (i) the depth of water in the cylinder when it is brought to rest, and (ii) the volume of water that remains in the cylinder if the speed is doubled. (See fig. 2.32.)

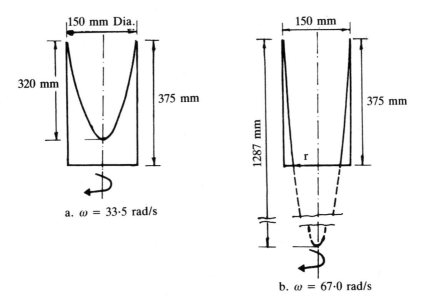

a. ω = 33·5 rad/s

b. ω = 67·0 rad/s

Figure 2.32 Rotating cylinder

Solution:

Angular velocity, ω = 33·5 rad/s

Height of the paraboloid (fig. 2.32a), $y = \omega^2 r^2/2g$

$$= (33·5 \times 0·075)^2/19·62$$

$$= 0·32 \text{ m}$$

Amount of water spilled out = volume of the paraboloid

$$= \frac{1}{2} \times \text{volume of circumscribing cylinder}$$

$$= \frac{1}{2} \pi (0·075)^2 \times 0·32$$

$$= 2·83 \times 10^{-3} \text{ m}^3$$

Original volume of water $= \pi(0·075)^2 \times 0·375$

$$= 6·63 \times 10^{-3} \text{ m}^3$$

∴ Remaining volume of water $= (6·63 - 2·83) \times 10^{-3}$

$$= 3·8 \times 10^{-3} \text{ m}^3$$

Hence depth of water at rest $= 3·8 \times 10^{-3}/\pi(0·075)^2$

$$= 0·215 \text{ m}$$

If the speed is doubled, $\omega = 67$ rad/s

$$\therefore \text{Height of paraboloid} = (67 \times 0\cdot075)^2/2g$$

$$= 1\cdot287 \text{ m}$$

The free surface in the vessel assumes the shape as show in fig. 2.32b, and we can write:

$$1\cdot287 - 0\cdot375 = \omega^2 r^2/2g$$

$$\therefore r = \sqrt{2g \times 0\cdot912/67^2} = 0\cdot063 \text{ m}$$

$\therefore$ Volume of water spilled out

$$= \frac{1}{2} \pi (0\cdot075)^2 \times 1\cdot287 - \frac{1}{2} \pi (0\cdot063)^2 \times 0\cdot912$$

$$= 5\cdot684 \times 10^{-3} \text{ m}^3$$

Hence volume of water left $= (6\cdot63 - 5\cdot684) \times 10^{-3}$

$$= 0\cdot946 \times 10^{-3} \text{ m}^3.$$

Recommended reading

1. Davis, C.V. and Sorensen, K.E. (Editors) (1993) *Handbook of applied hydraulics*, 4th edn. New York: McGraw-Hill.
2. Rouse, H. (1946) *Elementary mechanics of fluids*. Chichester: Wiley.
3. Streeter, V.L. and Wylie, E.B. (1985) *Fluid mechanics*, 8th edn. New York: McGraw-Hill.

Problems

1. (a) A large storage tank contains a salt solution of variable density given by $\rho = 1050 + kh$ in kg/m^3, where $k = 50\backslash$ kg/m^4, at a depth h metres below the free surface. Calculate the pressure intensity at the bottom of the tank holding 5 m of the solution.

   (b) A Bourdon type pressure gauge is connected to a hydraulic cylinder activated by a piston of 20 mm diameter. If the gauge balances a total mass of 10 kg placed on the piston, determine the gauge reading in metres of water.

2. A closed cylindrical tank 4 m high is partly filled with oil of density 800 kg/m^3 to a depth of 3 m. The remaining space is filled with air under pressure. A U-tube containing mercury (relative density 13·6) is used to measure the air pressure, with one end open to atmosphere. Find the gauge pressure at the base of the tank when the mercury deflection in the open limb of the U-tube is (i) 100 mm above, and (ii) 100 mm below the level in the other limb.

3.   A manometer consists of a glass tube, inclined at 30° to horizontal, connected to a metal cylinder standing upright. The upper end of the cylinder is connected to a gas supply under pressure. Find the pressure in millimetres of water when the manometer fluid of relative density 0·8 reads a deflection of 80 mm along the tube. Take the ratio, r, of the diameters of the cylinder and the tube as 64. What value of r would you suggest so that the error due to disregarding the change in level in the cylinder will not exceed 0·2%?

4.   In order to measure the pressure difference between two points in a pipeline carrying water, an inverted U-tube is connected to the points and air under atmospheric pressure is entrapped in the upper portion of the U-tube. If the manometer deflection is 0·8 m and the downstream tapping is 0·5 m below the upstream point, find the pressure difference between the two points.

5.   A high pressure gas pipeline is connected to a macromanometer consisting of four U-tubes in series with one end open to atmosphere and a deflection of 500 mm of mercury (relative density 13·6) has been observed. If water is entrapped between the mercury columns of the manometer and the relative density of the gas is $1·2 \times 10^{-3}$, calculate the gas pressure in N/mm^2, the centre line of the pipeline being at a height of 0·50 m above the top mercury level.

6.   A dock gate is to be reinforced with three identical horizontal beams. If the water stands to depth of 5 m and 3 m on either side, find the positions of the beams, measured above the floor level, so that each beam will carry an equal load, and calculate the load on each beam per unit length.

7.   A storage tank of a sewage treatment plant is to discharge excess sewage into the sea through a horizontal rectangular culvert 1 m deep and 1.3 m wide. The face of the discharge end of the culvert is inclined at 40° to the vertical and the storage level is controlled by a flap-gate weighing 4·5 kN, hinged at the top edge and just covering the opening. When the sea water stands to the hinge level, to what height above the top of the culvert will the sewage be stored before a discharge occurs? Take the density of the sewage as $1000 \ kg/m^3$ and of the sea water as $1025 \ kg/m^3$.

8.   A radial gate, 2 m long, hinged about a horizontal axis, closes the rectangular sluice of a control dam by the application of a counter-weight W (see fig. 2.33).

Determine (i) the total hydrostatic thrust and its location on the gate when the storage depth is 4 m, and (ii) for the gate to be stable, the counter-weight W. Explain what will happen if the storage increases beyond 4 m.

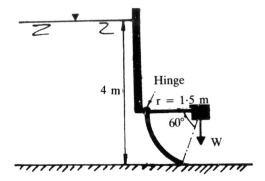

Figure 2.33 Hinged radial gate

9.   A sector gate of radius 3 m and length 4 m retains water as shown in fig. 2.34.

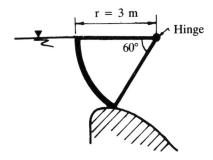

Figure 2.34 Sector or tainter gate

Determine the magnitude, direction and location of the resultant hydrostatic thrust on the gate.

10.   The profile of the inner face of a dam is a parabola with equation y = 0·30 x^2 (see fig. 2.35). The dam retains water to a depth of 30 m above the base. Determine the hydrostatic thrust on the dam per unit length, its inclination to the vertical and the point at which the line of action of this thrust intersects the horizontal base of the dam.

11.   A homogeneous wooden cylinder of circular section, relative density 0·7, is required to float in oil of density 900 kg/m³. If d and h are the diameter and height of the cylinder respectively, establish the upper limiting value of the ratio h/d for the cylinder to float with its axis vertical.

12.   A conical buoy floating in water with its apex downwards has a diameter d and a vertical height h. If the relative density of the material of the buoy is s, prove that for stable equilibrium,

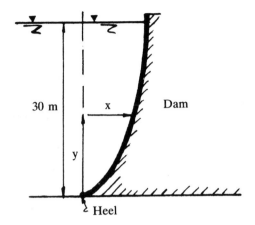

Figure 2.35 Parabolic profile of the inner face of a dam

$$h/d < \frac{1}{2}\sqrt{\frac{s^{\frac{1}{3}}}{1 - s^{\frac{1}{3}}}}$$

13.   A cyclindrical buoy weighing 20 kN is to float in sea water whose density is 1020 kg/m³. The buoy has a diameter of 2 m and is 2·5 m high. Prove that it is unstable.

   If the buoy is anchored with a chain attached to the centre of its base, find the tension in the chain to keep the buoy in vertical position.

14.   A floating platform for offshore drilling purposes is in the form of a square floor supported by 4 vertical cylinders at the corners. Determine the location of the centroid of the assembly in terms of the side L of the floor and the depth of submergence h of the cylinders, so as to float in neutral equilibrium under a uniformly distributed loading condition.

15.   A platform constructed by joining two 10 m long wooden beams as shown in fig. 2.36 is to float in water. Examine the stability of a single beam and of the platform and determine their stability moments. Neglect the weight of the connecting pieces and take the density of wood as 600 kg/m³.

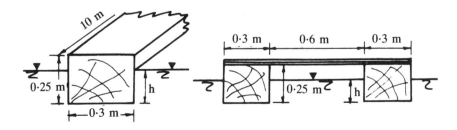

a. Single beam   b. Platform

Figure 2.36 Floating platform

16.   A rectangular barge 10 m wide and 20 m long is 5 m deep and weighs 6 MN when loaded without any ballast. The barge has two compartments each 4 m wide and 20 m long, symmetrically placed about its central axis, and each containing 1 MN of water ballast. The water surface in each compartment is free to move. The centre of gravity without ballast is 3·0 m above the bottom and on the geometrical centre of the plan. (i) Calculate the metacentric height for rolling, and (ii) if 100 kN of the deck load is shifted 5 m laterally find the approximate heel angle of the barge.

17,   A U-tube acceleration meter consists of two vertical limbs connected by a horizontal tube of 400 mm long parallel to the direction of motion. Calculate the level difference of the liquid in the U-tube when it is subjected to a horizontal uniform acceleration of 6 m/s^2.

18.   An open rectangular tank 4 m long and 3 m wide contains water up to a depth of 2 m. Calculate the slope of the free surface of water when the tank is accelerated at 2 m/s^2, (i) up a slope of 30°, and (ii) down a slope of 30°.

19.   Prove that, in the forced vortex motion (fluids subjected to rotation externally) of a liquid, the rate of increase of the pressure, p, with respect to the radius, r, at a point in liquid is given by $dp/dr = \rho\omega^2 r$, in which ω is the angular velocity of the liquid and ρ is its mass density. Hence calculate the thrust of the liquid on the top of a closed vertical cylinder of 450 mm diameter, completely filled with water under a pressure of 10 N/cm^2, when the cylinder rotates about its axis at 240 rpm.

Chapter 3
Fluid Flow Concepts and Measurements

C. Nalluri

3.1 Kinematics of fluids

The kinematics of fluids deal with space-time relationships for fluids in motion. In the Lagrangian method of describing the fluid motion one is concerned to trace the paths of the individual fluid particles (elements) and to find their velocities, pressures, etc., with the passage of time. The co-ordinates of a particle $A(x,y,z)$ at any time, t, (fig. 3.1a) are dependent on its initial co-ordinates (a,b,c) at the instant t_0 and can be written as functions of a,b,c and t, i.e.

$$x = \phi_1 (a,b,c,t)$$

$$y = \phi_2 (a,b,c,t)$$

$$z = \phi_3 (a,b,c,t)$$

The path traced by the particles over a period of time is known as the pathline. Due to the diffusivity phenomena of fluids and their flows, it is difficult to describe the motion of individual particles of a flow field with time. More appropriate for describing the fluid motion is to know the flow characteristics such as velocity and pressure, of a particle or group of particles at a chosen point in the flow field at any particular time; such a description of fluid flow is known as Eulerian method.

In any flow field, velocity is the most important characteristic to be

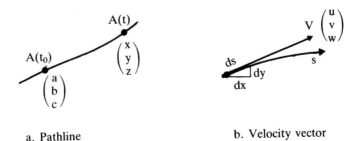

a. Pathline　　　　　　　　　　b. Velocity vector

Figure 3.1　Descriptions of fluid flow

identified at any point. The velocity vector at a point in the flow field is a function of s and t and can be resolved into u, v and w components, representing velocities in the x, y and z directions respectively; these components are functions of x,y,z and t and written as (fig. 3.1b):

$$u = f_1 (x,y,z,t)$$

$$v = f_2 (x,y,z,t)$$

$$w = f_3 (x,y,z,t)$$

defining the vector V at each point in the space, at any instant t. A continuous curve traced tangentially to the velocity vector at each point in the flow field is known as the streamline.

3.2 Steady and unsteady flows

The flow parameters such as velocity, pressure and density of a fluid flow are independent of time in a steady flow whereas they depend on time in unsteady flows. For example, this can be written as:

$$(\partial V/\partial t)_{x_0, y_0, z_0} = 0 \text{ for steady flow} \tag{3.1a}$$

$$\text{and } (\partial V/\partial t)_{x_0, y_0, z_0} \neq 0 \text{ for unsteady flow} \tag{3.1b}$$

At a point, in reality these parameters are generally time dependent but often remain constant on average over a time period T. For example, the average velocity $\bar{u}$ can be written as:

$$\bar{u} = \frac{1}{T} \int_{t}^{t+T} u \, dt \qquad \text{where } u = u(t) = \bar{u} \pm u' (t),$$

u' being the velocity fluctuation from mean, with time t; such velocities are called temporal mean velocities.

In steady flow, the streamline has a fixed direction at every point and is therefore fixed in space. A particle always moves tangentially to the streamline and hence in steady flow the path of a particle is a streamline.

3.3 Uniform and non-uniform flows

A flow is uniform if its characteristics at any given instant remain the same at different points in the direction of flow; otherwise it is termed as non-uniform flow. Mathematically this can be expressed as:

$$(\partial V/\partial s)_{t_0} = 0 \text{ for uniform flow} \tag{3.2a}$$

$$\text{and } (\partial V/\partial s)_{t_0} \neq 0 \text{ for non-uniform flow} \tag{3.2b}$$

The flow through a long uniform pipe at a constant rate is steady uniform flow and at a varying rate is unsteady uniform flow. Flow through a diverging pipe at a constant rate is steady non-uniform flow and at a varying rate is unsteady non-uniform flow.

3.4 Rotational and irrotational flows

If the fluid particles within a flow have rotation about any axis, the flow is called rotational and if they do not suffer rotation, the flow is in irrotational motion. The non-uniform velocity distribution of real fluids close to a boundary causes particles to deform with a small degree of rotation whereas, the flow is irrotational if the velocity distribution is uniform across a section of the flow field.

3.5 One, two and three dimensional flows

The velocity component transverse to the main flow direction is neglected in one dimensional flow analysis. Flow through a pipe may usually be characterised as one dimensional. In two dimensional flow, the velocity vector is a function of two co-ordinates and the flow conditions in a straight, wide river may be considered as two dimensional. Three dimensional flow is the most general type of flow in which the velocity vector varies with space and is generally complex.

Thus in terms of the velocity vector $V(s,t)$, we can write:

$$V = f(x,t) \quad - \text{ one dimensional flow} \tag{3.3a}$$

$$V = f(x,y,t) \quad - \text{ two dimensional flow} \tag{3.3b}$$

$$V = f(x,y,z,t) - \text{ three dimensional flow} \tag{3.3c}$$

3.6 Streamtube and continuity equation

A streamtube consists of a group of streamlines whose bounding surface is made up of these several streamlines. Since the velocity at any point along a streamline is tangential to it, there can be no flow across the surface of a streamtube and therefore, the streamtube surface behaves like a boundary of a pipe across which there is no flow. This concept of the streamtube is very useful in deriving the continuity equation.

Considering an elemental streamtube of the flow (fig. 3.2), we can state:

mass entering the tube/second = mass leaving the tube/second

since there is no mass flow across the tube (principle of mass conservation).

$$\therefore \rho_1 V_1 dA_1 = \rho_2 V_2 dA_2$$

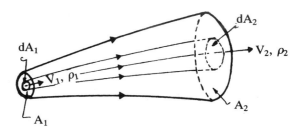

Figure 3.2 Streamtube

where V_1 and V_2 are the steady average velocities at the entrance and exit of the elementary streamtube of cross-sectional areas dA_1 and dA_2 and ρ_1 and ρ_2 are the corresponding densities of entering and leaving fluid.

Therefore, for a collection of such streamtubes along the flow:

$$\bar{\rho}_1 \bar{V}_1 A_1 = \bar{\rho}_2 \bar{V}_2 A_2 \tag{3.4}$$

where $\bar{\rho}_1$ and $\bar{\rho}_2$ are the average densities of fluid at the entrance and exit, and $\bar{V}_1$ and $\bar{V}_2$ are the average velocities over the entire entrance and exit sections of areas A_1 and A_2 of the flow tube.

For incompressible steady flow, equation 3.4 reduces to the one dimensional continuity equation:

$$A_1 \bar{V}_1 = A_2 \bar{V}_2 = Q \tag{3.5}$$

and Q is the volumetric rate of flow called discharge, expressed in m³/s ($:L^3T^{-1}$), often referred to as cumecs.

3.7 Accelerations of fluid particles

In general, the velocity vector V of a flow field is a function of space and time, written as:

$$V = f(s,t)$$

which shows that the fluid particles experience accelerations due to (a) change in velocity in space (convective acceleration), and (b) change in velocity in time (local or temporal acceleration).

(a) Tangential acceleration

If V_s, in the direction of motion $= f(s,t)$

$$dV_s = \frac{\partial V_s}{\partial s} ds + \frac{\partial V_s}{\partial t} dt$$

$$\text{or } dV_s/dt = \frac{ds}{dt}\frac{\partial V_s}{\partial s} + \frac{\partial V_s}{\partial t}$$

$$= V_s \frac{\partial V_s}{\partial s} + \frac{\partial V_s}{\partial t} \tag{3.6}$$

dV_s/dt being the total tangential acceleration equal to the sum of tangential convective and tangential local accelerations.

(b) Normal acceleration

The velocity vectors of the particles negotiating curved paths (fig. 3.3a) may experience both change in direction and magnitude.

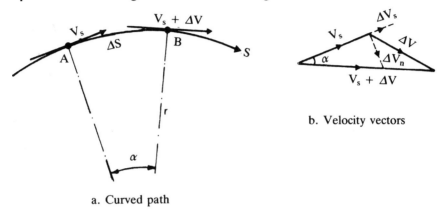

b. Velocity vectors

a. Curved path

Figure 3.3 Curved motion

Along a flow line of radius of curvature, r, the velocity vector V_s at A changes to $V_s + \Delta V$ at B. The vector change ΔV can be resolved into two components, one along the vector V_s and the other normal to the vector V_s. The tangential change in velocity vector ΔV_s produces tangential convective acceleration whereas the normal component ΔV_n produces normal convective acceleration,

$$\Delta V_n/\Delta t \left(= \frac{\Delta V_n}{\Delta s} \frac{\Delta s}{\Delta t} \right).$$

From similar triangles (fig. 3.3b):

$$\Delta V_n/V_s = \Delta s/r$$

$$\text{or } \Delta V_n/\Delta s = V_s/r$$

∴ The total normal acceleration can now be written as:

$$dV_n/dt = \frac{V_s^2}{r} + \frac{\partial V_n}{\partial t} \tag{3.7}$$

$\partial V_n/\partial t$ being the local normal acceleration.

Examples of streamline patterns and their corresponding types of acceleration in steady flows (:$\partial V/\partial t = 0$) are shown in fig. 3.4.

Fluid flows between straight parallel boundaries (fig. 3.4a) do not experience any kind of accelerations whereas between straight converging (fig. 3.4b) or diverging boundaries the flow suffers tangential convective acceleration or decelerations.

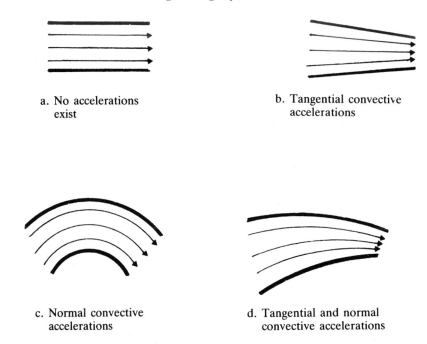

a. No accelerations
   exist

b. Tangential convective
   accelerations

c. Normal convective
   accelerations

d. Tangential and normal
   convective accelerations

Figure 3.4 Streamline patterns and types of acceleration

Flow in a concentric curved bend (fig. 3.4c) experiences normal convective accelerations while in a converging (fig. 3.4d) or diverging bend both tangential and normal convective accelerations or decelerations exist.

3.8 Two kinds of fluid flow

Fluid flow may be classified as laminar or turbulent. In laminar flow, the fluid particles move along smooth layers, one layer gliding over an adjacent layer. Viscous shear stresses dominate in this kind of flow in which the shear stress and velocity distribution are governed by Newton s law of viscosity (equation 1.1). In turbulent flows, which occur most commonly in engineering practice, the fluid particles move in erratic paths causing instantaneous fluctuations in the velocity components. These turbulent fluctuations cause an exchange of momentum setting up additional shear stresses of large magnitudes. An equation of the form similar to Newton's law of viscosity (equation 1.1) may be written for turbulent flow replacing μ by η. The coefficient η, called the eddy viscosity, depends upon the fluid motion and the density.

The type of a flow is identified by the Reynolds number, $R_e = \rho VL/\mu$, where ρ and μ are the density and viscosity of the fluid and V is the flow velocity and L is a characteristic length such as the pipe diameter (D) in the case of a pipe flow. Reynolds number represents the ratio of inertial forces to the viscous forces that exist in the flow field and is dimensionless.

The flow through a pipe is always laminar if the corresponding Reynolds number ($R_c = \rho VD/\mu$) is less than 2000 and for all practical purposes the flow may be assumed to pass through a transition to full turbulent flow in the range of Reynolds numbers from 2000 to 4000.

3.9 Dynamics of fluid flow

The study of fluid dynamics deals with the forces responsible for fluid motion and the resulting accelerations. A fluid in motion experiences, in addition to gravity, pressure forces, viscous and turbulent shear resistances, boundary resistance, and forces due to surface tension and compressibility effects of the fluid. The presence of such a complex system of forces in real fluid flow problems makes the analysis very complicated.

However, a simplifying approach to the problem may be made by assuming the fluid to be ideal or perfect i.e. non-viscous or frictionless and incompressible. Water has a relatively low viscosity and is practically incompressible and is found to behave like an ideal fluid. The study of ideal fluid motion is a valuable background information to encounter the problems of civil engineering hydraulics.

3.10 Energy equation for an ideal fluid flow

Consider an elemental streamtube in motion along a streamline (fig. 3.5) of an ideal fluid flow.

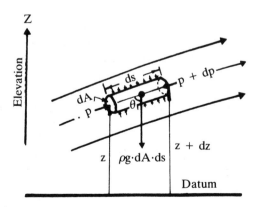

Figure 3.5 Euler's equation of motion

The forces responsible for its motion are the pressure forces, gravity and accelerating force due to change in velocity along the streamline. All frictional forces are assumed to be zero and the flow is irrotational i.e. uniform velocity distribution across streamlines.

By Newton's second law of motion along the streamline (Force = mass × acceleration):

$$p \, dA - (p + dp) \, dA - \rho g \, dA \, ds \cos \theta = \rho \, dA \, ds \, \frac{dV}{dt}$$

or $$- dp - \rho g \, ds \cos \theta = \rho \, ds \, \frac{dV}{dt}$$

The tangential acceleration (along streamline) for steady flow,

$$\frac{dV}{dt} = V \frac{dV}{ds} \quad \text{(equation 3.6)}$$

and $\cos \theta = dz/ds$ (fig. 3.5)

$$\therefore \; - dp - \rho g \, dz = \rho V \, dV$$

or $dz + dp/\rho g + d(V)^2/2g = 0$ (3.8)

Equation 3.8 is the Euler equation of motion applicable to steady state, irrotational flow of an ideal and incompressible fluid.

On integration along the streamline, we get:

$$z + \frac{p}{\rho g} + \frac{V^2}{2g} = \text{Constant} \tag{3.9a}$$

$$\text{or } z_1 + \frac{p_1}{\rho g} + \frac{V_1^2}{2g} = z_2 + \frac{p_2}{\rho g} + \frac{V_2^2}{2g} \tag{3.9b}$$

The three terms on the left-hand side of equation 3.9a have the dimension of length and the sum can be interpreted as the total energy of a fluid element of unit weight. For this reason equation 3.9b, known as Bernoulli's equation, is sometimes called the energy equation for steady ideal fluid flow along a streamline between two sections 1 and 2.

Bernoulli's theorem states that the total energy at all points along a steady continuous streamline of an ideal incompressible fluid flow is constant and is written as:

$$z + \frac{p}{\rho g} + \frac{V^2}{2g} = \text{Constant}$$

where z = elevation, p = pressure and V = average (uniform) velocity of the fluid at a point in the flow under consideration.

The first term, z, is the elevation or potential energy per unit weight of fluid with respect to an arbitrary datum, z N m/N (or metres) of the fluid, called elevation or potential head. The second term, $p/\rho g$, represents the work done in pushing a body of fluid by fluid pressure and is known as pressure energy per unit weight of fluid. The work done over a volume $\dot{V}$ is $p\,\dot{V}$ and $\dot{V} = W/\rho g$, where W is the corresponding weight of fluid, giving the pressure energy per unit weight as $p/\rho g$ N m/N (or metres) of the fluid, called pressure head. The third term, $V^2/2g$, is the kinetic energy per unit weight of fluid (K.E. $= \frac{1}{2}mV^2$ and mass m $= W/g$) in N m/N (or metres) of the fluid, known as velocity head.

The units of the total energy can be written as N m/N of fluid (or metres) of fluid in which case it is known as total head.

3.11 Modified energy equation for real fluid flows

Bernoulli's equation can be modified in the case of real incompressible fluid flow (i) by introducing a loss term in the equation 3.9b which would take into account the energy expended in overcoming the frictional resistances caused by viscous and turbulent shear stresses and other resistances due to changes of section, valves, fittings, etc., and (ii) by correcting the velocity energy term for true velocity distribution. The frictional losses depend upon the type of flow; in a laminar pipe flow they vary directly with the viscosity, the length and the velocity and inversely with the square of the diameter whereas in turbulent flow they vary directly with the length, square of the velocity and inversely with the diameter. The turbulent losses also depend upon the roughness of the interior surface of the pipe wall and the fluid properties of density and viscosity.

Therefore, for real incompressible fluid flow, we can write:

$$z_1 + \frac{p_1}{\rho g} + \frac{\alpha_1 V_1^2}{2g} = z_1 + \frac{p_2}{\rho g} + \frac{\alpha_2 V_2^2}{2g} + \text{Losses} \tag{3.10a}$$

where α is the velocity (kinetic) energy correction factor.

Note (a) A general energy equation from the principles of conservation of energy can be derived for a fluid flow taking into account the mass, momentum and heat transfer and the thermal energy due to friction in real fluid. For a steady flow situation between two sections of a flow field, an energy equation of the form

$$z_1 + p_1/\rho_1 g + \alpha_1 V_1^2/2g + E_m =$$
$$z_2 + p_2/\rho_2 g + \alpha_2 V_2^2/2g + J[(I_2 - I_1) + q] \tag{3.10b}$$

can be written where E_m is the external energy supplied by some machine, I is the internal energy, q is the heat energy transferred to the surroundings of the fluid and J is the mechanical equivalent of heat; this equation reduces to equation 3.10a in the case of a real incompressible fluid flow without the supply of external energy, the loss term being $J[(I_2 - I_1) + q]$. Thus the Bernoulli's equation is a specific case of energy equation.

Note (b) Veloctiy energy correction factor, α:

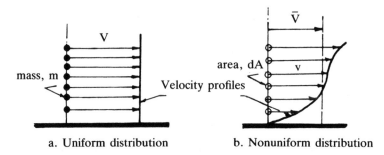

a. Uniform distribution   b. Nonuniform distribution

Figure 3.6   Velocity (kinetic) energy correction factor

Total kinetic energy over the section =

Σ kinetic energies of individual particles of mass, m.

In the case of uniform velocity distribution (fig. 3.6a), each particle moves with a velocity V and its kinetic energy is $\frac{1}{2}mV^2$.

$\therefore$ Total kinetic energy at the section $= \dfrac{1}{2}$ (m + m + m + ...) V^2

$$= \frac{1}{2} (W/g) V^2$$

$$= V^2/2g \text{ per unit weight of fluid.}$$

In the case of non-uniform velocity distribution (fig. 3.6b), the particles move with different velocities.

Mass of individual elements passing through an elementary area dA

$= \rho \, dA \, v$

$\therefore$ Kinetic energy of individual mass element

$= \dfrac{1}{2} \rho \, dA \, v \, v^2$

and hence total kinetic energy over the section

$= \displaystyle\int_A \frac{1}{2} \rho \, v^3 \, dA$

$= \alpha \dfrac{1}{2} \rho \, A\bar{V} \, \bar{V}^2$ or $\alpha \dfrac{1}{2} \rho \, A\bar{V}^3$, $\bar{V}$ being the average velocity at the section.

$\therefore \alpha = \dfrac{1}{A} \displaystyle\int_A (v/\bar{V})^3 \, dA$ \hfill (3.11)

(For turbulent flows α lies between 1.03 and 1.3 and for laminar flows α is 2·0.)

α is commonly referred to as the Coriolis coefficient.

3.12 Separation and cavitation in fluid flow

Consider a rising main (fig. 3.7a) of uniform pipeline.

At any point, by Bernoulli's equation:

Total energy $= z + p/\rho g + v^2/2g = $ Constant

For a given discharge the velocity is the same at all sections (uniform diameter) and hence we have: $z + p/\rho g = $ Constant.

As the elevation z increases, the pressure p in the system decreases and if p becomes vapour pressure of the fluid, the fluid tends to boil liberating dissolved gases and air bubbles. With further liberation of gases the bubbles tend to grow in size eventually blocking the pipe section thus allowing the discharge to take place intermittently. This phenomenon is known as separation and greatly reduces the efficiency of the system.

If the tiny air bubbles formed at the separation point are carried to a high pressure region (fig. 3.7b) by the flowing fluid, they collapse extremely abruptly or implode producing a violent hammering action on any boundary surface on which the imploding bubbles come in contact and cause pitting and vibration to the system which is highly undesirable. The whole phenomenon is called cavitation and should be avoided while designing any hydraulic system.

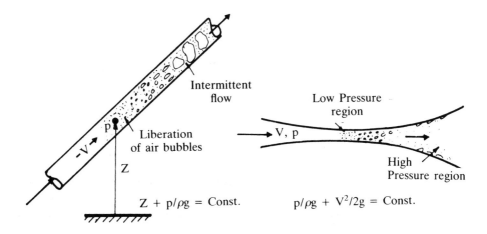

a. Rising main of uniform diameter   b. Horizontal converging-diverging pipe

Figure 3.7 Separation and cavitation phenomena

3.13 Impulse-momentum equation

Momentum of a body is the product of its mass and velocity (kg m/s: MLT^{-1}) and Newton's second law of motion states that the resultant external force acting on any body in any direction is equal to the rate of change of momentum of the body in that direction.

In x direction, this can be expressed as:

$$F_x = \frac{d}{dt}(M_x)$$

$$\text{or } F_x dt = d(M_x) \tag{3.12}$$

Equation 3.12 is known as implulse-momentum equation and can be written as:

$$F_x dt = m \, dv_x \tag{3.13}$$

where m is the mass of the body and dv is the change in velocity in the direction considered; F dt is called the impulse of applied force F.

Momentum correction factor (β)

In the case of non-uniform velocity distribution (fig. 3.6b), the particles move with different velocities across a section of the flow field.

$\therefore$ Total momentum of the flow = Σ momenta of individual elements of

mass, m and can be written as : $\int_A \rho \, dA \, v \, v = \beta \rho \, A \, \bar{V} \, \bar{V}$

where $\bar{V}$ is the average velocity at the section.

$$\therefore \beta = \frac{1}{A} \int_A \left(\frac{v}{\bar{V}}\right)^2 dA \tag{3.14}$$

(For turbulent flows β is seldom greater than 1.1 and for laminar flows β is 1·33.)

β is commonly referred to as the Boussinesq coefficient.

3.14 Energy losses in sudden transitions

Flow through a sudden expansion experiences separation from the boundary to some length downstream of the flow. In these regions of separation turbulent eddies form with a consequence of pressure loss dissipating in the form of heat energy.

Referring to fig. 3.8a (the pressure against the angular area $A_2 - A_1$ is experimentally found to be the same as the pressure p_1, just before the entrance).

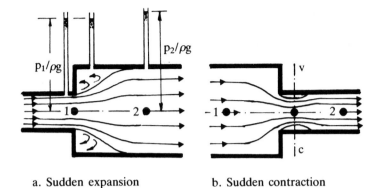

a. Sudden expansion   b. Sudden contraction

Figure 3.8   Energy losses in sudden transitions

Energy equation: $p_1/\rho g + V_1^2/2g = p_2/\rho g + V_2^2/2g + loss$ (i)

Momentum equation: Net force on the control volume between 1 and 2
= rate of change of momentum

$p_1 A_1 + p_1 (A_2 - A_1) - p_2 A_2 = \rho Q (V_2 - V_1)$ (ii)

Continuity equation: $A_1 V_1 = A_2 V_2 = Q$ (iii)

∴ The head or energy loss between 1 and 2 (from equations (i), (ii) and (iii)),

$h_L = (V_1 - V_2)^2/2g$ (3.15)

Referring to fig. 3.8b, the head loss is mainly due to sudden enlargement of flow from vena-contracta to section 2 and therefore, the contraction loss can be written as (from equation 3.15):

$h_L = (V_c - V_2)^2/2g$ (iv)

where V_c is velocity at vena-contracta v − c.

By continuity, $A_c V_c = A_2 V_2 = Q$

∴ $V_c = (A_2/A_c) V_2 = V_2/C_c$

where C_c is the coefficient of contraction (= A_c/A_2).
∴ Equation (iv) reduces to:

$h_L = (1/C_c - 1)^2 V_2^2/2g = k V_2^2/2g$ (3.16)

where k is a function of the contraction ratio A_2/A_1.

3.15 Flow measurement through pipes

Application of continuity, energy and momentum equations to a given system of fluid flow makes velocity and volume measurements possible.

(a) Venturi meter and orifice meter

A pressure differential is created along the flow by providing either a gradual (venturi meter) or sudden (orifice plate meter) constriction in the pipeline, and is related to flow velocities and discharge by the energy and continuity principles. (See fig. 3.9.)

Bernoulli's equation between inlet section and constriction:

$$p_1/\rho g + v_1^2/2g = p_2/\rho g + v_2^2/2g \text{ neglecting losses}$$

$$\therefore \frac{p_1 - p_2}{\rho g} = h = \frac{v_1^2 - v_2^2}{2g} \tag{i}$$

Continuity equation gives: $a_1 v_1 = a_2 v_2 = Q$ (ii)

From (i) and (ii)

$$v_1 = a_2 \sqrt{\frac{2gh}{a_1^2 - a_2^2}}$$

$$\text{and} \therefore Q = a_1 v_1 = \frac{a_1 a_2 \sqrt{2gh}}{\sqrt{a_1^2 - a_2^2}} \tag{3.17}$$

$$\text{or } Q = a_1 \sqrt{\frac{2gh}{k^2 - 1}} \text{ where } k = a_1/a_2 \tag{3.18}$$

Equation 3.18 is an ideal equation obtained by neglecting all losses.

The actual discharge is, therefore, written by introducing a coefficient C_d in equation 3.18

$$\text{Discharge, } Q = C_d a_1 \sqrt{\frac{2gh}{k^2 - 1}} \tag{3.19}$$

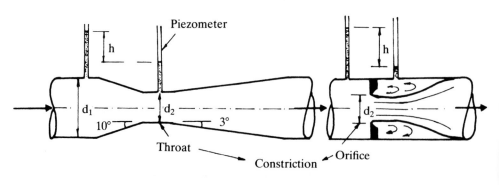

a. Venturi meter   b. Orifice meter

Figure 3.9 Discharge measurement through pipes

The numerical value of C_d, the coefficient of discharge, will depend upon the ratio a_1/a_2, type of transition, velocity and viscosity of the flowing fluid.

The gradual transitions of the venturi meter (fig. 3.9a) between its inlet and outlet induce least amount of losses and the value of its C_d lies between 0.96 and 0.99 for turbulent flows.

The transition in the case of an orifice plate meter (fig. 3.9b) is sudden and hence the flow within the meter experiences greater losses due to contraction and expansion of the jet through the orifice. Its discharge coefficient has a much lower value (0.6 to 0.63) as the area a_2 in equation 3.17 refers to the orifice and not to the contracted jet.

The reduction in the constriction diameter causes velocity to increase, and correspondingly a large pressure differential is created between inlet and constriction, thus enabling greater accuracy in its measurement. High velocities at the constriction cause low pressures in the system and if these fall below the vapour pressure limit of the fluid, cavitation sets in which is highly undesirable. Therefore, the selection of the ratio d_2/d_1 is to be considered carefully. This ratio may be kept between $\frac{1}{3}$ and $\frac{3}{4}$ and a more common value is $\frac{1}{2}$.

(b) Pitot tube
A Pitot tube in its simplest form is an L-shaped tube held against the flow as shown in fig. 3.10, creating a stagnation point in the flow.

The stagnation pressure at point 2 (velocity is zero),

$$p_2/\rho g = p_1/\rho g + v^2/2g \text{ by Bernoulli's equation.}$$

∴ h, the rise in water level or pressure differential between 1 and 2

$$(p_2 - p_1)/\rho g = v^2/2g$$

or v, the velocity $= \sqrt{2gh}$ \hfill (3.20)

The actual velocity will be slightly less than the velocity given by equation 3.20 and it is modified by introducing a coefficient, K (usually between 0·95 and 1·0) as:

$$v = K \sqrt{2gh} \hfill (3.21)$$

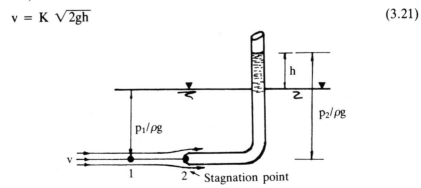

Figure 3.10 Pitot tube

3.16 Flow measurement through orifices and mouthpieces

(a) Small orifice

If the head, h, causing flow through an orifice of diameter, d, is constant (small orifice: h $\gg$ d) as shown (fig. 3.11), by Bernoulli's equation:

$$h + p_1/\rho g + v_1{}^2/2g = 0 + p_2/\rho g + v_2{}^2/2g + \text{losses}$$

With $p_1 = p_2$ (both atmospheric), assuming $v_1 \simeq 0$ and ignoring losses we get

$$v_2{}^2/2g = h$$

or the velocity through the orifice, $v_2 = \sqrt{2gh}$ (3.22)

Equation 3.22 is called Torricelli's theorem and the velocity is called the theoretical velocity.

The actual velocity $= C_v \sqrt{2gh}$ where C_v is the coefficient of velocity defined as:

$$C_v = \text{actual velocity/theoretical velocity} \qquad (3.23)$$

The jet area is much less than the area of the orifice due to contraction and the corresponding coefficient of contraction is defined as:

$$C_c = \text{area of jet/area of orifice, a} \qquad (3.24)$$

At a section very close to the orifice, known as the vena-contracta, the velocity is normal to the cross-section of the jet and hence the discharge can be written as:

$$Q = \text{area of jet} \times \text{velocity of jet (at vena-contracta)}$$

$$= C_c \, a \times C_v \, \sqrt{2gh}$$

$$= C_d \, a \, \sqrt{2gh} \qquad (3.25)$$

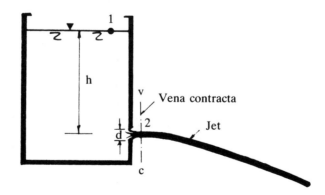

Figure 3.11   Small orifice (h $\gg$ d)

where C_d is called the coefficient of discharge and defined as:

C_d = actual discharge/theoretical discharge, a $\sqrt{2gh}$

$= C_c C_v$ (3.26)

Some typical orifices and mouthpieces (short pipe lengths attached to orifice) and their coefficients, C_c, C_v and C_d are shown in fig. 3.12.

(b) Large rectangular orifice (see fig. 3.13)
As the orifice is large the velocity across the jet is no longer constant; however, if we consider a small area, b dh, at a depth, h,

the velocity through this area = $\sqrt{2gh}$ (equation 3.22)

∴ The actual discharge through the strip area,

dq = C_d × area of strip × velocity through the strip

   = C_d b dh $\sqrt{2gh}$

∴ Total discharge through the entire opening, (h from H_2 to H_1)

$$Q = \int dq = C_d \, b \, \sqrt{2g} \int_{H_2}^{H_1} h^{1/2} \, dh$$

$$= \frac{2}{3} C_d \, \sqrt{2g} \, b \, (H_1^{3/2} - H_2^{3/2}) \qquad (3.27)$$

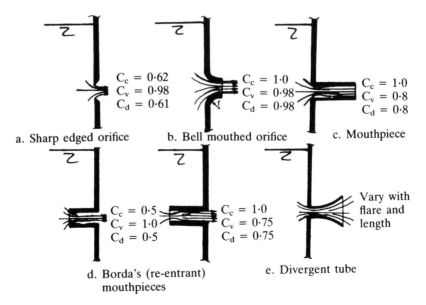

a. Sharp edged orifice   b. Bell mouthed orifice   c. Mouthpiece

$C_c = 0.62$
$C_v = 0.98$
$C_d = 0.61$

$C_c = 1.0$
$C_v = 0.98$
$C_d = 0.98$

$C_c = 1.0$
$C_v = 0.8$
$C_d = 0.8$

d. Borda's (re-entrant) mouthpieces   e. Divergent tube

$C_c = 0.5$
$C_v = 1.0$
$C_d = 0.5$

$C_c = 1.0$
$C_v = 0.75$
$C_d = 0.75$

Vary with flare and length

Figure 3.12 Hydraulic coefficients for some typical orifices and mouthpieces

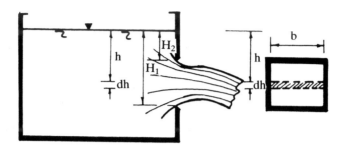

Figure 3.13 Large rectangular orifice

Modification of equation 3.27

(a) If V_a is the velocity of approach, the head responsible for the strip velocity is $h + \alpha\,V_a^2/2g$ (fig. 3.14a) and hence the strip velocity is $\sqrt{2g(h + \alpha\,V_a^2/2g)}$, α being the kinetic energy correction factor (Coriolis coefficient).

∴ Discharge through the strip, $dq = C_d\, b\, dh\,\sqrt{2g(h + \alpha\,V_a^2/2g)}$

and the total discharge,

$$Q = \int dq = \frac{2}{3} C_d\,\sqrt{2g}\ b\ [(H_1 + \alpha\,V_a^2/2g)^{3/2} - (H_2 + \alpha\,V_a^2/2g)^{3/2}]$$

(3.28)

(b) Side wall of the tank inclined at an angle β (see fig. 3.14b):

The effective strip area $= b\,dh/\cos\beta$

$$\therefore \text{Discharge, } Q = \int dq = \int_{H_2}^{H_1} C_d\,\frac{b\,dh}{\cos\beta}\,\sqrt{2g(h + \alpha\,V_a^2/2g)}$$

$$= \frac{2}{3} C_d\,\sqrt{2g}\,\frac{b}{\cos\beta}\,[(H_1 + \alpha\,V_a^2/2g)^{3/2} -$$
$$(H_2 + \alpha\,V_a^2/2g)^{3/2}]\qquad(3.29)$$

(c) Submerged orifice (see fig. 3.14c):
It can be shown by the Bernoulli's equation that the velocity across the jet is constant and equal to $\sqrt{2gH}$ or $\sqrt{2g(H + \alpha\,V_a^2/2g)}$ if V_a is considered.

∴ The discharge through a submerged orifice,

$$Q = C_d \times \text{area of orifice} \times \text{velocity}$$

$$= C_d\,A\,\sqrt{2g(H + \alpha\,V_a^2/2g)}\qquad(3.30)$$

Note: Discharge under varying head: Since the head causing flow is varying, the discharge through the orifice varies with time.

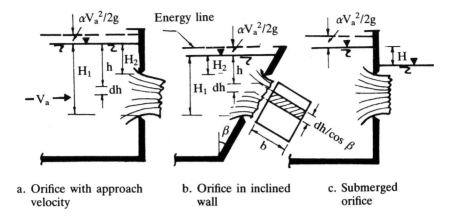

a. Orifice with approach
velocity

b. Orifice in inclined
wall

c. Submerged
orifice

Figure 3.14 Velocity of approach — large rectangular orifice

If h is the head at any instant t (see fig. 3.15), the velocity through the orifice at that instant $= \sqrt{2gh}$

Let the water level drop down by a small amount, dh, in a time dt.

We can write: Volume reduced = Volume escaped through the orifice

$$- A\, dh = C_d\, a\, \sqrt{2gh}\, dt \quad \text{(dh is negative)}$$

$$\therefore\ dt = -\ \frac{A}{C_d\, a\, \sqrt{2g}}\ \frac{dh}{h^{1/2}} \tag{3.31}$$

$\therefore$ Time taken to lower the water level from H_1 to H_2:

$$T = \int dt = \frac{2A(H_1^{1/2} - H_2^{1/2})}{C_d\, a\, \sqrt{2g}} \tag{3.32}$$

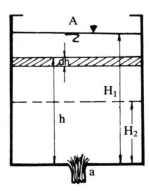

Figure 3.15 Time of emptying a tank

3.17 Flow measurement in channels

Notches and weirs are regular obstructions placed across open streams over which the flow takes place. The head over the sill of such an obstruction is related to the discharge through energy principles. A weir or a notch may be regarded as a special form of large orifice with the free water surface below its upper edge. Thus equation 3.27 with $H_2 = 0$, for example, gives the discharge through a rectangular notch. In general, the discharge over such structures can be written as

$$Q = K H^n \tag{3.33}$$

where K and n depend on the geometry of notch.

(a) Rectangular notch

Considering a small strip area of the notch at a depth, h, below free water surface (see fig. 3.16), the total head responsible for the flow is written as:

$h + \alpha V_a^2/2g$, α being the energy correction factor.

∴ The velocity through the strip $= \sqrt{2g(h + \alpha V_a^2/2g)}$

and discharge, $dq = C_d b\, dh \sqrt{2g(h + \alpha V_a^2/2g)}$

∴ The total discharge,

$$Q = \int dq = C_d \sqrt{2g}\, b \int_0^H (h + \alpha V_a^2/2g)^{1/2}\, dh$$

$$= \frac{2}{3} C_d \sqrt{2g}\, b\, [(H + \alpha V_a^2/2g)^{3/2} - (\alpha V_a^2/2g)^{3/2}] \tag{3.34}$$

The discharge coefficient C_d largely depends upon the shape, contraction of the nappe, sill height, head causing flow, sill thickness, etc.

As the effective width of the notch for the flow is reduced by the presence

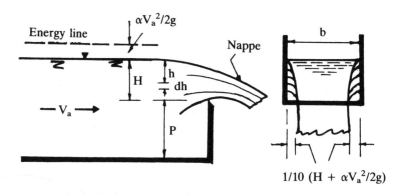

Figure 3.16 Rectangular notch with end contractions

of end contractions, each contraction being one-tenth of the total head (experimental result), equation 3.34 is modified as: (in S.I. units)

$$Q = 1 \cdot 84 \, [b - 0 \cdot 1n \, (H + \alpha \, V_a^2/2g)] \, [(H + \alpha \, V_a^2/2g)^{3/2} - (\alpha \, V_a^2/2g)^{3/2}]$$

(3.35)

taking an average value of $C_d = 0 \cdot 623$. This is known as Francis formula, n being the number of end contractions.

Note: (a) Bazin formula:

$$Q = (0 \cdot 405 + 0 \cdot 003/H_1) \, \sqrt{2g} \, b \, (H_1)^{3/2}$$

(3.36)

where $H_1 = H + 1 \cdot 6 \, (V_a^2/2g)$

   (b) Rehbock formula:

$$Q = [1 \cdot 78 + 0 \cdot 245(H_e/P)] \, b(H_e)^{3/2}$$

(3.37)

where $H_e = H + 0 \cdot 0012$ m, and P the height of sill, the coefficient of discharge, C_d being:

$$C_d = 0 \cdot 602 + 0 \cdot 083 \, (H_e/P)$$

(3.38)

(b) Triangular or V-notch
A similar approach to determine the discharge over a triangular notch of an included angle θ, results in the equation

$$Q = \frac{8}{15} C_d \, \sqrt{2g} \, \tan \frac{\theta}{2} \left[\left(H + \frac{\alpha \, V_a^2}{2g} \right)^{5/2} - \left(\frac{\alpha \, V_a^2}{2g} \right)^{5/2} \right]$$

(3.39a)

If the approach velocity V_a is neglected, equation 3.39a reduces to

$$Q = \frac{8}{15} C_d \, \sqrt{2g} \, \tan \frac{\theta}{2} \, H^{5/2}$$

(3.39b)

(c) Cipolletti weir
This is a trapezoidal weir with 14° side slopes (1 horizontal : 4 vertical). The discharge over such a weir may be computed by using the formula for a suppressed (no end contractions) rectangular weir with equal sill width.

A trapezoidal notch may be considered as one rectangular notch of width b and two half V-notches (apex angle $\frac{1}{2} \theta$) and the discharge equation

written as: $Q = \frac{2}{3} C_d \, \sqrt{2g} \, (b - 0 \cdot 2H) \, H^{3/2} + \frac{8}{15} C_d \, \sqrt{2g} \, \tan \frac{1}{2} \theta \, H^{5/2}$

(3.40)

Equation 3.40 reduces to that for a suppressed rectangular weir (weir with no end contractions) if the reduction in discharge due to the presence of end contractions is compensated by the increase provided by the presence of two half V-notches.

$$\therefore \text{ We can write: } \frac{2}{3} C_d \sqrt{2g} \times 0.2H \times H^{3/2} = \frac{8}{15} C_d \sqrt{2g} \tan \frac{1}{2} \theta H^{5/2}$$

Assuming C_d constant throughout, we get

$$\tan \frac{1}{2} \theta = \frac{1}{4} \text{ or } \frac{1}{2} \theta = 14°2'$$

(d) Proportional or Sutro weir

In general, the discharge through any type of weir may be expressed as $Q \propto H^n$. A weir with $n = 1$, i.e. the discharge is proportional to the head over the weir's crest, is called a proportional weir (fig. 3.17).

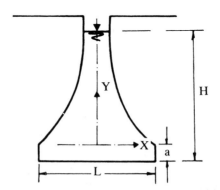

Figure 3.17   Sutro or proportional weir

Sutro's analytical approach resulted in the relationship,

$$x \propto y^{-1/2}$$

for the proportional weir profile and to overcome the practical limitation (as $y \rightarrow 0$, $x \rightarrow \infty$) he proposed the weir shape in the form of hyperbolic curves of the equation:

$$\frac{2x}{L} = \left[1 - \frac{2}{\pi} \tan^{-1} \sqrt{y/a}\right] \tag{3.31}$$

where a and L are the height and width of the rectangular aperture forming the base of the weir.

The discharge, $Q = C_d L(2ga)^{1/2} (H - a/3)$ \hfill (3.42)

The proportional weir is a very useful device, for example, in chemical dosing and sampling, irrigation outlets, etc.

(e) Ogee spillway

Excess flood flows behind dams are normally discharged by providing spillways. The profile of an Ogee spillway conforms to the shape of a sharp crested weir (see fig. 3.18) at a design head, H_d. A discharge equation similar to that of the weir, but with a higher discharge coefficient, C_{do}, (since the reference sill level for the spillway is slightly shifted), written in the form

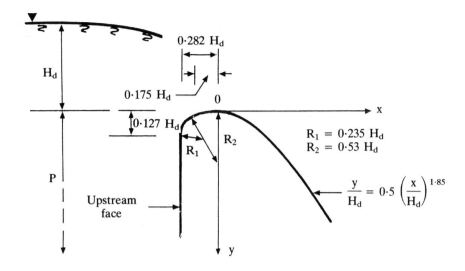

Figure 3.18 Cross-section of an Ogee spillway

$$Q = \frac{2}{3} C_{do} \ B \ \sqrt{2g} \ H_{de}^{3/2} \tag{3.43}$$

is applicable, in which $H_{de} = H_d + V_a^2/2g$, V_a being the velocity of approach. For spillways of $P/H_{de} \geq 3$ the value of $C_{do} \approx 0{\cdot}75$; fig. 3.19 shows the variation of C_{do} with P/H_{de}. For heads other than the design head the discharge coefficient varies as the underside of the nappe no longer conforms to the spillway profile; fig. 3.20 shows the variation of C_d/C_{do} with H_e/H_{de}, H_e being any other energy head with a corresponding discharge coefficient, C_d. For larger values of P/H the approach velocity, V_a, may be negligible leading to $H_e \approx H$.

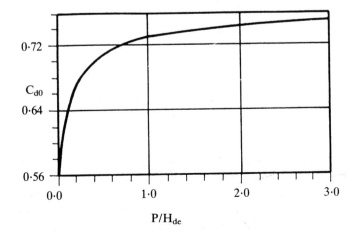

Figure 3.19 Variation of C_{do} with P/H_{de}

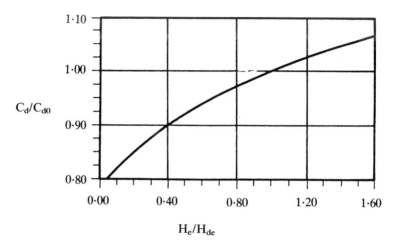

Figure 3.20 Variation of C_d/C_{d0} with H_e/H_{de}

(f) Other forms of flow measuring devices
Open channel flows may also be measured by broad crested weir and venturi
flume (see Chapter 8) and some special structures like Crump weir (see
Novak *et al.*[7], for example).

Effect of submergence of flow measuring structures

If the water level (H_2) downstream of a measuring device is below the sill
level, the discharge is said be modular (free flow, Q_f) and the above equations
are valid to compute the free flows. When the downstream water level is
above the sill level, the structure is said to be drowned and the discharge
(non-modular or drowned flow) is affected, i.e. reduced. The non-modular
flow, Q_s, is given by the equation

$$Q_s = Q_f \left[1 - \left(\frac{H_2}{H_1} \right)^m \right]^{0.385} \tag{3.44}$$

where m is the exponent of H_1 (upstream water level above sill) in the weir
equations; m = 1·5 for rectangular weir and m = 2·5 for triangular weir.

Worked examples

Example 3.1
A pipeline of 300 mm diameter carrying water at an average velocity of
4·5 m/s branches into two pipes of 150 mm and 200 mm diameters. If the

average velocity in the 150 mm pipe is $\frac{5}{8}$ of the velocity in the main pipeline, determine the average velocity of flow in the 200 mm pipe and the total flow rate in the system in l/s. (See fig. 3.21.)

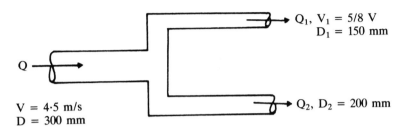

$$Q_1, V_1 = 5/8 \ V$$
$$D_1 = 150 \ mm$$

$$Q_2, D_2 = 200 \ mm$$

$$V = 4\cdot5 \ m/s$$
$$D = 300 \ mm$$

Figure 3.21   Branching pipeline

Solution:

Discharge, $Q = AV = Q_1 + Q_2$ by continuity

$$\therefore \ AV = A_1 V_1 + A_2 V_2$$

$$\frac{1}{4} \pi \ (0\cdot3)^2 \times 4\cdot5 = \frac{1}{4} \pi \ (0\cdot15)^2 \times \frac{5}{8} \times 4\cdot5 + \frac{1}{4} \pi \ (0\cdot2)^2 \times V_2$$

or $V_2 = 8\cdot54$ m/s

and total flow rate, $Q = \frac{1}{4} \pi \ (0\cdot3)^2 \times 4\cdot5$

$$= 0\cdot318 \ m^3/s$$

$$= 318 \ l/s.$$

Example 3.2

A storage reservoir supplies water to a pressure turbine (fig. 3.22) under a head of 20 m. When the turbine draws 500 l/s of water the head loss in the 300 mm diameter supply line amounts to 2·5 m. Determine the pressure intensity at the entrance to the turbine. If a negative pressure of 30 kN/m² exists at the 600 mm diameter section of the draft tube 1·5 m below the supply line, estimate the energy absorbed by the turbine in kW neglecting all frictional losses between the entrance and exit of the turbine. Hence find the output of the turbine assuming an efficiency of 85%.

Solution:

Referring to fig. 3.19, by Bernoulli's equation (between points 1 and 2):

$$z_1 + p_1/\rho g + V_1^2/2g = z_2 + p_2/\rho g + V_2^2/2g + \text{Loss} \qquad \text{(i)}$$

With section 2 as datum equation (i) becomes ($p_1 = 0$ and $V_1 = 0$)

$$20 = p_2/\rho g + V_2^2/2g + 2\cdot5$$

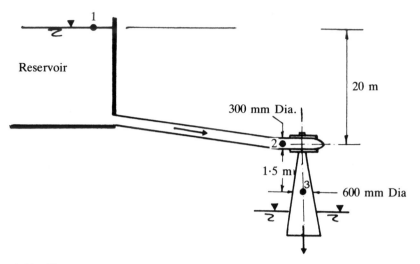

Figure 3.22 Flow through a hydraulic turbine

By the continuity equation $Q = A_2 V_2 = A_3 V_3$

$\therefore$ Average velocities, $V_2 = Q/A_2 = 0.5/\dfrac{\pi}{4}(0.3)^2 = 7.07$ m/s

and $V_3 = Q/A_3 = 0.5/\dfrac{\pi}{4}(0.6)^2 = 1.77$ m/s

$\therefore p_2/\rho g = 20 - 2.5 - (7.07)^2/2g$

$= 14.95$ m of water

or $p_2 = \rho g \times 14.95 = 9.81 \times 14.95$

$= 146.95$ kN/m^2

Between sections 2 and 3 we can write:

$$z_2 + p_2/\rho g + V_2{}^2/2g = z_3 + p_3/\rho g + V_3{}^2/2g + E_t + \text{Losses} \qquad \text{(ii)}$$

where E_t is the energy absorbed by the machine/unit weight of water flowing.
Assuming no losses between 2 and 3, equation (ii) reduces to

$1.5 + 14.95 + (7.07)^2/2g = -30 \times 10^3/\rho g + (1.77)^2/2g + E_t$

$\therefore E_t = 1.5 + 14.95 + 2.55 + 3.06 - 0.16$

$= 21.9$ N m/N

Weight of water flowing through the turbine/s,

$W = \rho g\, Q$

$= 10^3 \times 9.81 \times 0.5$

$= 4.905$ kN/s

∴ Total energy absorbed by the machine

$$= E_t \times W = 21{\cdot}9 \times 4{\cdot}905$$
$$= 107{\cdot}42 \text{ kW}$$

Hence its output = efficiency × input

$$= 0{\cdot}85 \times 107{\cdot}42$$

$$= 91{\cdot}31 \text{ kW.}$$

Example 3.3
A 500 mm diameter vertical water pipeline discharges water through a constriction of 250 mm diameter (fig. 3.23). The pressure difference between the normal and constricted sections of the pipe is measured by an inverted U-tube. Determine (i) the difference in pressure between these two sections when the discharge through the system is 600 l/s, and (ii) the manometer deflection, h, if the inverted U-tube contains air.

Solution:
Discharge, $Q = 600$ l/s $= 0{\cdot}6$ m³/s

$$\therefore V_a = 0{\cdot}6/\frac{\pi}{4} (0{\cdot}5)^2 = 3{\cdot}056 \text{ m/s}$$

$$\text{(by continuity)}$$
$$\text{and } V_b = 0{\cdot}6/\frac{\pi}{4} (0{\cdot}25)^2 = 7{\cdot}54 \text{ m/s}$$

By Bernoulli's equation between aa and bb (assuming no losses):

$$0{\cdot}50 + p_a/\rho g + (3{\cdot}056)^2/2g = 0 + p_b/\rho g + (7{\cdot}54)^2/2g$$
$$\text{or } (p_a - p_b)/\rho g = [(7{\cdot}54)^2 - (3{\cdot}056)^2]/2g - 0{\cdot}50$$

$$= 2{\cdot}42 - 0{\cdot}50$$

$$= 1{\cdot}92 \text{ m of water} \tag{i}$$

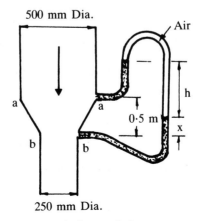

Figure 3.23 **Flow through a vertical constriction**

$$\therefore \ p_a - p_b = 10^3 \times 9 \cdot 81 \times 1 \cdot 92$$

$$= 18 \cdot 8 \ \text{kN/m}^2$$

Manometer equation:

$$p_a/\rho g - (h + x - 0 \cdot 50) + x = p_b/\rho g$$

$$\therefore \ (p_a - p_b)/\rho g = h - 0 \cdot 50 = 1 \cdot 92 \quad \text{from equation (i)}$$

$$\text{or } h = 1 \cdot 92 + 0 \cdot 50$$

$$= 2 \cdot 42 \ \text{m.}$$

Example 3.4

A drainage pump having a tapered suction pipe, discharges water out of a sump. The pipe diameters at the inlet and at the upper end are 1 m and 0·5 m respectively. The free water surface in the sump is 2 m above the centre of the inlet and the pipe is laid at a slope of 1 (vertical) : 4 (along pipeline). The pressure at the top end of the pipe is 0·25 m of mercury below atmosphere and it is known that the loss of head due to friction between the two sections is 1/10 of the velocity head at the top section. Compute the discharge in l/s through the pipe if its length is 20 m. (See fig. 3.24.)

Solution:

By the continuity equation: $Q = a_2 v_2 = a_3 v_3$ ⠀⠀⠀⠀⠀⠀⠀⠀⠀⠀⠀(i)

By Bernoulli's equation between (1), (2) and (3):

$$2 + 0 + 0 = 0 + p_2/\rho g + v_2{}^2/2g = 20 \times \frac{1}{4} + p_3/\rho g + v_3{}^2/2g + (1/10)v_3{}^2/2g$$
⠀⠀⠀⠀⠀⠀⠀⠀⠀⠀⠀⠀⠀⠀⠀⠀⠀⠀⠀⠀⠀⠀⠀⠀⠀⠀⠀⠀⠀⠀⠀⠀⠀⠀⠀⠀⠀(ii)

assuming the velocity in the sump at (1) as zero and a datum through (2).

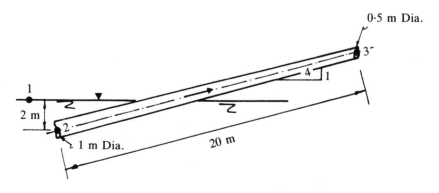

Figure 3.24 Flow through the suction pipe of a pump

The pressure at the top end, $p_3/\rho g$ = 0·25 m of mercury below
atmosphere

$$= -0·25 \times 13·6$$

$$= -3·4 \text{ m of water}$$

$$\therefore \ 1·1 \times v_3^2/2g = 2 - 5 + 3·4 = 0·4$$

$$\text{or } v_3^2/2g = 0·4/1·1 = 0·364 \text{ m}$$

$$v_3 = 2·67 \text{ m/s}$$

Hence discharge, $Q = a_3 \times v_3 = \dfrac{1}{4} \pi \ (0·5)^2 \times 2·67$

$$= 0·524 \text{ m}^3/\text{s}$$

$$= 524 \text{ l/s.}$$

Example 3.5
A jet of water issues out from a fire hydrant nozzle fitted at a height of 3 m
from the ground at an angle of 45° with the horizontal. If the jet under a
particular flow condition strikes the ground at a horizontal distance of 15 m
from the nozzle, find (i) the jet velocity, and (ii) the maximum height the jet
can reach and its horizontal distance from the nozzle. Neglect air resistance.
(See fig. 3.25.)

Solution:
In the horizontal direction, acceleration a = 0

$$\therefore \ V_1 \cos \theta_1 = V \cos \theta = \text{Constant} \tag{i}$$

and in time t, horizontal distance covered $x_1 = V \cos \theta \times t$

$$\text{or } t = x_1/V \cos \theta \tag{ii}$$

and vertical distance $y_1 = V \sin \theta \times t - \dfrac{1}{2} gt^2$ (since a = $-g$)

$$\therefore \ y_1 = V \sin \theta \times \frac{x_1}{V \cos \theta} - \frac{1}{2} g \left(\frac{x_1}{V \cos \theta} \right)^2 \quad \text{from (ii)}$$

$$= x_1 \tan \theta - \frac{1}{2} g \ x_1^2 \sec^2 \theta/V^2 \tag{iii}$$

Co-ordinates of the point where the jet strikes the ground are:

$$y = -3 \text{ m} \quad \text{and} \quad x = 15 \text{ m}$$

$$\therefore \ \text{from (iii) } V = 11·07 \text{ m/s}$$

Highest point P: velocity vector is horizontal and is V cos θ from (i)

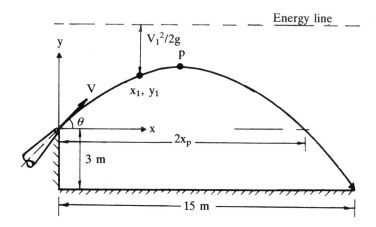

Figure 3.25   Jet dynamics

∴ In the vertical direction, initial velocity $= V \sin \theta$

final velocity $= 0$

giving $0 - V^2 \sin^2 \theta = -2g\, y_{max}$ (since $a = -g$)

or $y_{max} = V^2 \sin^2 \theta / 2g = 3\cdot12$ m

we can also write: $0 = V \sin \theta - g\, t_p$

or $t_p = V \sin \theta / g$ (iv)

and horizontal distance $x_p = V \cos \theta \times t_p$

$= V^2 \sin2\, \theta / 2g$

$= 6\cdot24$ m.   (v)

Note: Total horizontal distance traversed by the jet

$= 2\, x_p = V^2 \sin2\, \theta / g$ (vi)

Example 3.6

A 500 mm diameter siphon pipeline discharges water from a large reservoir.
Determine (i) the maximum possible elevation of its summit for a discharge
of 2·15 m³/s without the pressure becoming less than 20 kN/m² absolute,
and (ii) the corresponding elevation of its discharge end. Take atmospheric
pressure as 1 bar and neglect all losses.

Solution:

Consider the three points, A, B and C along the siphon system as shown in
fig. 3.26.

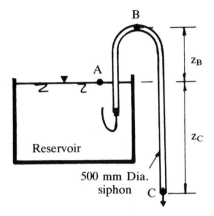

Figure 3.26 Siphon pipeline

Discharge, $Q = av = 2 \cdot 15 \text{ m}^3/\text{s}$

$\therefore$ Velocity, $v = 2 \cdot 15 / \frac{1}{4} \pi (0 \cdot 5)^2 = 10 \cdot 95 \text{ m/s}$

and $v^2/2g = 6 \cdot 11 \text{ m}$

Atmospheric pressure $= 1 \text{ bar} = 10^5 \text{ N/m}^2$

$\qquad\qquad\qquad = \text{pressures at A and C}$

Minimum pressure at $B = 20 \text{ kN/m}^2$ absolute (given)

By Bernoulli's equation between A and B (reservoir water surface as datum):

$$0 + \frac{10^5}{\rho g} + 0 = z_B + \frac{20 \times 10^3}{\rho g} + 6 \cdot 11$$

$$\therefore z_B = 10^5/\rho g - \frac{20 \times 10^3}{\rho g} - 6 \cdot 11 = 2 \cdot 04 \text{ m}$$

Between A and C (with exit end as datum):

$$z_C + p_A/\rho g + 0 = 0 + p_C/\rho g + 6 \cdot 11$$

$$\therefore z_C = 6 \cdot 11 \text{ m} \quad (p_A = p_C = \text{atmospheric pressure})$$

Hence the exit end is to be $6 \cdot 11$ m below the reservoir level.

Example 3.7
A horizontal bend in a pipeline conveying 1 cumec of water gradually reduces from 600 mm to 300 mm in diameter and deflects the flow through an angle of 60°. At the larger end the pressure is 170 kN/m². Determine the magnitude and direction of the force exerted on the bend. Assume $\beta = 1 \cdot 0$. (See fig. 3.27.)

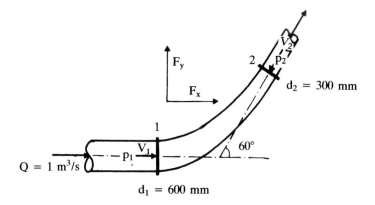

Figure 3.27   Forces on a converging bend

Solution:

Discharge, $Q = 1$ m³/s $= A_1 V_1 = A_2 V_2$: continuity equation

$$\therefore V_1 = 1/\frac{1}{4} \pi (0\cdot6)^2 = 3\cdot54 \text{ m/s}$$

and $V_2 = 1/\frac{1}{4} \pi (0\cdot3)^2 = 14\cdot15$ m/s

Energy equation neglecting friction losses:

$$p_1/\rho g + V_1^2/2g = p_2/\rho g + V_2^2/2g$$

Pressure at 1, $p_1 = 170 \times 10^3$ N/m²

$$\therefore p_2/\rho g = \frac{170 \times 10^3}{10^3 \times 9\cdot81} + \frac{(3\cdot54)^2}{19\cdot62} - \frac{(14\cdot15)^2}{19\cdot62}$$

or $p_2 = 7\cdot62 \times 10^4$ N/m²

Momentum equation: Gravity forces are zero along the horizontal plane and the only forces acting on the fluid mass are pressure and momentum forces.

Let F_x and F_y be the two components of the total force, F, exerted by the bent boundary surface on the fluid mass; these are considered positive if F_x is left to right and F_y upwards.

In x-direction:

$$p_1 A_1 + F_x - p_2 A_2 \cos \theta = \rho Q (V_2 \cos \theta - V_1)$$

and in y-direction:

$$0 + F_y - p_2 A_2 \sin \theta = \rho Q (V_2 \sin \theta - 0)$$

$$\therefore F_x = 10^3 \times 1\ (14\cdot15\cos 60° - 3\cdot54) + 7\cdot62 \times 10^4 \times \frac{1}{4}\pi\ (0\cdot3)^2 \cos 60°$$

$$- 17 \times 10^4 \times \frac{1}{4}\pi\ (0\cdot6)^2$$

$$= -4\cdot2 \times 10^4\ \text{N (negative sign indicates } F_x \text{ is right to left)}$$

$$\text{and } F_y = 10^3 \times 1\ (14\cdot15 \sin 60°) + 7\cdot62 \times 10^4 \times \frac{1}{4}\pi\ (0\cdot3)^2 \sin 60°$$

$$= 1\cdot7 \times 10^4\ \text{N (upwards)}.$$

According to Newton's third law of motion, the forces R_x and R_y exerted by the fluid on the bend will be equal and opposite to F_x and F_y.

$$\therefore R_x = -F_x = 4\cdot2 \times 10^4\ \text{N (left to right)}$$

$$\text{and } R_y = -F_y = -1\cdot7 \times 10^4\ \text{N (downwards)}$$

$\therefore$ Resultant force on the bend,

$$R = \sqrt{R_x^2 + R_y^2}$$

$$= 4\cdot53 \times 10^4\ \text{N} \quad \text{or} \quad 45\cdot3\ \text{kN}$$

acting at an angle, $\theta = \tan^{-1}\ (R_y/R_x)$

$$= 22°\ \text{to the x-direction}.$$

Example 3.8
Derive an expression for the normal force on a plate inclined at $\theta°$ to the jet.

A 150 mm × 150 mm square metal plate, 10 mm thick, is hinged about a horizontal edge. If a 10 mm diameter horizontal jet of water impinging 50 mm below the hinge keeps the plate inclined at 30° to the vertical, find the velocity of the jet. Take the specific weight of the metal as 75 kN/m³.
Referring to fig. 3.28a, force in the normal direction to the plate,

$$F = (\text{mass} \times \text{change in velocity normal to the plate) of jet}$$

$$F = \rho aV\ [V \cos (90 - \theta) - 0]$$

$$= \rho aV^2 \sin \theta\ \text{N}$$

Now referring to fig. 3.28b,

$$F = \rho\ aV^2 \sin 60°$$

$$= 10^3 \times \frac{1}{4}\pi\ (0\cdot01)^2 \times \sin 60° \times V^2 = 6\cdot8 \times 10^{-2}\ V^2\ \text{N}$$

Weight of the plate, $W = 0\cdot150 \times 0\cdot150 \times 0\cdot010 \times 75\,000$

$$= 16\cdot87\ \text{N}$$

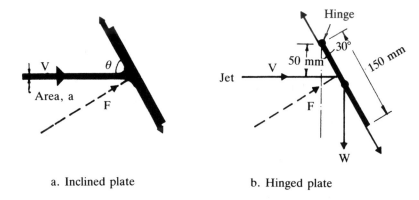

a. Inclined plate   b. Hinged plate

Figure 3.28 Forces on flat plates

Taking moments about the hinge,

$$F \times 50 \sec 30° = W \times 75 \sin 30°$$

$$\text{or } 6{\cdot}8 \times 10^{-2} \, V^2 \times 50 \sec 30° = 16{\cdot}87 \times 75 \sin 30°$$

$$\therefore V = 12{\cdot}7 \text{ m/s}$$

Example 3.9
Estimate the energy (head) loss along a short length of pipe suddenly
enlarging from a diameter of 350 mm to 700 mm and conveying 300 litres
per second of water. If the pressure at the entrance of the flow is 10^5 N/m²,
find the pressure at the exit of the pipe. What would be the energy loss if
the flow were to be reversed with a contraction coefficient of 0·62?

Solution:
Case of sudden expansion:

$$Q = 0{\cdot}3 \text{ m}^3\text{/s} = a_1 v_1 = a_2 v_2$$

$$\therefore v_1 = 3{\cdot}12 \text{ m/s and } v_2 = 0{\cdot}78 \text{ m/s and hence } h_L = (3{\cdot}12 - 0{\cdot}78)^2/2g$$

$$= 0{\cdot}28 \text{ m of water}$$

$$\text{pressure } p_1 = 10^5 \text{ N/m}^2$$

By energy equation:

$$p_1/\rho g + v_1^2/2g = p_2/\rho g + v_2^2/2g + (v_1 - v_2)^2/2g$$

$$10{\cdot}2 + 0{\cdot}5 = p_2/\rho g + 0{\cdot}03 + 0{\cdot}28$$

$$\therefore p_2/\rho g = 10{\cdot}39 \text{ m} \quad \text{or} \quad p_2 = 1{\cdot}02 \times 10^5 \text{ N/m}^2$$

Case of sudden contraction:

$$h_L = (1/C_c - 1)^2 \, v^2/2g \quad \text{where v is the velocity in the smaller pipe}$$
$$= (1/0{\cdot}62 - 1)^2 \, (3{\cdot}12)^2/2g$$
$$= 0{\cdot}186 \text{ m of water.}$$

Example 3.10
A venturi meter is introduced in a 300 mm diameter horizontal pipeline carrying water under a pressure of 150 kN/m^2. The throat diameter of the meter is 100 mm and the pressure at the throat is 400 mm of mercury below atmosphere. If 3% of the differential pressure is lost between inlet and throat, determine the flow rate in the pipeline.

Solution:
Bernoulli's equation between inlet and throat:

$$p_1/\rho g + v_1^2/2g = p_2/\rho g + v_2^2/2g + 0{\cdot}03 \, (p_1/\rho g - p_2/\rho g)$$
$$\therefore \, 0{\cdot}97 \, (p_1 - p_2)/\rho g = (v_2^2 - v_1^2)/2g$$
$$p_1 = 150 \times 10^3 \text{ N/m}^2 = 15{\cdot}29 \text{ m of water}$$
$$p_2 = -400 \text{ mm of mercury} = -0{\cdot}4 \times 13{\cdot}6 \text{ m of water}$$
$$= -5{\cdot}44 \text{ m of water}$$
$$\therefore \, (p_1 - p_2)/\rho g = 15{\cdot}29 - (-5{\cdot}44)$$
$$= 20{\cdot}73 \text{ m}$$

and hence $(v_2^2 - v_1^2)/2g = 0{\cdot}97 \times 20{\cdot}73$
$$= 20{\cdot}11 \text{ m} \tag{i}$$

From continuity equation:

$$a_1 v_1 = a_2 v_2$$
$$\therefore \, v_1 = (a_2/a_1)v_2 = (d_2/d_1)^2 v_2$$
$$= (10/30)^2 v_2 = (1/9)v_2 \tag{ii}$$

From (i) and (ii):

$$v_2^2 \, (1 - 1/81)/2g = 20{\cdot}11$$

$$\text{or } v_2 = \sqrt{\frac{2g \times 20{\cdot}11}{1 - 1/81}} = 19{\cdot}89 \text{ m/s}$$

$$\therefore \, \text{Flow rate, } Q = a_2 v_2 = \frac{1}{4} \pi \, (0{\cdot}1)^2 \times 19{\cdot}98$$

$$= 0{\cdot}157 \text{ m}^3/\text{s or } 157 \text{ l/s.}$$

Example 3.11
A 50 mm × 25 mm venturi meter with a coefficient of discharge of 0·98 is to be replaced by an orifice meter having a coefficient of discharge of 0·6. If both meters are to give the same differential mercury manometer reading for a discharge of 10 l/s, determine the diameter of the orifice.

Solution:
Discharge through venturi meter = discharge through orifice meter

$$k = a_1/a_2 = (50/25)^2 = 4 \text{ for the venturi meter}$$

and k_o for the orifice meter = $(50/d_o)^2$ where d_o is the diameter of orifice

$$\therefore Q = 0·01 = 0·98 \times \frac{1}{4} \pi (0·05)^2 \sqrt{\frac{2gh}{4^2 - 1}}$$

$$= 0·6 \times \frac{1}{4} \pi (0·05)^2 \sqrt{\frac{2gh}{k_o^2 - 1}}$$

$$\text{or } \sqrt{k_o^2 - 1} = \frac{0·6 \sqrt{15}}{0·98}$$

$$\therefore k_o = \left(\frac{50}{d_o}\right)^2 = 2·57 \quad \text{or} \quad d_o = \frac{50}{\sqrt{2·57}} = 31·2 \text{ mm}$$

Example 3.12
A Pitot tube was used to measure the quantity of water flowing in a pipe of 300 mm diameter. The stagnation pressure at the centre line of the pipe is 250 mm of water more than the static pressure. If the mean velocity is 0·78 times the centre line velocity and the coefficient of the Pitot tube is 0·98, find the rate of flow in l/s.

Solution:
The centre line velocity in the pipe, $v = K \sqrt{2gh}$

$$= 0·98 \sqrt{2 \times 9·81 \times 0·25}$$

$$= 2·17 \text{ m/s}$$

$$\therefore \text{ mean velocity of flow} = 0·78 \times 2·17$$

$$= 1·693 \text{ m/s}$$

Hence the discharge, $Q = av$

$$= \frac{1}{4} \pi (0·3)^2 \times 1·693$$

$$= 0·12 \text{ m}^3/\text{s or } 120 \text{ l/s}.$$

Example 3.13
A large rectangular orifice 0·40 m wide and 0·60 m deep placed with the upper edge in a horizontal position 0·90 m vertically below the water surface in a vertical side wall of a large tank, is discharging to atmosphere. Calculate the rate of flow through the orifice if its discharge coefficient is 0·65.

Solution:
The discharge rate when $b = 0·4$ m, $H_1 = 0·90 + 0·60 = 1·5$ m, $H_2 = 0·90$ m and $C_d = 0·65$ from equation 3.27,

$$Q = \frac{2}{3} \times 0·65 \times \sqrt{2g} \times 0·40 \times [(1·5)^{3/2} - (0·9)^{3/2}]$$

$$= 0·755 \text{ m}^3/\text{s}.$$

Example 3.14
A vertical circular tank 1·25 m diameter is fitted with a sharp edged circular orifice 50 mm diameter in its base. When the flow of water into the tank was shut off, the time taken to lower the head from 2 m to 0·75 m was 253 seconds. Determine the rate of flow in l/s, through the orifice under a steady head of 1·5 m.

Solution:

$T = 253$ seconds, $H_1 = 2$m, $H_2 = 0·75$ m, $a = \frac{1}{4} \pi (0·05)^2 = 1·96 \times 10^{-3}$ m^2 and $A = \frac{1}{4} \pi (1·25)^2 = 1·228$ m^2

$\therefore$ From equation 3·32, $C_d = 0·61$

Hence steady discharge under a head of 1·5 m $= C_d a \sqrt{2gH}$

$= 0·61 \times 1·96 \times 10^{-3} \times \sqrt{2g} \times (1·5)^{1/2}$

$= 0·0065$ m^3/s or 6·5 l/s.

Example 3.15
Determine the discharge over a sharp crested weir 4·5 m long with no lateral contractions, the measured head over the crest being 0·45 m. The width of the approach channel is 4·5 m and the sill height of the weir is 1 m.

Solution:
Equation 3.35 is rewritten as:

$$Q = 1·84 \, b \, [(H + V_a^2/2g)^{3/2} - (V_a^2/2g)^{3/2}] \tag{i}$$

for a weir with no lateral contractions (suppressed weir) and $\alpha = 1$.

$$\text{Equation (i) reduces to: } Q = 1·84 \, b \, (H)^{3/2} \tag{ii}$$

neglecting velocity of approach as a first approximation.

$$\therefore \text{ from (ii) } Q = 1{\cdot}84 \times 4{\cdot}5 \times (0{\cdot}45)^{3/2}$$

$$= 2{\cdot}5 \text{ m}^3/\text{s}$$

Now velocity of approach, $V_a = 2{\cdot}5/4{\cdot}5 \; (1 + 0{\cdot}45)$

$$= 0{\cdot}383 \text{ m/s}$$

$$\text{and } V_a{}^2/2g = 7{\cdot}48 \times 10^{-3} \text{ m}$$

$$\therefore Q = 1{\cdot}84 \times 4{\cdot}5 \; [(0{\cdot}45 + 0{\cdot}00748)^{3/2} - (0{\cdot}00748)^{3/2}]$$

$$= 2{\cdot}556 \text{ m}^3/\text{s}.$$

Example 3.16
The discharge over a triangular notch can be written as:

$$Q = (8/15) \; C_d \; \sqrt{2g} \; \tan \frac{1}{2} \, \theta \; H^{5/2}$$

If an error of 1% in measuring H is introduced determine the corresponding error in the computed discharge.

A right angled triangular notch is used for gauging the flow of a laboratory flume. If the coefficient of discharge of the notch is 0·593 and an error of 2 mm is suspected in observing the head, find the percentage error in computing an estimated discharge of 20 l/s.

Solution:
We can write $Q = K \; H^{5/2}$

$$\therefore \; dQ = 5/2 \; K \; H^{3/2} \; dH$$

$$\text{and } dQ/Q = \frac{5/2 \; K \; H^{3/2} \; dH}{K \; H^{5/2}}$$

$$= 5/2 \; \frac{dH}{H} \qquad\qquad\qquad \text{(i)}$$

$\therefore$ If dH/H is 1%, the error in the discharge, $dQ/Q = 2{\cdot}5\%$ from (i)

$$Q = 0{\cdot}02 = 8/15 \times 0{\cdot}593 \times \sqrt{2g} \times 1 \times H^{5/2}$$

$$H^{5/2} = 1{\cdot}4275 \times 10^{-2} \text{ or } H = 0{\cdot}183 \text{ m or } 183 \text{ mm}$$

$$\text{and } dQ/Q = (2{\cdot}5) \; (dH/H)$$

$$= 2{\cdot}5 \times 2/183 = 2{\cdot}73\%.$$

Example 3.17
If the velocity distribution of a turbulent flow in an open channel is given by a power law,

$$\frac{v}{v_{max}} = \left(\frac{y}{y_0}\right)^{1/7}$$

where v is the velocity at a distance y from the bed and v_{max} is the maximum velocity in the channel with a flow depth of y_0, determine the average velocity and the energy (α) and momentum (β) correction factors; assume the flow to be two-dimensional.

Solution:
If the mean velocity of flow is V, the discharge per unit width of the channel is

$$y_0 V = q = \int_0^{y_0} v\,dy$$

which gives $q = (7/8)v_{max}y_0$.
Therefore $V = q/y_0 = (7/8)v_{max}$.

The kinetic energy correction factor α, given by equation 3.11, can be written as

$$\alpha = \frac{1}{A}\int_A \left(\frac{v}{V}\right)^3 dA = \frac{1}{y_0 V^3}\int_0^{y_0}\left[v_{max}\left(\frac{y}{y_0}\right)^{1/7}\right]^3 dy$$

Replacing v_{max} {= (8/7)V} and integrating we obtain $\alpha = 1\cdot045$. The momentum correction factor, β, given by equation 3.14 can be written as

$$\beta = \frac{1}{A}\int_A \left(\frac{v}{V}\right)^2 dA = \frac{1}{y_0 V^2}\int_0^{y_0}\left[v_{max}\left(\frac{y}{y_0}\right)^{1/7}\right]^2 dy$$

Again replacing v_{max} and integrating we obtain $\beta = 1\cdot016$.
   Note: The energy and momentum correction factors α and β for open channels may be computed by the equations

$$\alpha = 1 + 3\varepsilon^2 - 2\varepsilon^3 \text{ and } \beta = 1 + \varepsilon^2 \text{ where } \varepsilon = (v_{max}/V) - 1.$$

If the velocity distributions are not described by any equation and if the measured data is available, α and β values may be computed by graphical methods; plots of $\int vdy$, $\int v^3dy$ and $\int v^2dy$ will give V, α and β respectively.

Example 3.18
An Ogee spillway of large height is to be designed to evacuate a flood discharge of 200 m³/s under a head of 2 m. The spillway is spanned by piers to support a bridge deck above. The clear span between piers is limited to 6 m. Determine the number of spans required in order to pass the flood discharge with the head not exceeding 2 m; assume the pier contraction coefficient $k_p = 0\cdot01$ and the abutment contraction coefficient $k_a = 0\cdot10$.

Solution:
The flow between the piers and abutments is contracted, thus reducing the spillway width for the flow to B_e. Each pier has two end contractions and abutment one; and hence the effective width is given by

$$B_e = B - 2(nk_p + k_a)H_e, \text{ n being the number of piers.}$$

The pier contraction coefficient depends on the shape of its nose ($k_p = 0$ for pointed nose and $k_p = 0.02$ for square nose), whereas the abutment contraction coefficient may be as high as 0.2 for square abutment, reducing to zero for rounded abutment. If the velocity of approach, V_a, is not negligible, a trial and error procedure is to be used for the discharge computations; for large heights (P), $V_a \approx 0$ and hence $H_e \approx H$. Here assuming $V_a \approx 0$ we can write equation 3.43 as

$$Q = 200 = \frac{2}{3} \, C_{do} \, \sqrt{(2g)} \, \{6(n+1) - 2(n \times 0.01 + 0.10)2.0\}2^{3/2}$$

which gives n = 4.36 with $C_{do} = 0.75$ (P/H > 3). Hence provide five piers. Thus the clear span of the spillway (for flow) = 36 m. From the discharge equation we can now compute the corresponding head for this flow. In fact the spillway is capable of discharging a larger flood flow at the specified design head of 2 m. A stage (head) – discharge relationship can be established by using appropriate discharge coefficients (read from fig. 3.20).

Recommended reading

1. BS 1042, Part I (1964) *Methods for the measurement of fluid flow in pipes.* London: British Standards Institution.
2. BS 3680, Part 4A (1981) *Methods of measurement of liquid flow in open channels.* London: British Standards Institution.
3. Prandtl, L. (1952) *Essentials of fluid dynamics.* Glasgow: Blackie.
4. Rouse, H. (1959) *Advanced mechanics of fluids.* Chichester: Wiley.
5. Shames, I.H. (1992) *Mechanics of fluids,* 3rd edn. Maidenhead: McGraw-Hill.
6. Webber, N.B. (1971) *Fluid mechanics for civil engineers.* London: Chapman.
7. Novak, P., Moffat, A.I.B., Nalluri, C. and Narayanan, R. (1997) *Hydraulic Structures,* 2nd edn. London: E&FN Spon.

Problems

1.   A tapered nozzle is so shaped that the velocity of flow along its axis changes from 1.5 m/s to 15 m/s in a length of 1.35 m. Determine the magnitude of the convective acceleration at the beginning and end of this length.

2.   The spillway section of a dam ends in a curved shape (known as the bucket) deflecting water away from the dam. The radius of this bucket is 5 m and when the spillway is discharging 5 cumecs of water per metre length of crest, the average thickness of the sheet of water over the bucket is 0.5 m. Compare the resulting normal or centripetal acceleration with the acceleration due to gravity.

3.   The velocity distribution of a real fluid flow in a pipe is given by the equation $v = V_{max} (1 - r^2/R^2)$, where V_{max} is velocity at the centre of the pipe, R is pipe radius, and v is the velocity at radius, r, from the centre of the pipe. Show that the kinetic energy correction factor for this flow is 2.

4.   A pipe carrying oil of relative density 0·8 changes in diameter from 150 mm to 450 mm, the pressures at these sections being 90 kN/m^2 and 60 kN/m^2 respectively. The smaller section is 4 m below the other and if the discharge is 145 l/s determine the energy loss and the direction of flow.

5.   Water is pumped from a sump (see fig. 3.29) to a higher elevation by installing a hydraulic pump with the data:

Discharge of water = 6·9 m^3/min
Diameter of suction pipe = 150 mm
Diameter of delivery pipe = 100 mm
Energy supplied by the pump = 25 kW.

(i) Determine the pressure in kN/m^2 at points A and B neglecting all losses.
(ii) If the actual pressure at B is 25 kN/m^2 determine the total energy loss in kW between the sump and the point B.

6.   A fire-brigade man intends to reach a window 10 m above the ground with a fire stream from a nozzle of 40 mm diameter held at a height of 1·5 m above the ground. If the jet is discharging 1000 l/min, determine the maximum distance from the building at which the fireman can stand to hit the target. Hence find the angle of inclination with which the jet issues from the nozzle.

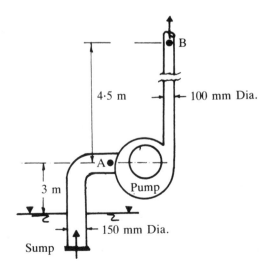

Figure 3.29   Flow through a hydraulic pump

7.　A pipeline 600 mm diameter conveying oil of relative density 0·85 at the rate of 2 cumecs has a 90° bend in a horizontal plane. The pressure at the inlet to the bend is 2 m of oil. Find the magnitude and direction of the force exerted by the oil on the bend. If the ends of the bend are anchored by tie-rods at right angles to the pipeline, determine tension in each tie-rod.

8.　The diameter of pipe bend is 300 mm at inlet and 150 mm at outlet and the flow is turned through 120° in a vertical plane. The axis at inlet is horizontal and the centre of the outlet section is 1·5 m below the centre of the inlet section. The total volume of fluid contained in the bend is $8·5 \times 10^{-2}$ m^3. Neglecting friction, calculate the magnitude and direction of the force exerted on the bend by water flowing through it at a rate of 0·225 m^3/s when the inlet pressure is 140 kN/m^2.

9.　A sluice gate is used to control the flow of water in a horizontal rectangular channel, 6 m wide. The gate is lowered so that the stream flowing under it has a depth of 800 mm and a velocity of 12 m/s. The depth upstream of the sluice gate is 7 m. Determine the force exerted by the water on the sluice gate assuming uniform velocity distribution in the channel and neglecting frictional losses.

10.　A jet of water 50 mm diameter strikes a curved vane at rest with a velocity of 30 m/s and is deflected through 135° from its original direction. Neglecting friction, compute the resultant force on the vane in magnitude and direction.

11.　A horizontal rectangular outlet downstream of a dam, 2·5 m high and 1·5 m wide discharges 70 m^3/s of water on to a concave concrete floor of 12 m radius and 6 m long, deflecting the water away from the outlet to dissipate energy. Calculate the resultant thrust the fluid exerts on the floor.

12.　A venturi meter is to be fitted to a pipe of 250 mm diameter where the pressure head is 6 m of water and the maximum flow 9 m^3/min. Find the smallest diameter of the throat to ensure that the pressure head does not become negative.

13.　(a) Determine the diameter of throat of a Venturi meter to be introduced in a horizontal section of a 100 mm diameter main so that deflection of a differential mercury manometer connected between the inlet and throat is 600 mm when the discharge is 20 l/s of water. The discharge coefficient of the meter is 0·95.
(b) What difference will it make to the manometer reading if the meter is introduced in a vertical length of the pipeline, with water flowing upwards and the distance from inlet to throat of the meter is 200 mm?

14.　A Pitot tube placed in front of a submarine moving horizontally in sea 16 m below the water surface, is connected to the two limbs of a U-tube mercury manometer. Find the speed of the submarine for a manometer deflection of 200 mm. Relative densities of mercury and sea water are 13·6 and 1·026 respectively.

15.   In an experiment to determine the hydraulic coefficients of a 25 mm diameter sharp-edged orifice, it was found that the jet issuing horizontally under a head of 1 m travelled a horizontal distance of 1·5 m from vena-contracta in the course of a vertical drop of 612 mm from the same point. Further, the impact force of the jet on a flat plate held normal to it at vena-contracta was measured as 5.5 N. Determine the three coefficients of the orifice assuming an impact coefficient of unity.

16.   A swimming pool with vertical sides is 25 m long and 10 m wide. Water at the deep end is 2·5 m and shallow end 1 m. If there are two outlets each 500 mm diameter, one at each of the deep and shallow ends, find the time taken to empty the pool. Assume the discharge coefficients for both the outlets as 0·8.

17.   A convergent-divergent nozzle is fitted to the vertical side of a tank containing water to a height of 2 m above the centre line of the nozzle. Find the throat and exit diameters of the nozzle, if it discharges 7 l/s of water into the atmosphere, assuming that (i) the pressure head in the throat is 2·5 m of water absolute, (ii) atmospheric pressure is 10 m of water, (iii) there is no hydraulic loss in the convergent part of the nozzle, and (iv) the head loss in the divergent part is one-fifth of exit velocity head.

18.   When water flows through a right-angled V-notch, show that the discharge is given by $Q = kH^{5/2}$, in which k is a dimensional constant and H is the height of water surface above the bottom of the notch. (i) What are the dimensions of k if H is in metres and Q in m^3/s? (ii) Determine the head causing flow when the discharge through this notch is 1·42 l/s. Take $C_d = 0·62$. (iii) Find the accuracy with which the head in (ii) must be measured if the error in the estimation of discharge is not to exceed $1\frac{1}{2}\%$.

19.   (a) What is meant by a 'suppressed' weir? Explain the precautions that you would take in using such a weir as discharge measuring structure. (b) A suppressed weir with two ventilating pipes is installed in a laboratory flume with the following data:

Width of flume = 1000 mm
Height of weir (P) = 300 mm
Diameter of ventilating pipes = 30 mm
Pressure difference between the two sides of the nappe = 1 N/m^2
Head over sill (h) = 150 mm
Density of air = 1·25 kg/m^3
Coefficient of discharge, C_d = 0·611 + 0·075 (h/P)

Assuming a smooth entrance to the ventilating pipes and neglecting the velocity of approach, find the air demand in terms of percentage of water discharge.

20.   State the advantages of a triangular weir over a rectangular one, for measuring discharges.

The following observations of head and the corresponding discharge were made in a laboratory to calibrate a 90° V-notch.

Head (mm)	50	75	100	125	150
Discharge (l/s)	0·81	2·24	4·76	8·03	12·66

Determine K and n in the discharge equation, $Q = K H^n$ (H in m, Q in m³/s) and hence find the value of the coefficient of discharge.

21.   A reservoir has an area of 8·5 ha and is provided with a weir 4·5 m long ($C_d = 0·6$). Find how long will it take for the water level above the sill to fall from 0·60 m to 0·30 m.

22.   A submerged weir of 1 m height spans the entire width of a rectangular channel 7 m wide. Find the discharge when the depths of flow on the upstream and downstream sides of the weir are 1·8 m and 1·25 m. Use Francis formula for the free discharge. Consider velocity of approach and assume the energy correction factor (Coriolis coefficient), $\alpha = 1·1$.

Chapter 4
Flow of Incompressible Fluids in Pipelines

R.E. Featherstone

4.1 Resistance in circular pipelines flowing full

A fluid moving through a pipeline is subjected to energy losses from various sources. A continuous resistance is exerted by the pipe walls due to the formation of a boundary layer in which the velocity decreases from the centre of the pipe to zero at the boundary. In steady flow in a uniform pipeline the boundary shear stress τ_o is constant along the pipe, since the boundary layer is of constant thickness, and this resistance results in a uniform rate of total energy or head degradation along the pipeline. The total head loss along a specified length of pipeline is commonly referred to as the 'head loss due to friction' and denoted by h_f. The rate of energy loss or energy gradient $S_f = \dfrac{h_f}{L}$.

The hydraulic grade line shows the elevation of the pressure head along the pipe. In a uniform pipe the velocity head, $\dfrac{\alpha V^2}{2g}$, is constant and the energy grade line is parallel with the hydraulic grade line (fig. 4.1). Applying the Bernoulli equation to sections 1 and 2,

$$Z_1 + \frac{p_1}{\rho g} + \frac{\alpha V_1^2}{2g} = Z_2 + \frac{p_2}{\rho g} + \frac{\alpha V_2^2}{2g} + h_f$$

and since $V_1 = V_2$, $Z_1 + \dfrac{p_1}{\rho g} = Z_2 + \dfrac{p_2}{\rho g} + h_f$ (4.1)

In steady uniform flow the motivating and drag forces are exactly balanced. Equating between sections 1 and 2:

$$(p_1 - p_2)A + \rho g \, AL \sin \theta = \tau_o \, PL \tag{4.2}$$

where A = area of cross-section, P the wetted perimeter and τ_o the boundary shear stress.

Rearranging equation (4.2) and noting that $L \sin \theta = Z_1 - Z_2$;

$$\frac{p_1 - p_2}{\rho g} + Z_1 - Z_2 = \frac{\tau_o \, PL}{\rho g \, A}$$

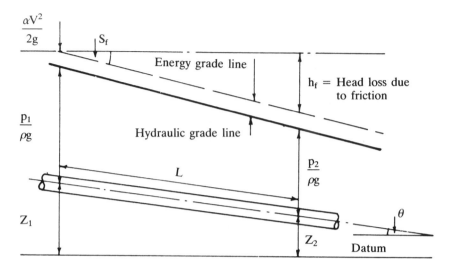

Figure 4.1   Pressure head and energy gradients in full, uniform pipe flow

and, from equation (4.1), $h_f = \dfrac{p_1 - p_2}{\rho g} + Z_1 - Z_2$

whence $h_f = \dfrac{\tau_o \, PL}{\rho g \, A}$

or $\tau_o = \rho g \, R \dfrac{h_f}{L} = \rho g \, R \, S_f$ (4.3)

where R (hydraulic radius) $= \dfrac{A}{P}$ $(= D/4$ for a circular pipe of diameter D).

The head loss due to friction in steady uniform flow is given by the Darcy-Weisbach equation:

$$h_f = \frac{\lambda \, LV^2}{2g \, D}$$ (4.4)

the derivation of which is to be found in any standard textbook. λ is a non-dimensional coefficient which, for turbulent flow, can be shown to be a function of k/D, the relative roughness, and the Reynolds number, Re $= \dfrac{VD}{\nu}$. k is the effective roughness size of the pipe wall. For laminar flow, (Re $\leqslant$ 2000), h_f can be obtained theoretically in the form of the Hagen-Pouiseuille equation:

$$h_f = \frac{32 \, \mu \, LV}{\rho g \, D^2}$$ (4.5)

Thus in equation (4.4) $\lambda = \dfrac{64}{Re}$ for laminar flow.

In the case of turbulent flow experimental work on smooth pipes by Blasius (1913) yielded the relationship

$$\lambda = \frac{0 \cdot 3164}{Re^{\frac{1}{4}}} \qquad (4.6)$$

Later work by Prandtl and Nikuradse on smooth and artificially roughened pipes revealed three zones of turbulent flow:
(i)   smooth turbulent zone in which the friction factor λ, a function of Reynolds number only and expressed by

$$\frac{1}{\sqrt{\lambda}} = 2 \log \frac{Re \sqrt{\lambda}}{2 \cdot 51} \qquad (4.7)$$

(ii)   transitional turbulent zone in which λ is a function of both k/D and Re

(iii)   rough turbulent zone in which λ is a function of k/D only and expressed by:

$$\frac{1}{\sqrt{\lambda}} = 2 \log \frac{3 \cdot 7 D}{k} \qquad (4.8)$$

Equations (4.7) and (4.8) are known as the Kármán-Prandtl equations. Colebrook and White (1939) found that the function resulting from addition of the rough and smooth equations (4.7) and (4.8) in the form:

$$\frac{1}{\sqrt{\lambda}} = -2 \log \left[\frac{k}{3 \cdot 7 D} + \frac{2 \cdot 51}{Re \sqrt{\lambda}} \right] \qquad (4.9)$$

fitted observed data on commercial pipes over the three zones of turbulent flow. Further background notes on the development of the form of the Kármán-Prandtl equations are given in Chapter 7. The Colebrook-White equation, (4.9) was first plotted in the form of a $\lambda - Re$ diagram by Moody (1944) (fig. 4.2) and hence is generally referred to as the 'Moody diagram'. This was presented originally with a logarithmic scale of λ. Figure 4.2 has been drawn, from computation of equation (4.9), with an arithmetic scale of λ for more accurate interpolation.

Combining the Darcy-Weisbach and Colebrook-White equations, (4.4) and (4.9), yields an explicit expression for the V:

$$V = -2 \sqrt{2g \, D \, S_f} \, \log \left[\frac{k}{3 \cdot 7 D} + \frac{2 \cdot 51 \, v}{D \sqrt{2g \, D \, S_f}} \right] \qquad (4.10)$$

This equation forms the basis of the *Charts for the hydraulic design of channels and pipes* (4th edition) produced by the Hydraulics Research Station[3]. A typical chart is reproduced as fig. 4.3.

Due to the implicit form of the Colebrook-White equation a number of approximations in explicit form in λ have been proposed.
Moody produced the following formulation:

$$\lambda = 0 \cdot 0055 \left[1 + \left(20\,000 \, \frac{k}{D} + \frac{10^6}{Re} \right)^{1/3} \right] \qquad (4.11)$$

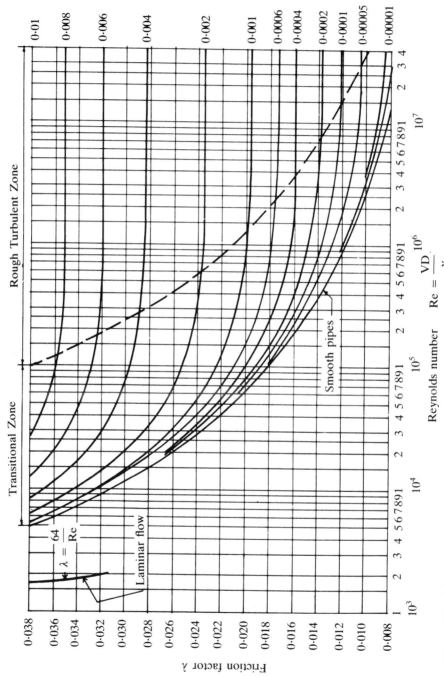

Figure 4.2 **Moody diagram**

This is claimed to give values of λ within $\pm 5\%$ for Reynolds numbers between 4×10^3 and 1×10^7 and for k/D up to $0 \cdot 01$.

More recently Barr[1] proposed the following form based partly on an approximation to the logarithmic smooth turbulent element in the Colebrook-White function by White:

$$\frac{1}{\sqrt{\lambda}} = -2 \log \left(\frac{k}{3 \cdot 7\ D} + \frac{5 \cdot 1286}{Re^{0 \cdot 89}} \right) \tag{4.12}$$

Further development by Barr[7] led to an even closer approximation which was expressed as:

$$\frac{1}{\sqrt{\lambda}} = -2 \log \left[\frac{k}{3 \cdot 7\ D} + \frac{5 \cdot 02 \log (Re/4 \cdot 518 \log (Re/7))}{Re\ (1 + Re^{0 \cdot 52}/29\ (D/k)^{0 \cdot 7})} \right] \tag{4.13}$$

Typical percentage errors in λ given by (4.13) compared with the solution of the Colebrook-White function are:

$\dfrac{k}{D}$	Reynolds number		
	3×10^4	3×10^5	3×10^6
10^{-3}	$-0 \cdot 12$	$0 \cdot 00$	$-0 \cdot 07$
10^{-4}	$-0 \cdot 16$	$-0 \cdot 07$	$+0 \cdot 03$

The values given by (4.13) should be sufficiently accurate for most purposes but substitution of these values once into the right hand side of the Colebrook-White function produces λ values with a maximum discrepancy of $+0 \cdot 04\%$.

4.2 Resistance to flow in non-circular sections

In order to use the same form of resistance equations such as the Darcy (4.4) and Colebrook-White (4.9) it is convenient to treat the non-circular section as an equivalent hypothetical circular section yielding the same hydraulic gradient at the same discharge.

The 'transformation' is achieved by expressing the diameter D in terms of the hydraulic radius $R = \left(\dfrac{A}{P} \right)$ and since for circular pipes $R = \dfrac{D}{4}$, equations (4.4) and (4.9) become

$$h_f = \frac{\lambda\ LV^2}{8g\ R} \tag{4.14}$$

and $$\frac{1}{\sqrt{\lambda}} = -2 \log \left[\frac{k}{14 \cdot 8\ R} + \frac{2 \cdot 51\ \nu}{4V\ R\sqrt{\lambda}} \right] \tag{4.15}$$

Discharge Q (l/s) *for pipes flowing full*

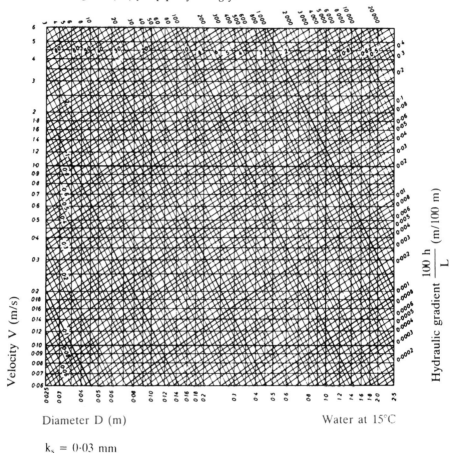

$k_s = 0.03$ mm

Figure 4.3   Extract from *Charts for the hydraulic design of channels and pipes*. Repro-
                 duced by permission of the Controller HMSO, courtesy Hydraulics
                 Research Station, Wallingford, England. Crown copyright.

Due to the fact that in the actual non-circular section the boundary shear stress
is not constant around the wetted perimeter, whereas it is in the equivalent
circular section, the 'transformation' is not exact but experiments have shown
that the error is small.

It is important to note that the equivalent circular pipe does not have the
same area as the actual conduit; their hydraulic radii are equal.

4.3 Local losses

In addition to the spatially continuous head loss due to friction, local head
losses occur at changes of cross-section, at valves and at bends. These local

losses are sometimes referred to as 'minor' losses since in long pipelines their effect may be small in relation to the friction loss. However the head loss at a control valve has a primary effect in regulating the discharge in a pipeline.

Typical values for circular pipelines

$$\text{Head loss at abrupt contraction} = K_c \frac{V_2{}^2}{2g}$$

where V_2 = mean velocity in downstream section of diameter D_2; D_1 = upstream diameter.

$\dfrac{D_2}{D_1}$	0	0·2	0·4	0·6	0·8	1·0
K_c	0·5	0·45	0·38	0·28	0·14	0

Note that the value of $K_c = 0·5$ relates to the abrupt entry from a tank into a circular pipeline.

$$\text{Head loss at abrupt enlargement} = \frac{V_2{}^2}{2g}\left(\frac{A_2}{A_1} - 1\right)^2$$

$$\text{Head loss at } 90° \text{ elbow} = 1·0\,\frac{V^2}{2g}$$

$$\text{Head loss at } 90° \text{ smooth bend} = \frac{V^2}{2g}$$

$$\text{Head loss at a valve} = K_v\,\frac{V^2}{2g}$$

where K_v depends upon the type of valve and percentage of closure. See also reference 6.

The following examples demonstrate the application of the above theory and equations to the analysis and design of pipelines.

Worked examples

Example 4.1
Crude oil of density 925 kg/m³ and absolute viscosity 0·065 Ns/m² at 20°C is pumped through a horizontal pipeline 100 mm in diameter, at a rate of 10 l/s. Determine the head loss in each kilometre of pipeline and the shear stress at the pipe wall. What power is supplied by the pumps per kilometre length?

Solution:
Determine if the flow is laminar.

Area of pipe = 0·00786 m²

Mean velocity of oil = 1·27 m/s

$$\text{Reynolds number} \left(\text{Re} = \frac{VD}{\nu} \right) = \frac{1 \cdot 27 \times 0 \cdot 1 \times 925}{0 \cdot 065}$$

$$\text{Re} = 1810$$

Thus the flow may therefore be assumed to be laminar:

$$\text{Hence } \lambda = \frac{64}{\text{Re}} = 0 \cdot 0354.$$

$$\text{Friction head loss/km} = \frac{\lambda \, LV^2}{2g \, D} = \frac{0 \cdot 0354 \times 1000 \times 1 \cdot 27^2}{19 \cdot 62 \times 0 \cdot 1}$$

$$= 29 \cdot 2 \text{ m}$$

Boundary shear stress $\tau_o = \rho g \, R \, S_f$

$$\tau_o = 925 \times 9 \cdot 81 \times \frac{0 \cdot 1}{4} \times \frac{29 \cdot 2}{1000}$$

$$\tau_o = 6 \cdot 62 \text{ N/m}^2$$

$$\text{Power consumed} = \rho g \, Q \, h_f$$

$$= 925 \times 9 \cdot 81 \times 0 \cdot 01 \times 29 \cdot 2 \text{ watts}$$

$$= 2 \cdot 65 \text{ kW/km}$$

Note that if the outlet end of the pipeline were elevated above the head of oil at inlet the pumps would have to deliver more power to overcome the static lift. This is dealt with more fully in Chapter 6 which covers pumps.

Example 4.2
A uniform pipeline, 5000 m long, 200 mm in diameter and roughness size 0·03 mm, conveys water at 15°C between two reservoirs, the difference in water level between which is maintained constant at 50 m. In addition to the entry loss of $0 \cdot 5 \dfrac{V^2}{2g}$ a valve produces a head loss of $\dfrac{10 \, V^2}{2g}$. $\alpha = 1 \cdot 0$.

Determine the steady discharge between the reservoirs using
(a) the Colebrook-White equation
(b) the Moody diagram
(c) the Hydraulics Research Station Charts, and
(d) an explicit function for λ. (See fig. 4.4.)

Solution:
Apply the Bernoulli equation to A and B

$$H = \frac{0 \cdot 5 \, V^2}{2g} + \frac{V^2}{2g} + \frac{10 \, V^2}{2g} + \frac{\lambda \, LV^2}{2g \, D} \qquad \text{(i)}$$

| Gross head | entry loss | velocity head | valve head loss | friction head loss |

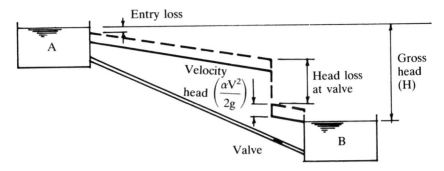

Figure 4.4 Energy losses in uniform pipeline

(a) Colebrook-White equation: $\dfrac{1}{\sqrt{\lambda}} = -2 \log \left[\dfrac{k}{3 \cdot 7 \, D} + \dfrac{2 \cdot 51}{Re \sqrt{\lambda}} \right]$ (ii)

The solution to the problem is obtained by solving (i) and (ii) simultaneously. However, direct substitution of λ from (i) into (ii) yields a complex implicit function in V which can only be evaluated by trial or graphical interpolation.

A simpler computational procedure is obtained if terms other than the friction head loss in equation (i) are initially ignored; in other words the gross head is assumed to be totally absorbed in overcoming friction. Then equation 4.10 can be used to obtain an approximate value of V.

i.e. $V = -2 \sqrt{2g \, D \dfrac{h_f}{L}} \, \log \left[\dfrac{k}{3 \cdot 7 \, D} + \dfrac{2 \cdot 51 \, v}{D \sqrt{2g \, D \dfrac{h_f}{L}}} \right]$ (iii)

Writing $h_f = H = 50$ m; $\dfrac{h_f}{L} = 0 \cdot 01$

whence $V = -2 \sqrt{19 \cdot 62 \times 0 \cdot 2 \times 0 \cdot 01}$

$\log \left[\dfrac{0 \cdot 03 \times 10^{-3}}{3 \cdot 7 \times 0 \cdot 2} + \dfrac{2 \cdot 51 \times 1 \cdot 13 \times 10^{-6}}{0 \cdot 2 \sqrt{19 \cdot 62 \times 0 \cdot 2 \times 0 \cdot 01}} \right]$

$V = 1 \cdot 564$ m/s

The terms other than friction loss in equation (i) can now be evaluated:

$h_m = 11 \cdot 5 \, \dfrac{V^2}{2g} = 1 \cdot 435$ m

(where h_m denotes the sum of the minor head loss).
A better estimate of h_f is thus $h_f = 50 - 1 \cdot 435 = 48 \cdot 565$ m.
Whence from equation (iii) $V = 1 \cdot 544$ m/s.

Repeating until successive values of V are sufficiently close yields

$$V = 1 \cdot 541 \text{ m/s and } Q = 48 \cdot 41 \text{ l/s with } h_f = 48 \cdot 61 \text{ m and } h_m = 1 \cdot 39 \text{ m}$$

Convergence is usually rapid since the friction loss usually predominates.

(b) The use of the Moody chart, fig. 4.2, involves the determination of the Darcy friction factor. In this case the minor losses need not be neglected initially. However the solution is still iterative and an estimate of the mean velocity is needed.

$$\text{Estimate } V = 2 \cdot 0 \text{ m/s}; \quad Re = \frac{2 \times 0 \cdot 2}{1 \cdot 13 \times 10^{-6}} = 3 \cdot 54 \times 10^5$$

$$\text{Relative roughness,} \quad \frac{k}{D} = 0 \cdot 00015$$

From the Moody chart $\lambda = 0 \cdot 015$

$$\text{Rearranging (i);} \qquad V = \sqrt{\frac{2g \, H}{11 \cdot 5 + \dfrac{\lambda \, L}{D}}} \qquad\qquad \text{(iv)}$$

And a better estimate of mean velocity is given by

$$V = \sqrt{\frac{19 \cdot 62 \times 50}{11 \cdot 5 + \dfrac{0 \cdot 015 \times 5000}{0 \cdot 2}}} = 1 \cdot 593 \text{ m/s}$$

$$\text{Revised } Re = \frac{1 \cdot 593 \times 0 \cdot 2}{1 \cdot 13 \times 10^{-6}} = 2 \cdot 82 \times 10^5$$

whence $\lambda = 0 \cdot 016$ and equation (iv) yields: $V = 1 \cdot 54$ m/s

Further change in λ due to the small change in V will be undetectable in the Moody diagram.
Thus, accept $V = 1 \cdot 54$ m/s; $Q = 48 \cdot 41$ l/s.

(c) The solution by the use of the Hydraulics Research Station Charts is basically the same as method (a) except that values of V are obtained directly from the chart instead of from equation (4.10).
   Making the initial assumption that $h_f = H$, the hydraulic gradient $(^m/100 \text{ m}) = 1 \cdot 0$.

Entering fig. 4.3 with $D = 0 \cdot 2$ m and $\dfrac{h_f}{100} = 1 \cdot 0$ yields $V = 1 \cdot 55$ m/s.

The minor loss term $\dfrac{11 \cdot 5 \, V^2}{2g} = 1 \cdot 41$ m and a better estimate of h_f is therefore $0 \cdot 972$ m/100 m.
   Whence from fig. 4.3 $V = 1 \cdot 5$ m/s and $Q = 47 \cdot 1$ l/s. Note the loss of fine accuracy due to the graphical interpolation in fig. 4.3.

(d) Using equation (4.12), $\dfrac{1}{\sqrt{\lambda}} = -\, 2 \log \left(\dfrac{k}{3\cdot 7\ D} + \dfrac{5\cdot 1286}{Re^{0\cdot 89}} \right)$ (v)

Assuming $V = 2\cdot 0$ m/s, $\lambda = 0\cdot 0156$ (from (v))

Using equation (iv) $V = \sqrt{\dfrac{2g\ H}{11\cdot 5 + \dfrac{\lambda\ L}{D}}}$

$V = 1\cdot 563$ m/s; $\lambda = 0\cdot 0161$ (from (v))

whence $V = 1\cdot 54$ m/s

Thus, accept $V = 1\cdot 54$ m/s which is essentially identical with that obtained using the other methods.

Example 4.3 (Pipes in series)
Reservoir A delivers to reservoir B through two uniform pipelines AJ : JB of diameters 300 mm and 200 mm respectively. Just upstream of the change in section, which is assumed gradual, a controlled discharge of 30 l/s is taken off.

Length of AJ = 3000 m; length of JB = 4000 m; effective roughness size of both pipes = 0·015 mm; gross head = 25·0m. Determine the discharge to B, neglecting the loss at J. (See fig. 4.5.)

Solution:
Apply the energy equation between A and B.

$$H = \frac{0\cdot 5\ V_1^2}{2g} + h_{f,1} + h_{f,2} + \frac{V_2^2}{2g}$$

i.e. $H = \dfrac{0\cdot 5\ V_1^2}{2g} + \dfrac{\lambda_1 L_1 V_1^2}{2g\ D_1} + \dfrac{\lambda_2 L_2 V_2^2}{2g\ D_2} + \dfrac{V_2^2}{2g}$ (i)

Since λ_1 and λ_2 are initially unknown the simplest method of solution is to input a series of trial values of Q_1.

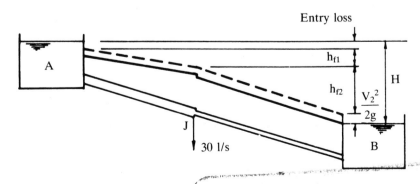

Figure 4.5   Energy losses in pipes in series with outflow

Since $Q_2 = Q_1 - 30$ (l/s), the corresponding values of Reynolds number can be calculated and hence λ_1 and λ_2 can be obtained from the Moody diagram, fig. 4.2. The total head loss H, corresponding with each trial value of Q_1 is then evaluated directly from equation (i). From a graph of H v. Q_1 the value of Q_1 corresponding with H = 25 m can be read off.

$$\frac{k_1}{D_2} = 0.00005; \qquad \frac{k_2}{D_2} = 0.000075$$

Q_1 (l/s)	50	60	80
V_1 (m/s)	0.707	0.849	1.132
V_2 (m/s)	0.637	0.955	1.591
Re_1 (× 10^5)	1.88	2.25	3.00
Re_2 (× 10^5)	1.13	1.69	2.81
λ_1	0.0164	0.016	0.0156
λ_2	0.0184	0.018	0.016
H(m)	11.82	22.67	51.66

From the graph (fig. 4.6) $Q_1 = 62.5$ l/s, whence $Q_2 = 32.5$ l/s.
Note: This problem could also be solved by the 'quantity balance' method of pipe network analysis (see Chapter 5).

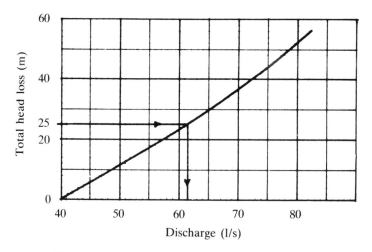

Figure 4.6   Head loss vs. discharge

Example 4.4 (Head loss in pipe with uniform lateral outflow)
Determine the total head loss due to friction over a 100 m length of a 200 mm diameter pipeline of roughness size 0.03 mm which receives an inflow of 150 l/s and releases a uniform lateral outflow of 1.0 l/s per metre. (See fig. 4.7.)

Theory: Note that the pressure head $\left(\dfrac{p}{\rho g}\right)_x$ at any section is not simply $h_1 - h_{f.x}$ since momentum effects occur along the pipe due to the continual withdrawal of water. In addition the velocity head decreases along the pipe. Thus applying the energy equation to 1 and section X

$$\frac{p_1}{\rho g} + \frac{V_1^2}{2g} = \frac{p_x}{\rho g} + PD_x + h_{f.x} + \frac{V_x^2}{2g}$$

where PD_x is the increase in pressure head due to the change in momentum between 1 and X. However the present example will deal only with the evaluation of the friction head loss $h_{f.x}$.

The flowrate in the pipe at section X is

$$Q_x = Q_1 - qx$$

The hydraulic gradient at x is

$$\frac{dh_f}{dx} = \frac{\lambda_x\, Q_x^2}{2g\, DA^2} = B\,\lambda_x Q_x^2; \qquad \text{where } B = \frac{1}{2g\, DA^2}$$

$$\frac{dh_f}{dx} = B\,\lambda_x\,[Q_1 - qx]^2$$

and the total head loss due to friction between the inlet and outlet is

$$h_f = B \int_o^L \lambda_x\,[Q_1 - qx]^2\, dx \tag{i}$$

Now λ_x is given by

$$\frac{1}{\sqrt{\lambda_x}} = -\,2 \log\left[\frac{k}{3{\cdot}7\,D} + \frac{2{\cdot}51\ v}{V_x D\ \sqrt{\lambda_x}}\right]$$

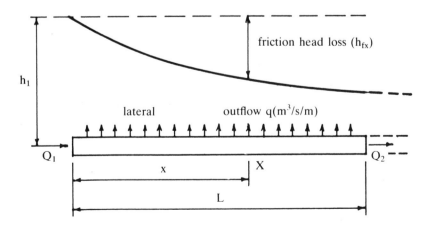

Figure 4.7 Head loss in pipe with uniform lateral outflow

Thus, an exact analytical solution to (i) is not possible but it could be evaluated approximately by summation over finite intervals δx.

However, if we take a constant value of λ_x, based on the average of the inlet and outlet values an approximate, explicit solution is obtained. Thus the solution to (i) is

$$h_f = B \bar{\lambda} \left[Q_1^2 x - q Q_1 x^2 + \frac{q^2 x^3}{3} \right]_0^L$$

$$= B \bar{\lambda} L \left[Q_1 - q L Q_1 + \frac{q^2 L^2}{3} \right] \tag{ii}$$

$$\frac{k}{D} = \frac{0 \cdot 03}{200} = 0 \cdot 00015$$

$$Q_1 = 150 \text{ l/s}; \qquad Q_2 = 50 \text{ l/s}$$

$$V_1 = 4 \cdot 775 \text{ l/s}; \qquad V_2 = 1 \cdot 59 \text{ l/s}$$

$$Re_1 = 8 \cdot 45 \times 10^5; \quad Re_1 = 2 \cdot 82 \times 10^5$$

whence $\lambda_1 = 0 \cdot 014;$ $\qquad \lambda_2 = 0 \cdot 016$ (from Moody diagram)

Taking $\bar{\lambda} = 0 \cdot 015$ and substituting into (ii):

$$h_f = 4 \cdot 195 \text{ m}.$$

Alternatively, by calculating the head loss in each 10 m interval and summating, the variation of λ along the pipe can be included, and a more accurate result should be obtained.

Then, using equation (ii) with subscripts 1 and 2 indicating the upstream and downstream ends of each section and with $L = 10$ m the table shows the head loss in each section.

x (m)	λ_1	λ_2	Δh_f (m)
10	0·014	0·014	0·760
20	0·014	0·014	0·659
30	0·014	0·0144	0·573
40	0·0144	0·0148	0·499
50	0·0148	0·0152	0·427
60	0·0152	0·0152	0·355
70	0·0152	0·0154	0·287
80	0·0154	0·0156	0·225
90	0·0156	0·016	0·173
100	0·016	0·0164	0·127

$$h_f = \Sigma \Delta h_f = 4 \cdot 086 \text{ m}$$

Example 4.5 (Flow between tanks where the level in the lower tank is dependent upon discharge)

A constant head tank delivers water through a uniform pipeline to a tank, at a lower level, from which the water discharges over a rectangular weir. Pipeline length 20·0 m, diameter 100 mm, roughness size 0·2 mm. Length of weir crest 0·25 m, discharge coefficient 0·6, crest level 2·5 m below water level in header tank. Calculate the steady discharge and the head of water over the weir crest. (See fig. 4.8.)

Solution:

For pipeline, $H = \dfrac{1 \cdot 5 \ V^2}{2g} + \dfrac{\lambda \ LV^2}{2g \ D} = (2 \cdot 5 - h)$ (i)

or $H = \dfrac{Q^2}{2g \ A^2}\left(1 \cdot 5 + \dfrac{\lambda \ L}{D}\right) = (2 \cdot 5 - h)$ (ii)

Discharge over weir: $Q = \dfrac{2}{3} C_D \sqrt{2g} \ B \ h^{3/2}$ (iii)

i.e. $Q = \dfrac{2}{3} \times 0 \cdot 6 \times \sqrt{19 \cdot 62} \times 0 \cdot 25 \times h^{3/2}$

$= 0 \cdot 443 \ h^{3/2}$

i.e. $h = \left(\dfrac{Q}{0 \cdot 443}\right)^{2/3}$ (iv)

Then in (ii) $\dfrac{Q^2}{2g \ A^2}\left(1 \cdot 5 + \dfrac{\lambda \ L}{D}\right) = 2 \cdot 5 - \left(\dfrac{Q}{0 \cdot 443}\right)^{2/3}$

or $\dfrac{Q^2}{2g \ A^2}\left(1 \cdot 5 + \dfrac{\lambda \ L}{D}\right) + \left(\dfrac{Q}{0 \cdot 443}\right)^{2/3} = 2 \cdot 5$ (v)

Since λ is unknown this equation can be solved by trial or interpolation i.e. inputting a number of trial Q values and evaluating the left-hand side of equation (v):

$H_1 = \dfrac{Q^2}{2g \ A^2}\left(1 \cdot 5 + \dfrac{\lambda \ L}{D}\right) + \left(\dfrac{Q}{0 \cdot 443}\right)^{2/3}$

For the same values of Q, the corresponding values of h are evaluated from equation (iv).

For each trial value of Q, the Reynolds number is calculated and the

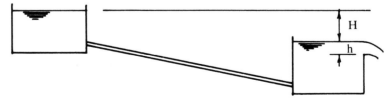

Figure 4.8 Flow through pipeline between two reservoirs, with outflow from receiving reservoir

friction factor obtained from the Moody diagram, for $\frac{k}{D} = 0{\cdot}0002$. See table below.

whence $Q = 0{\cdot}0213$ m³/s (21·3 l/s) when $H_1 = 2{\cdot}5$ m

and $h = 0{\cdot}132$ m.

Q m³/s	Re	λ	H_1 (m)	h(m)
0·010	1·13 × 10⁵	0·0250	0·617	0·08
0·015	1·69 × 10⁵	0·0243	1·287	0·105
0·018	2·03 × 10⁵	0·0241	1·810	0·118
0·020	2·25 × 10⁵	0·0241	2·215	0·126
0·022	2·48 × 10⁵	0·0240	2·655	0·135

Example 4.6 (Pipes in parallel)
A 200 mm diameter pipeline, 5000 m long and of effective roughness 0·03 mm delivers water between reservoirs the minimum difference in water level between which is 40 m.
(a) Taking only friction, entry and velocity head losses into account, determine the steady discharge between the reservoirs.
(b) If the discharge is to be increased to 50 l/s without increase in gross head determine the length of 200 mm diameter pipeline of effective roughness 0·015 mm to be fitted in parallel. Consider only friction losses.

Solution:
(a) Using the technique of Example 4.2:

$$40 = \frac{\lambda\, LV^2}{2g\, D} + \frac{1{\cdot}5\, V^2}{2g}$$

yields $Q = 43{\cdot}52$ l/s.

(b) (See fig 4.9.)

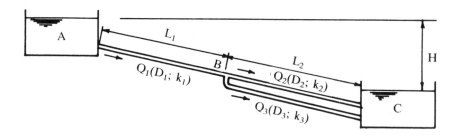

Figure 4.9 Pipes in parallel

In the general case where local losses, (h_m) occur in each branch:

$$H = \frac{0 \cdot 5 \; V_1{}^2}{2g} + h_{m,1} + h_{f,1} + h_{m,B} + h_{f,2} + h_{m,2} + \frac{V_2{}^2}{2g} \qquad \text{(i)}$$

$$\text{also } H = \frac{0 \cdot 5 \; V_1{}^2}{2g} + h_{m,1} + h_{f,1} + h_{m,B} + h_{f,3} + h_{m,3} + \frac{V_3{}^2}{2g} \qquad \text{(ii)}$$

where $h_{m,B}$ = head loss at junction.
Note that the head loss along branch 2 is equal to that along branch 3.

The local losses can be expressed in terms of the velocity heads. Thus equation (i) or (ii) can be solved simultaneously with the continuity equation at B

i.e. $\quad Q_1 = Q_2 + Q_3$ $\qquad\qquad$ (iii)

and $h_{L,2} = h_{L,3}$ $\qquad\qquad$ (iv)

and using the Colebrook-White equation (or Moody chart) for λ.
If friction losses predominate equation (i) reduces to

$$H = h_{f,1} + h_{f,2}$$

i.e. $H = \dfrac{\lambda_1 L_1 Q_1{}^2}{2g \; D_1 A_1{}^2} + \dfrac{\lambda_2 L_2 Q_2{}^2}{2g \; D_2 A_2{}^2}$ $\qquad\qquad$ (v)

Equation (iv) becomes: $\dfrac{\lambda_2 L_2 Q_2{}^2}{2g \; D_2 A_2{}^2} = \dfrac{\lambda_3 L_3 Q_3{}^2}{2g \; D_3 A_3{}^2}$ $\qquad\qquad$ (vi)

Since $D_2 = D_3$ and $L_2 = L_3$ we have

$$\lambda_2 Q_2{}^2 = \lambda_3 Q_3{}^2 \qquad\qquad \text{(vii)}$$

Also $Q_3 = 0 \cdot 05 - Q_2$

Equation (vii) can be solved by trial, of Q_2, and using the Moody diagram to obtain the corresponding λ values.

For example: $\dfrac{k_2}{D_2} = 0 \cdot 00015; \quad \dfrac{k_3}{D_3} = 0 \cdot 000075$

Try $Q_2 = 0 \cdot 022$ m^3/s; $\qquad Q_3 = 0 \cdot 028$ m^3/s

$\qquad$ Re$_2 = 1 \cdot 24 \times 10^5$; $\qquad$ Re$_3 = 1 \cdot 58 \times 10^5$

$\qquad \lambda_2 = 0 \cdot 0182$; $\qquad\qquad \lambda_3 = 0 \cdot 017$

$\lambda_2 Q_2{}^2 = 8 \cdot 81 \times 10^{-6}$; $\quad \lambda_3 Q_3{}^2 = 1 \cdot 33 \times 10^{-5}$

By adjusting Q_2 and repeating, equation (vii) is satisfied when $\lambda_2 Q_2{}^2 \simeq 1 \cdot 10 \times 10^{-5}$
Now equation (v) can be solved:

$$Q_1 = 0 \cdot 05 \text{ m}^3/\text{s}; \quad \text{Re}_1 = 2 \cdot 816 \times 10^5; \quad \frac{k_1}{D_1} = 0 \cdot 00015$$

whence $\lambda_1 = 0.0161$; $\dfrac{\lambda_1 Q_1^2}{2g\,DA_1^2} = 0.01039$

Substituting into (v), $40 = 0.01039 \times L_1 + \dfrac{1.10 \times 10^{-5}\,(5000 - L_1)}{19.62 \times 0.2 \times 0.03142^2}$

whence $L_1 = 3355$ m

and $L_2 = 1645$ m

or duplicated length $= 1645$ m.

Example 4.7 (Design of a uniform pipeline)
A uniform pipeline of length 20 km is to be designed to convey water at a minimum rate of 250 l/s from an impounding reservoir to a service reservoir, the minimum difference in water level between which is 160 m. Local losses including entry loss and velocity head total $\dfrac{10\ V^2}{2g}$.

(a) Determine the diameter of standard commercially available lined spun iron pipeline which will provide the required flow when in new condition ($k = 0.03$ mm).
(b) Calculate also the additional head loss to be provided by a control valve such that with the selected pipe size installed the discharge will be regulated exactly to 250 l/s.
(c) An existing pipeline in a neighbouring scheme, conveying water of the same quality, has been found to lose 5 per cent of its discharge capacity, annually, due to wall deposits (which are removed annually).
  (i) Check the capacity of the proposed pipeline after one year of use assuming the same percentage reduction; and
  (ii) determine the corresponding effective roughness size.

Solution:
(a):

$$160 = \frac{\lambda\,LV^2}{2g\,D} + \frac{10\ V^2}{2g} \tag{i}$$

Neglecting minor losses in the first instance

$$h_f = H = \frac{\lambda\,LV^2}{2g\,D} \tag{ii}$$

$$\therefore \ \frac{1}{\sqrt{\lambda}} = -2\log\left[\frac{k}{3.7\,D} + \frac{2.51}{Re\sqrt{\lambda}}\right] \tag{iii}$$

Combining (ii) and (iii)

$$V = -2\sqrt{2g\,D\,\frac{h_f}{L}}\ \log\left[\frac{k}{3.7\,D} + \frac{2.51\ v}{D\sqrt{2g\,D\,\dfrac{h_f}{L}}}\right] \tag{iv}$$

Substituting $h_f = 160$ in equation (iv), and calculating the corresponding discharge capacity for a series of standard pipe diameters; (and noting that there is no need to correct for the reduction due to minor losses each time since there is a considerable percentage increase in capacity between adjacent pipe sizes) the following table is produced

D mm	150	200	250	300	350	400
Q l/s	20·3	43·6	78·6	127·3	191·1	271·5

Thus a 400 mm diameter pipeline is required.
Now check the effect of minor losses

$$Q = 271·5 \text{ l/s}; \quad V = 2·16 \text{ m/s}; \quad h_m = \frac{10 \, V^2}{2g} = 2·38 \text{ m}$$

$h_f = 157·6$ and revised $Q = 269·4$ l/s;

the 400 mm diameter is satisfactory.
(b): To calculate the head loss at a valve to control the flow to 250 l/s calculate the hydraulic gradient corresponding with this discharge:

$$V = 1·99 \text{ m/s}$$

h_f may be obtained by trial in equation (iv) until the right-hand side $= 1·99$

Thus, $h_f = 137$ m

$$\text{Minor head loss, } h_m = \frac{10 \, V^2}{2g} = 2·0 \text{ m}$$

Thus, valve loss $= 160 - 137 - 2·0 = 21·0$ m

Alternatively using the Moody chart:

$$k/D = 0·03/400 = 0·000075; \quad Re = \frac{1·99 \times 0·4}{1·13 \times 10^{-6}} = 7·04 \times 10^5$$

$$\lambda = 0·0136; \quad h_f = \frac{0·0136 \times 20\,000 \times 1·99^2}{19·62 \times 0·4} = 137·25 \text{ m}$$

Adopting $h_f = 137$ m; $\quad 10 \, V^2/2g = 2·0$ m

Additional loss by valve $= 160 - (137 + 2·0) = 21$ m.
(c): (i) 5% annual reduction in capacity.
         Capacity at end of 1 year $= 0·95 \times 269·4 = 255·9$ l/s; pipe will be satisfactory if cleaned each year.
     (ii) To calculate the effective roughness size after 1 year's operation, use equation (iv)

$$\text{i.e. } \frac{k}{3 \cdot 7 \, D} = \text{antilog} \left(- \frac{V}{\sqrt{2g \, D \, h_f/L}} \right) - \frac{2 \cdot 51 \, v}{D \sqrt{2g \, D \, h_f/L}} \tag{v}$$

$$Q = 255 \cdot 9 \text{ l/s;} \quad V = 2 \cdot 036 \text{ m/s}$$

$$h_f = 160 - \frac{10 \, V^2}{2g} = 157 \cdot 89 \text{ m}$$

whence from (v), $k = 0 \cdot 0795$ mm.

Example 4.8 (Effect of booster pump in pipeline)
In the gravity supply system illustrated in Example 4.6, as an alternative to the duplicated pipeline, calculate the head to be provided by a pump to be installed on the pipeline and the power delivered by the pump. (See fig. 4.10.)

Solution:

$$L = 5000 \text{ m;} \quad D = 200 \text{ mm;} \quad k = 0.03 \text{ mm;}$$

$$H = 40 \text{ m;} \quad Q = 50 \text{ l/s;}$$

H_m = manometric head to be delivered by the pump.

Total head $= H_m + H$

$$H + H_m = \frac{1 \cdot 5 \, V^2}{2g} + \frac{\lambda \, LV^2}{2g \, D}$$

$$V = 1 \cdot 59 \text{ m/s;} \quad Re = 2 \cdot 83 \times 10^5; \quad \frac{k}{D} = 0 \cdot 00015$$

whence $\lambda = 0 \cdot 0162$ from Moody chart.

whence $H + H_m = 52 \cdot 48$ m

and $H_m = 12 \cdot 48$ m

Hydraulic power delivered $= \dfrac{\rho g \, Q \, H_m}{1000}$ kW

$$P = 9 \cdot 81 \times 0 \cdot 05 \times 12 \cdot 48$$

$$P = 6 \cdot 12 \text{ kW}$$

Note that the power consumed Pc will be greater than this.

$$Pc = \frac{P}{\eta}$$

Where η = overall efficiency of the pump and motor unit.
(Pump/pipeline combinations are dealt with in more detail in Chapter 6.)

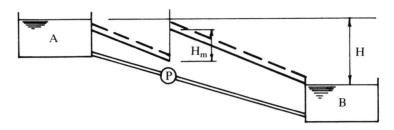

Figure 4.10 Pipeline with booster pump

Example 4.9 (Resistance in non-circular conduit)
A rectangular culvert to be constructed in reinforced concrete is being designed to convey a stream through a highway embankment. For short distances upstream and downstream of the culvert the existing stream channel will be improved to become rectangular and 6 m wide. The proposed culvert having a bed slope of 1 : 500 and length 100 m is 4 m wide and 2 m deep and is assumed to have an effective roughness of 0·6 mm. The design discharge is 40 m³/s at which flow the depth in the stream is 3·0 m. Water temperature = 4°C. Entry and exit from the culvert will be taken to be abrupt (although in the final design transitions at entry and exit would probably be adopted). Determine the depth at the entrance to the culvert at a flow of 40 m³/s. (See fig. 4.11.)

Solution:
Referring to fig. 4.11:
Apply the energy equation to 1 and 2

$$Z + y_1 + \frac{V_1^2}{2g} = y_3 + \frac{V_3^2}{2g} + \text{entry loss} + \text{friction loss in culvert} + \text{exit loss}$$

(i)

The entry loss coefficient, k_c, may be less than 0·5 which is commonly adopted for entry from a reservoir. Assuming that the loss at the contraction is similar to that for concentric pipes, k_c will depend on the ratio of upstream and downstream areas of flow, derived from the table in section 4.3 as follows:

A_2/A_1	0·0	0·04	0·16	0·36	1·0
k_c	0·5	0·45	0·38	0·28	0·0

Assume y_1 = 4·0 m (say) then A_2/A_1 = 8/24 = 0·3

whence k_c = 0·3

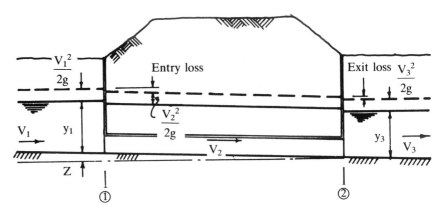

Figure 4.11   Flow resistance in non-circular conduit

$$\text{Discharge} = 40\!\cdot\!0 \ \text{m}^3/\text{s}; \quad V_2 = \frac{40}{8} = 5\!\cdot\!0 \ \text{m/s}$$

$$\text{entry loss} = 0\!\cdot\!3 \ \frac{V_2^{\,2}}{2g} = 0\!\cdot\!38 \ \text{m}$$

The exit loss (expansion) is expressed as $\dfrac{(V_2 - V_3)^2}{2g}$

$$V_3 = \frac{40}{18} = 2\!\cdot\!22 \ \text{m/s, wnence exit loss} = 0\!\cdot\!39 \ \text{m}$$

Friction head loss in culvert: referring to section 4.2, the resistance in the duct can be calculated by 'transforming' the cross-section into an equivalent circular section by equating the hydraulic radii. For the culvert $R = \dfrac{A}{P} = \dfrac{8}{12} \ \text{m}$ and the equivalent diameter, D_e, is therefore 2·67 m (= 4R).
Note that either the Colebrook–White equation (4.9) or its graphical form (fig. 4.2) can now be used with D = 2·67 m.

Kinematic viscosity of water at 4°C = $1\!\cdot\!568 \times 10^{-6} \ \text{m}^2/\text{s}$

$$V_2 = 5 \ \text{m/s}; \quad \text{Re} = \frac{5 \times 2\!\cdot\!67}{1\!\cdot\!568 \times 10^{-6}} = 8\!\cdot\!5 \times 10^6$$

$$\frac{k}{D} = \frac{0\!\cdot\!6 \times 10^{-3}}{2\!\cdot\!67} = 0\!\cdot\!000225$$

From the Moody chart $\lambda = 0\!\cdot\!014$

$$h_f = \frac{0\!\cdot\!014 \times 100 \times 5^2}{19\!\cdot\!62 \times 2\!\cdot\!67} = 0\!\cdot\!668 \ \text{m (say 0·67 m)}$$

$$\frac{V_3^{\,2}}{2g} = \frac{2\!\cdot\!22^2}{19\!\cdot\!62} = 0\!\cdot\!25 \ \text{m}$$

In (i) $0.2 + y_1 + \dfrac{V_1^2}{2g} = 3.0 + 0.25 + 0.38 + 0.67 + 0.39$

i.e. $0.2 + y_1 + \dfrac{Q^2}{2g\,(6y_1)^2} = 4.49$ m

whence, by trial $y_1 \qquad = 4.37$ m

Since this is close to the assumed value of y_1 the entry loss will not be significantly altered.

Example 4.10 (Pumped storage scheme — pipeline design)
The four pump-turbine units of a pumped storage hydro-electric scheme are each to be supplied by a high-pressure pipeline of length 2000 m. The minimum gross head (difference in level between upper and lower reservoirs) = 310 m and the maximum head = 340 m.
  The upper reservoir has a usable volume of 3.25×10^6 m^8 which could be released to the turbines in a minimum period of 4 hours.

Max. Power output required/turbine = 110 MW

Turbo-generator efficiency $\qquad$ = 80 per cent

Effective roughness of pipeline $\qquad$ = 0.6 mm

Taking minor losses in the pipeline, power station and draft tube to be 3.0 m,
(a) Determine the minimum diameter of pipeline to enable the maximum specified power to be developed.
(b) Determine the pressure head to be developed by the pump-turbines when reversed to act in the pumping mode to return a total volume of 3.25×10^6 m^3 to the upper reservoir uniformly during 6 hours in the off-peak period. (See fig. 4.12.)

Solution:
(a) Pipe capacity must be adequate to convey the required flow under MINIMUM head conditions:

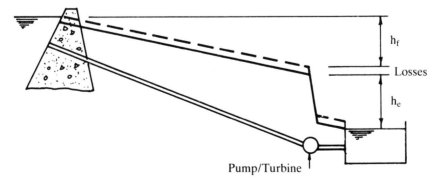

Figure 4.12 Pumped-storage power scheme in generating mode

Q max/unit = $3\cdot25 \times 10^6/4 \times 3600 = 56\cdot25$ m^3/s

Power generated: $P = \dfrac{\text{efficiency} \times \rho g\, Q\, \text{He}}{10^6}$ MW = 110 MW

where He = effective head at the turbines

i.e. $110 = \dfrac{0\cdot8 \times 1000 \times 9\cdot81 \times 56\cdot25 \times \text{He}}{10^6}$

whence He = 249·18 m

Total head loss = $310 - 249\cdot18 = 60\cdot82$ m

Head loss due to friction = 60·83 − minor losses

$\qquad\qquad\qquad\qquad = 60\cdot83 - 3\cdot0 = 57\cdot82$ m

$$h_f = \frac{\lambda\, LV^2}{2g\, D} \quad \text{and} \quad \frac{1}{\sqrt{\lambda}} = -\,2 \log\left[\frac{k}{3\cdot7\, D} + \frac{2\cdot51}{\text{Re}\sqrt{\lambda}}\right]$$

Since the hydraulic gradient is known but D is unknown it is preferable to use equation 4.10 in this case rather than use the Moody chart.

i.e. $$V = -\,2 \sqrt{2g\, D\, \frac{h_f}{L}} \, \log\left[\frac{k}{3\cdot7\, D} + \frac{2\cdot51\, \nu}{D\sqrt{2g\, D\, \dfrac{h_f}{L}}}\right]$$

and $Q = \dfrac{\pi\, D^2\, V}{4}$

Substituting values of D yields the corresponding discharge under the available hydraulic gradient.

D (m)	1·0	2·0	2·5	2·6	2·65
Q (m^3/s)	4·47	27·32	48·87	54·123	56·875

Hence required diameter = 2·65 metres.

(b) In pumping mode: Static lift = 340 m

$\qquad$ Q = $3\cdot25 \times 10^6/(6 \times 4 \times 3600) = 37\cdot616$ m^3/s

Since the diameter of the pipeline is known it is more straight-forward to use the Moody chart in this case.

$\qquad$ V = 6·82 m/s

$\qquad$ Re $= \dfrac{VD}{\nu} = 6\cdot82 \times 2\cdot65 \times 10^6$

$\qquad\qquad = 1\cdot8 \times 10^7$

$$\frac{k}{D} = 0.6 \times 10^{-3}/2.65 = 0.000226$$

$$\lambda = 0.0138$$

$$h_f = \frac{\lambda LV^2}{2g D} = \frac{0.0138 \times 2000 \times 6.82^2}{19.62 \times 2.65} = 24.69 \text{ m}$$

Total head on pumps $= 340 + 24.69 + 3.0$

$$= 367.69 \text{ m}.$$

Example 4.11
A high-head hydroelectric scheme consists of an impounding reservoir from which the water is delivered to four Pelton Wheel turbines through a low pressure tunnel, 10 000 m long, 4·0 m in diameter, lined with concrete, which splits into four steel pipelines (penstocks) 600 m long, 2·0 m in diameter each terminating in a single nozzle the area of which is varied by a spear valve. The maximum diameter of each nozzle is 0·8 m and the coefficient of velocity (C_v) is 0·98. The difference in level between reservoir and jets is 550 m. Roughness sizes of the tunnel and pipelines are 0·1 mm and 0·3 mm respectively.
(a) Determine the effective area of the jets for maximum power and the corresponding total power generated.
(b) A surge chamber is constructed at the downstream end of the tunnel. What is the difference in level between the water in the chamber and that in the reservoir under the condition of maximum power? (See fig. 4.13.)

Solution:
Let subscript T relate to tunnel and P to pipeline

$$H = \frac{0.5 \, Q_T^2}{2g \, A_T^2} + \frac{\lambda_T L_T}{2g \, D_T} \frac{Q_T^2}{A_T^2} + \frac{\lambda_P L_P Q_P^2}{2g \, D_P A_P^2} + \frac{\alpha_J V_J^2}{2g} + h_{1.n} \qquad (i)$$

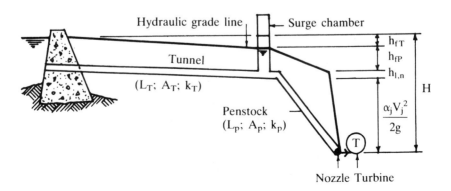

Figure 4.13 Tunnel and penstock

where $h_{1,n}$ = head loss in nozzle; V_J = velocity of jet issuing from nozzle. $h_{1,n}$ can be related to the coefficient of velocity C_v (see fig. 4.14).

$$h + \frac{\alpha_P V_P^2}{2g} = \frac{\alpha_J V_J^2}{2g} + h_{1,n} \tag{ii}$$

$$\text{and } V_J = C_v \sqrt{\frac{2g\left(h + \frac{\alpha_P V_P^2}{2g}\right)}{\alpha_J}} \tag{iii}$$

whence $h + \dfrac{\alpha_P V_P^2}{2g} = \dfrac{\alpha_J V_J^2}{C_v^2\,2g}$ and substituting in (ii),

$$h_{1,n} = \frac{\alpha_J V_J^2}{2g}\left(\frac{1}{C_v^2} - 1\right) \tag{iv}$$

Let a = area of each jet, N = number of jets (nozzles) per turbine, and since $Q_P = \dfrac{Q_T}{4}$ and $V_J = \dfrac{Q_P}{Na} = \dfrac{Q_T}{4Na}$

(i) becomes: $2g\,H = Q_T^2\left[\dfrac{0{\cdot}5 + \dfrac{\lambda_T L_T}{D_T}}{A_T^2}\right] + \dfrac{Q_T^2}{16}\left[\dfrac{\lambda_P L_P}{D_P A_P^2}\right]$

$$+ \frac{\alpha_J Q_T^2}{N^2 \times 16 \times C_v^2 \times a^2} \tag{v}$$

write $E = \left(\dfrac{0{\cdot}5 + \lambda_T L_T/D}{A_T^2}\right);\quad F = \dfrac{\lambda_P L_P}{16 D_P A_P^2};\quad G = \dfrac{\alpha_J}{16 N^2 C_v^2}$

(v) becomes: $2g\,H = Q_T^2\left(E + F + \dfrac{G}{a^2}\right) = Q_T^2\left(C + \dfrac{G}{a^2}\right)$

where C = E + F

$$\text{and } Q_T = \sqrt{\frac{2g\,H}{\left(C + \dfrac{G}{a^2}\right)}} \tag{vi}$$

Power of each jet $(P) = \rho g\,Q_J\,\dfrac{\alpha_J V_J^2}{2g}$

$$= \frac{\alpha_J \rho\, Q_T^3}{128 N^3 a^2}$$

and since $Q_J = \dfrac{Q_P}{N} = \dfrac{Q_T}{4N}$ and $V_J = \dfrac{Q_T}{4N}$

$$P = \frac{\alpha_J\, \rho\, Q_T^3}{128 N^3 a^2}$$

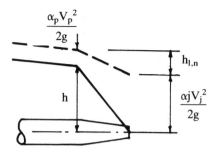

Figure 4.14 Pressure and velocity conditions at nozzle

substituting for Q_T from (vi), $P = \dfrac{\rho}{128N^3 a^2} \left[\dfrac{2g\ H}{\left(C + \dfrac{G}{a^2}\right)}\right]^{3/2}$

Thus $P \propto \dfrac{1}{a^2}\left(\dfrac{a^2}{Ca^2 + G}\right)^{3/2} \propto \dfrac{a^{2/3}}{(Ca^2 + G)}$

For max $\dfrac{dP}{da} = 0$ i.e. $-a^{2/3}(Ca^2 + G)^{-2} \times 2Ca + (Ca^2 + G)^{-1}\frac{2}{3}a^{-1/3} = 0$

whence $\dfrac{-3a^2\ C}{(Ca^2 + G)} + 1 = 0$

or a $= \sqrt{\dfrac{G}{2C}}$ (vii)

and $D_J = \sqrt{\dfrac{4a}{\pi}}$ (viii)

To evaluate λ_T, λ_P assume $V_T = V_P = 5$ m/s

$Re_T = 17 \cdot 6 \times 10^6;$ $(k/D)_T = \dfrac{0 \cdot 0003}{4} = 0 \cdot 000025$

$Re_P = 8 \cdot 8 \times 10^6;$ $(k/D)_P = \dfrac{0 \cdot 0003}{2} = 0 \cdot 00015$

$\lambda_T = 0 \cdot 0095$ $\lambda_P = 0 \cdot 013$

Noting that in this example $N = 1$ and taking $\alpha_J = 1 \cdot 0$

$E = 0 \cdot 1536;$ $F = 0 \cdot 0247;$ $C = 0 \cdot 1783;$ $G = 0 \cdot 065$

whence from (vii) and (viii), $D_J = 0 \cdot 737$ m

From (vi), $Q_T = 142 \cdot 0$ m³/s; $V_T = 11 \cdot 3$ m/s; $V_P = 11 \cdot 3$ m/s

Using revised estimates of V_T and $V_P = 11 \cdot 3$ m/s

$$\text{Re}_T = 40 \times 10^6; \quad \text{Re}_P = 20 \times 10^6$$

$$\lambda_T = 0\text{·}0092; \qquad \lambda_P = 0\text{·}013$$

whence $D_J = 0\text{·}742$ m

$$Q_T = 143\text{·}98 \text{ m}^3/\text{s}; \quad V_T = 11\text{·}46 \text{ m/s} = V_P$$

Power = 497·4 MW;   Head loss in tunnel = 157·24 m
= difference in elevation between water in reservoir and surge shaft.

Recommended reading

1. Barr, D.I.H. (Dec. 1975) *Two additional methods of direct solution of the Colebrook-White function.* Proc. Instn. Civ. Engrs, Part 2, Vol. 59, pp827−35.
2. Colebrook, C.F. (Feb. 1939) *Turbulent Flow in Pipes with Particular Reference to the Transition between the Smooth and Rough Pipe Laws.* J Inst CE, Vol. 11.
3. Hydraulics Research Station (1978) *Charts for the hydraulic design of channels and pipes.* 4th edition metric. Dept. of the Environment. London: H.M.S.O. 1978.
4. Moody, L.F. *Friction Factors for Pipe Flows.* Trans Am Soc Mech E, Vol. 66, p671.
5. Nikuradse, J. (1950) *Laws of Flow in Rough Pipes* (Translation of *Stromungsgesetze in rauhen Rohren VDI* — Forschungshaft 361 (1933)) Washington: NACA Tech Mem 1292.
6. Skeat, W.O. ed. (1969) *Manual of British Water Engineering* Practice, 4th edn. Institution of Water Engineers.
7. Barr, D.I.H. (Jun. 1981) *Solutions of the Colebrook-White function for the resistance to uniform turbulent flow.* Proc. Instn. Civ. Engrs, Part 2, Vol. 71, pp529−35.

Problems

(Note: Unless otherwise stated assume water temperature is 15°C $(v = 1\text{·}13 \times 10^{-6} \text{ m}^2/\text{s})$.)

1.   A pipeline 20 km long delivers water from an impounding reservoir to a service reservoir the minimum difference in level between which is 100 m. The pipe of uncoated cast iron (k = 0·3 mm) is 400 mm in diameter. Local losses, including entry loss and velocity head amount to $\dfrac{10 \; V^2}{2g}$.

(a) Calculate the minimum uncontrolled discharge to the service reservoir.
(b) What additional head loss would need to be created by a valve to regulate the discharge to 160 l/s?

2.   A long, straight horizontal pipeline of diameter 350 mm and effective roughness size 0·03 mm is to be constructed to convey crude oil of density 860 kg/m^3 and absolute viscosity 0·0064 Ns/m^2 from the oilfield to a port at a steady rate of 7000 m^3/day. Booster pumps, each providing a total head of 20 m with an overall efficiency of 60 per cent are to be installed at regular intervals. Determine the required spacing of the pumps and the power consumption of each.

3.   A service reservoir, A, delivers water through a trunk pipeline ABC to the distribution network having inlets at B and C.
Pipe AB: length = 1000 m;   diameter = 400 mm;   k = 0·06 mm.
Pipe BC: length = 600 m;   diameter = 300 mm;   k = 0·06 mm.
The water surface elevation in the reservoir is 110 m o.d. Determine the maximum permissible outflow at B such that the pressure head elevation at C does not fall below 90.0 m o.d. Neglect losses other than friction, entry and velocity head. Outflow at C = 160 l/s.

4.   (a) Determine the diameter of commercially available spun iron pipe (k = 0·03 mm) for a pipeline 10 km long to convey a steady flow of 200 l/s of water at 15°C from an impounding reservoir to a service reservoir under a gross head of 100 m. Allow for entry loss and velocity head. What is the unregulated discharge in the pipeline?
      (b) Calculate the head loss to be provided by a valve to regulate the flow to 200 l/s.

5.   Booster pumps are installed at 2 km intervals on a horizontal sewage pipeline of diameter 200 mm and effective roughness size, when new, of 0·06 mm. Each pump was found to deliver a head of 30 m when the pipeline was new. At the end of one year the discharge was found to have decreased by 10 per cent due to pipe wall deposits while the head at the pumps increased to 32 m. Considering only friction losses determine the discharge when the pipeline was in new condition and the effective roughness size after one year.

6.   An existing spun iron trunk pipeline 15 km long, 400 mm in diameter and effective roughness size 0·10 mm delivers water from an impounding reservoir to a service reservoir under a minimum gross head of 90 m. Losses in bends and valves are estimated to total $\dfrac{12\ V^2}{2g}$ in addition to the entry loss and velocity head.

(a) Determine the minimum discharge to the service reservoir.
(b) The impounding reservoir can provide a safe yield of 300 l/s. Determine the minimum length of 400 mm diameter uPVC pipeline (k = 0·03 mm) to be laid in parallel with the existing line so that a discharge equal to

the safe yield could be delivered under the available head. Neglect local losses in the new pipe and assume local losses of $\dfrac{12\,V^2}{2g}$ in the duplicated length of the original pipeline.

7.   A proposed small-scale hydro-power installation will utilise a single Pelton Wheel supplied with water by a 500 m long, 300 mm diameter pipeline of effective roughness 0·03 mm. The pipeline terminates in a nozzle ($C_v = 0·98$) which is 15 m below the level in the reservoir. Determine the nozzle diameter such that the jet will have the maximum possible power using the available head and determine the jet power.

8.   Oil of absolute viscosity 0·07 Ns/m^2 and density 925 kg/m^2 is to be pumped by a rotodynamic pump along a uniform pipeline 500 m long to discharge to atmosphere at an elevation of + 80 m o.d. The pressure head elevation at the pump delivery is 95 m o.d. Neglecting minor losses compare the discharges attained when the pipe, of roughness 0·06 mm, is (a) 100 mm, (b) 150 mm and state in each whether the flow is laminar or turbulent.

9.   A pipeline 10 km long is to be designed to deliver water from a river through a pumping station to the inlet tank of a treatment works. Elevation of delivery pressure head at pumping station = 50 m o.d.; elevation of water in tank = 30 m o.d. Neglecting minor losses, compare the discharges obtainable using

(a) a 300 mm diameter plastic pipeline which may be considered to be smooth
    (i) using the Colebrook-White equation
    (ii) using the Blasius equation;
(b) a 300 mm diameter pipeline with an effective roughness of 0·6 mm
    (i) using the Kármán-Prandtl rough law
    (ii) using the Colebrook-White equation.

10.   Determine the hydraulic gradient in a rectangular concrete culvert 1 m wide and 0.6 m high of roughness size 0·06 mm when running full and conveying water at a rate of 2·5 m^3/s.

Chapter 5
Pipe Network Analysis

R.E. Featherstone

5.1 Introduction

Water distribution network analysis provides the basis for the design of new systems and the extension of existing systems. Design criteria are that specified minimum flow rates and pressure heads must be attained at the outflow points of the network. The flow and pressure distributions across a network are affected by the arrangement and sizes of the pipes and the distribution of the outflows. Since a change of diameter in one pipe length will affect the flow and pressure distribution everywhere, network design is not an explicit process. Optimal design methods almost invariably incorporate the hydraulic analysis of the system in which the pipe diameters are systematically altered (see for example reference 3).

Pipe network analysis involves the determination of the pipe flow rates and pressure heads which satisfy the continuity and energy conservation equations. These may be stated:

(i) Continuity: The algebraic sum of the flow rates in the pipes meeting at a junction, together with any external flows, is zero.

$$\sum_{I=1}^{I=NP(J)} Q_{IJ} - F_J = 0, \ J = 1, NJ \tag{5.1}$$

in which Q_{IJ} is the flow rate in pipe IJ, at junction I, NP(J) is the number of pipes meeting at junction J, F_J is the external flow rate (outflow) at J and NJ is the total number of junctions in the network.

(ii) Energy conservation: The algebraic sum of the head losses in the pipes, together with any heads generated by in-line booster pumps, around any closed loop formed by pipes is zero.

$$\sum_{J=1}^{J=NP(I)} h_{L,I,J} - Hm,_{IJ} = 0, \ I = 1, NL \tag{5.2}$$

in which $h_{L,I,J}$ is the head loss in pipe J of loop I and $Hm,_{IJ}$ is the manometric head generated by a pump in line I, J.

When the equation relating energy losses to pipe flow rate are introduced into equations (5.1) or (5.2) systems of non-linear equations are produced. No method exists for the direct solution of such sets of equations and all methods of pipe network analysis are iterative. Pipe network analysis is therefore ideally suited to digital computer application but simple networks can be analysed with the aid of a pocket calculator.

The earliest systematic method of network analysis, due to Professor Hardy-Cross, and known as the head balance or 'loop' method is applicable to systems in which the pipes form closed loops. Assumed pipe flow rates, complying with the continuity requirement, equation (5.1), are successively adjusted, loop by loop, until in every loop equation (5.2) is satisfied within a specified small tolerance. In a similar, later, method, due to Cornish, assumed junction head elevations are systematically adjusted until equation (5.1) is satisfied at every junction within a small tolerance; it is applicable to both open and closed-loop networks. These methods are amenable to desk calculation but can be programmed for automatic digital computer analysis. However convergence is slow since the hydraulic parameter is adjusted at one element (either loop or junction) at a time. In later methods such as the Newton-Raphson and Linear Theory methods, systems of simultaneous linear equations, derived from equations (5.1), (5.2) and the head loss v. flow rate relationship, are formed, enabling corrections to the hydraulic parameters (flows or heads) to be made over the whole network simultaneously. Convergence is much more rapid but since a number of simultaneous linear equations, depending on the size of the network, have to be solved, the Newton-Raphson and Linear Theory methods are only realistically applicable to computer evaluation.

The majority of the worked examples in this chapter illustrate the use of equations (5.1) and (5.2) in systems which can be analysed by desk calculation using either the head balance or quantity balance methods. In addition to friction losses, the effect of local losses and booster pumps is shown. The networks illustrated have been analysed by computer but the intermediate steps in the computations have been reproduced, enabling the reader to follow the process as though it were by desk calculation; the numbers have been rounded to an appropriate number of decimal places. An example showing the Linear Theory method is given.

5.2 The head balance method ('Loop' method)

This method is applicable to closed-loop pipe networks. It is probably more widely applied to this type of network than the quantity balance method. The head balance method was originally devised by Professor Hardy-Cross and is often referred to as the Hardy-Cross method. Figure 5.1 represents the main pipes in a water distribution network.

The outflows from the system are generally assumed to occur at the nodes (junctions); this assumption results in uniform flows in the pipelines which simplifies the analysis.

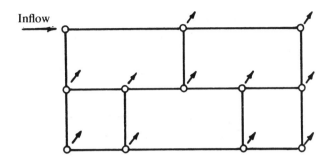

Figure 5.1 Closed loop type network

For a given pipe system, with known junction outflows the head balance method is an iterative procedure based on initially estimated flows in the pipes. At each junction these flows must satisfy the continuity criterion.

The head balance criterion is that the algebraic sum of the head losses around any closed loop is zero; the sign convention that clockwise flows (and the associated head losses) are positive is adopted.

The head loss along a single pipe is

$$h_L = KQ^2$$

If the flow is estimated with an error ΔQ

$$h_L = K(Q + \Delta Q)^2 = K[Q^2 + 2Q\Delta Q + \Delta Q^2]$$

Neglecting ΔQ^2, assuming ΔQ to be small:

$$h_L = K(Q^2 + 2Q\Delta Q)$$

Now round a closed loop $\Sigma h_L = 0$ and ΔQ is the same for each pipe to maintain continuity.

$$\Sigma h_L = \Sigma KQ^2 + 2\Delta Q \Sigma KQ = 0$$

$$\text{i.e. } \Delta Q = -\frac{\Sigma KQ^2}{2\Sigma KQ} = -\frac{\Sigma KQ^2}{2\Sigma \dfrac{KQ^2}{Q}}$$

$$\text{which may be written } \Delta Q = -\frac{\Sigma h}{2\Sigma h/Q}$$

where h is the head loss in a pipe based on the estimated flow Q.

5.3 The quantity balance method ('nodal' method)

Figure 5.2 shows a branched pipe system delivering water from the impounding reservoir A to the service reservoirs B, C and D. F is a known direct outflow from J.

If Z_J is the true elevation of the pressure head at J the head loss along each pipe can be expressed in terms of the difference between Z_J and the pressure head elevation at the other end.

For example: $h_{L,AJ} = Z_A - Z_J$.

Expressing the head loss in the form: $h_L = KQ^2$, N such equations can be written where N is the number of pipes.

i.e.
$$
\begin{bmatrix} Z_A - Z_J \\ Z_B - Z_J \\ \cdot \quad \cdot \\ \cdot \quad \cdot \end{bmatrix} = \begin{bmatrix} (SIGN) \; K_{AJ}(|Q_{AJ}|)^2 \\ (SIGN) \; K_{BJ}(|Q_{BJ}|)^2 \\ \cdot \quad \cdot \quad \cdot \\ \cdot \quad \cdot \quad \cdot \end{bmatrix}
$$

and in general $\quad \begin{bmatrix} Z_I - Z_J \end{bmatrix} = \begin{bmatrix} (SIGN) \; K_{IJ} \, (|Q_{IJ}|)^2 \end{bmatrix}$ \hfill (5.3)

(SIGN) is $+$ or $-$ according to the sign of $(Z_I - Z_J)$. Thus flows towards the junction are positive and flows away from the junction are negative. K_{Ij} is composed of the friction loss and minor loss coefficients.

The continuity equation for flow rates at J is:

$$\Sigma \, Q_{IJ} - F \, (= Q_{AJ} + Q_{BJ} + Q_{CJ} + Q_{DJ} - F) = 0 \tag{5.4}$$

Examination of equations, sets (5.3) and (5.4) shows that the correct value of Z_J will result in values of Q_{IJ}, calculated from set (5.3) which will satisfy equation (5.4).

Rearranging set (5.3) we have

$$[Q_{IJ}] = \left[(SIGN) \left(\frac{|Z_I - Z_J|}{K_{IJ}} \right)^{1/2} \right] \tag{5.5}$$

The value of Z_J can be found using an iterative method, by making an initial estimate of Z_J, calculating the pipe discharges from equation set (5.5) and testing the continuity condition in equation (5.4).

If $(\Sigma Q_{IJ} - F) \neq 0$ (with acceptable limits) a correction, ΔZ_J is made to Z_J and the procedure repeated until equation (5.4) is reasonably satisfied. A systematic correction for ΔZ_J can be developed: expressing the head loss

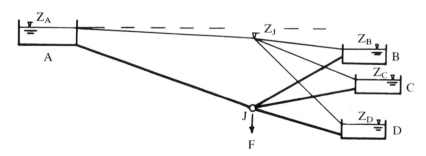

Figure 5.2 Branched-type pipe network

along a pipe as $h_L = KQ^2$, for a small error in the estimate Z_J, the correction ΔZ_J can be derived as:

$$\Delta Z_J = \frac{2(\Sigma Q_{IJ} - F_J)}{\displaystyle\sum \frac{Q_{IJ}}{h_{L,IJ}}}$$

Example 5.7 shows the procedure for networks with multiple unknown junction head elevations.

Evaluation of K_{IJ}:

$$K_{IJ} = \frac{\lambda L}{2g\, DA^2} + \frac{C_m}{2g\, A^2} \; (= K_f + K_m)$$

where C_m = sum of the minor loss coefficients. λ can be obtained from the Moody chart using an initially assumed value of velocity in the pipe (say 1 m/s). A closer approximation to the velocity is obtained when the discharge is calculated. For automatic computer analysis equation (5.5) should be replaced by the Darcy-Colebrook-White combination:

$$Q = - 2A \sqrt{2g\, D\, \frac{h_f}{L}} \; \log \left[\frac{k}{3 \cdot 7\, D} + \frac{2 \cdot 51\, v}{D \sqrt{2g\, D\, \dfrac{h_f}{L}}} \right] \qquad (5.6)$$

For each pipe $h_{f,IJ}$ (friction head loss) is initialised to $Z_I - Z_J$, Q_{IJ} calculated from equation (5.6) and $h_{f,IJ}$ re-evaluated from $h_{f,IJ} = (Z_I - Z_J) - K_m Q_{IJ}^2$. This subroutine follows the procedure of Example 4.2.

5.4 Newton Raphson method

The Newton Raphson method differs from the 'head' and 'quantity' balance methods in that it makes corrections to assumed heads or flow rates over the whole network simultaneously (see reference 4).

5.5 The linear theory method

The energy conservation equations when expressed in terms of flow rate produce a set of non-linear equations:

$$\sum_{J=1}^{NP(I)} K_{IJ} Q_{IJ}^2 = 0, \; I = 1, NL$$

in which NL is the number of closed loops.

By writing the head loss ($h_{L,p}$) in pipe, p, as $h_{L,p} = K'(p)Q(p)$ in which

$$K'(p) = K(p)\, Q_o(p)$$

where $Q_o(p)$ is the estimated, or current value of Q, the non-linear equations are transformed into linear equations. When the NL linearised loop equations are combined with NJ-1 independent continuity equations a system of NP linear equations is formed. Solving by Gaussian elimination, the values of $Q(p)$ are obtained. Since initial values of Q are estimated the procedure is repeated until successive $Q(p)$ values are sufficiently close. Like the Newton Raphson method, the Linear Theory method is not suited to hand calculation; it also converges much more quickly than the 'head' or 'quantity' balance methods. Example 5.8 illustrates the technique.

5.6 The gradient method

In addition to equations (5.1) to (5.6) the Gradient Method needs the following vector and matrixes definitions:

NT = Number of pipelines in the network

NN = Number of unknown piezometric head nodes

[A12] = 'Connectivity matrix' associated with each one of the nodes. Its dimension is NT × NN with only two non-zero elements in the ith row:

−1 in the column corresponding to the initial node of pipe i

1 in the column corresponding to the final node of pipe i

NS = Number of fixed head nodes.

[A10] = Topologic matrix: pipe to node for the NS fixed head nodes. Its dimension is NT × NS with a −1 value in rows corresponding to pipelines connected to fixed head nodes.

Thus, the head losses in each pipe between two nodes is:

$$[A11][Q] + [A12][H] = -[A10][H_o] \tag{5.7}$$

where

[A11] = Diagonal matrix of NT × NT dimension, defined as:

$$[A11] = \begin{bmatrix} \alpha_1 Q_1^{(n_1-1)} + \beta_1 + \frac{\gamma_1}{Q_1} & 0 & 0 & \cdots & 0 \\ 0 & \alpha_2 Q_2^{(n_2-1)} + \beta_2 + \frac{\gamma_2}{Q_2} & 0 & \cdots & 0 \\ 0 & 0 & \alpha_3 Q_3^{(n_3-1)} + \beta_3 + \frac{\gamma_3}{Q_3} & \cdots & 0 \\ 0 & 0 & 0 & \cdots & \alpha_{NT} Q_{NT}^{(n_{NT}-1)} + \beta_{NT} + \frac{\gamma_{NT}}{Q_{NT}} \end{bmatrix} \tag{5.8}$$

[Q] = Discharge vector with NT × 1 dimension

[H] = Unknown piezometric head vector with NN × 1 dimension

[H_o] = Fixed piezometric head vector with NS × 1 dimension

Equation (5.7) is an energy conservation equation. The continuity equation for all nodes in the network is:

$$[A21][Q] = [q] \tag{5.9}$$

where

[A21] = Transpose matrix of [A12]

[q] = Water consumption and water supply vector in each node with NN × 1 dimension

In matrix form, equations (5.7) and (5.9) are:

$$\begin{bmatrix} [A11] & [A12] \\ [A21] & [0] \end{bmatrix} \begin{bmatrix} [Q] \\ [H] \end{bmatrix} = \begin{bmatrix} -[A10][H_o] \\ [q] \end{bmatrix} \tag{5.10}$$

The upper part is *nonlinear*, which implies that equation (5.10) must use some *iterative* algorithm for its solution. Gradient method consists of a truncated Taylor expansion. Operating simultaneously on ([Q], [H]) field and applying the gradient operator, we can write:

$$\begin{bmatrix} [N][A11]' & [A12] \\ [A21] & [0] \end{bmatrix} \begin{bmatrix} [dQ] \\ [dH] \end{bmatrix} = \begin{bmatrix} [dE] \\ [dq] \end{bmatrix} \tag{5.11}$$

where

[N] = Diagonal matrix (n_1, n_2, ... n_{NT}) with NT × NT dimension

[A11]′ = NT × NT matrix defined as:

$$[A11]' = \begin{bmatrix} \alpha_1 Q_1^{(n_1-1)} & 0 & 0 & \dots & 0 \\ 0 & \alpha_2 Q_2^{(n_2-1)} & 0 & \dots & 0 \\ 0 & 0 & \alpha_3 Q_3^{(n_3-1)} & \dots & 0 \\ . & . & . & \dots & 0 \\ 0 & 0 & 0 & \dots & \alpha_{NT} Q_{NT}^{(n_{NT}-1)} \end{bmatrix} \tag{5.12}$$

In any iteration **i**, [dE] is the energy imbalance in each pipe and [dq] is the discharge imbalance in each node. These are given by:

$$[dE] = [A11][Q_i] + [A12][H_i] + [A10][H_o] \tag{5.13}$$

and

$$[dq] = [A21][Q_i] - [q] \tag{5.14}$$

The objective of the gradient method is to solve the system described by equation (5.11), taking into account that in each iteration:

$$[dQ] = [Q_{i+1}] - [Q_i] \tag{5.15}$$

and

$$[dH] - [H_{i+1}] - [H_i] \tag{5.16}$$

Using matrix algebra, it is possible to show that the solution to the system represented by equation (5.11) is:

$$[H_{i+1}] = \quad -\{[A21]([N][A11]')^{-1}[A12]\}^{-1}\{[A21]([N][A11]')^{-1} \\ ([A11][Q_i]) + [A10][H_o] - ([A21][Q_i]) - [q]\} \tag{5.17}$$

$$[Q_{i+1}] = \{ [I] - ([N][A11]') - [A11] \} [Q_i] - \{ ([N][A11]')^{-1}([A12]$$
$$[H_{i+1}] + [A10][H_o]) \} \qquad (5.18)$$

The method has the advantage of a faster convergence than the linear theory method as it uses smaller matrixes for a given network. Also, it does not need continuity balancing in each node to begin the process. The method is not suited to hand calculation. Example 5.9 illustrates the methodology.

Worked examples

Example 5.1
Neglecting minor losses in the pipes determine the flows in the pipes and the pressure heads at the nodes. (See fig. 5.3.)

Data:

Pipe	AB	BC	CD	DE	EF	AF	BE
Length (m)	600	600	200	600	600	200	200
Diameter (mm)	250	150	100	150	150	200	100

Roughness size of all pipes = 0·06 mm
Pressure head elevation at A = 70 m o.d.

Elevation of pipe nodes

Node	A	B	C	D	E	F
Elevation (m o.d.)	30	25	20	20	22	25

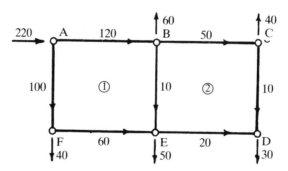

Figure 5.3   Two-looped network

Procedure:
1.   Identify loops. When using hand calculation the simplest way is to employ adjacent loops, e.g. Loop 1: ABEFA; Loop 2: BCDEB.
2.   Allocate estimated flows in the pipes. Only one estimated flow in each loop is required; the remaining flows follow automatically from the continuity condition at the nodes, e.g. since the total required inflow is 220 l/s, if Q_{AB} is estimated at 120 l/s then $Q_{AF} = 100$ l/s. The initial flows are shown in fig. 5.3.

3.   The head loss coefficient $K = \dfrac{\lambda\, L}{2g\, DA^2}$ ($h_L = KQ|Q|$) is evaluated for each pipe, λ being obtained from the λ v. Re diagram (fig. 4.2) corresponding with the flow in the pipe. Alternatively Barr's equation (4.12) may be used.

If the Reynolds numbers are fairly high ($\nleq 10^5$), it may be possible to proceed with the iterations using the initial λ values making better estimates as the solution nears convergence.

The calculations proceed in tabular form. Note that Q is written in l/s simply for convenience; all computations are based on Q in m^3/s. However h_L/Q could have been expressed in $\left(\dfrac{m}{l/s}\right)$ yielding ΔQ directly in l/s.

Loop	Pipe	k/D	Q(l/s)	Re($\times 10^5$)	λ	K	h_L (m)	$h_{L/Q}\left(\dfrac{m}{m^3/s}\right)$
	AB	0·00024	120·00	5·41	0·0157	797·0	11·48	95·64
1	BE	0·00060	10·00	1·31	0·0205	33877·0	3·39	338·77
	EF	0·00040	−60·00	4·51	0·0172	11229·1	−40·42	673·75
	FA	0·00030	−100·00	5·63	0·0162	836·6	−8·36	83·66
						Σ	−33·91	1191·82

$$\Delta Q = \frac{-\,\Sigma h}{2\Sigma\, h/Q} = \frac{-(-33.91)}{2 \times 1191.82} = 0.01423 = 14.23 \text{ l/s}$$

Loop	Pipe	Q(l/s)	Re($\times 10^5$)	λ	K	h_L (m)	$h_{L/Q}\left(\dfrac{m}{m^3/s}\right)$
	BC	50·0	3·76	0·0174	11359·7	28·40	567·98
2	CD	10·0	1·13	0·0205	33877·0	3·39	338·77
	DE	−20·0	1·50	0·0189	12338·9	−4·94	246·78
	EB	−24·23	2·73	0·0189	31232·9	−18·34	756·77
					Σ	8·51	1910·30

$\Delta Q = -2\cdot23$ l/s
(Note that the previously corrected value of flow in the 'common' pipe, EB, has been used in loop 2.)

Loop	Pipe	Q(l/s)	Re($\times 10^5$)	λ	K	h_L (m)	$h_{L/Q}$
	AB	134·23	6·05	0·0156	791·9	14·27	106·30
1	BE	26·46	2·98	0·0188	31067·7	21·75	822·05
	EF	−45·77	3·44	0·0175	11424·9	−23·93	522·92
	FA	−85·77	4·83	0·0164	846·9	−6·23	72·64
					Σ	5·86	1523·91

$\Delta Q = -1.92$ l/s

Proceed to loop 2 again and continuing in this way the solution is obtained within the required specified limit on Σh_L in any loop after several further iterations. The solution given is obtained for $\Sigma h_L < 0.01$ m but an acceptable result may be achieved with a larger tolerance.

Final values (Flows in direction of pipe identifier; e.g. A → B)

Pipe	Q(l/s)	h_L (m)
AB	131·55	13·70
BE	25·02	19·55
FE	48·45	26·67
AF	88·45	6·59
BC	46·53	24·74
CD	6·55	1·52
ED	23·47	6·69

Pressure heads

Node	Pressure head (m)
A	40·00
B	31·29
C	11·57
D	10·05
E	14·74
F	38·41

Example 5.2
In the network shown a valve in BC is partially closed to produce a local head loss of $10.0 \dfrac{V_{BC}^2}{2g}$. Analyse the flows in the network. (See fig. 5.4.)

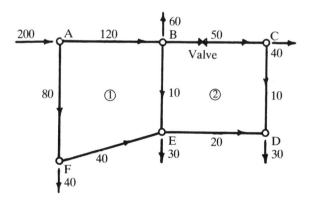

Figure 5.4 Pipe network with valve losses

Roughness of all pipes = 0·06 mm

Pipe	AB	BC	CD	DE	BE	EF	AF
Length (m)	500	400	200	400	200	600	300
Diameter (mm)	250	150	100	150	150	200	250

Solution:
The procedure is identical with that of the previous problem. K_{BC} is now composed of the valve loss coefficient and the friction loss coefficient.

With the initial assumed flows shown in the table below: Q_{BC} = 50 l/s; Re = $3·7 \times 10^5$; k/D = 0·0004; λ = 0·0174 (from Moody chart). Hence K_f = 7573 and K_m = 1632 and K_{BC} = 9205.

Loop	Pipe	k/D	Q(l/s)	Re($\times 10^5$)	λ	K	h_L (m)	$h_{L/Q}\left(\dfrac{m}{m^3/s}\right)$
	AB	0·00024	120·00	5·41	0·0157	664·2	9·56	79·70
	BE	0·0004	10·00	0·75	0·0208	4526·5	0·45	45·26
1	EF	0·0003	−40·00	2·25	0·0175	2711·2	−4·34	108·45
	FA	0·00024	−80·00	3·61	0·0163	413·7	−2·65	33·10
						Σ	3·03	266·51

ΔQ = −5·69 l/s

Loop	Pipe	k/D	Q(l/s)	Re($\times 10^5$)	λ	K	h_L (m)	$h_{L/Q}$
	BC	0·0004	50·00	3·75	0·0174	9205·2	23·01	460·26
	CD	0·0006	10·00	1·13	0·0205	33877·0	3·39	338·77
2	DE	0·0004	−20·00	1·50	0·0190	8226·0	−3·29	164·52
	EB	0·0004	−4·31	0·32	0·0242	5266·4	−0·10	22·70
						Σ	23·01	986·25

ΔQ = −11·67 l/s

Proceeding in this way the solution is obtained within a small limit on Σh_L in any loop:

Final values

Pipe	AB	BE	FE	FA	BC	CD	ED
Q(l/s)	111·52	16·48	48·48	88·48	35·05	4·95	34·95
h_L (m)	8·31	1·15	6·26	3·20	11·57	0·91	9·52

Example 5.3
If in the network shown in Example 5.2 a pump is installed in line BC boosting the flow towards C and the valve removed, analyse the network. Assume that the pump delivers a head of 10 m. (Note: In reality it would not be possible to predict the head generated by the pump since this will

depend upon the discharge. The head-discharge relationship for the pump e.g. $H = AQ^2 + BQ + C$ must therefore be solved for the discharge in the pipe at each iteration. However, for the purpose of illustration of the basic effect of a pump the head in this case is assumed known.) An example of a network analysis in which the pump head discharge curve is used is given in Chapter 6 (Example 6.8). Consider length BC. (See fig. 5.5)

The net loss of head along BC ($Z_B - Z_C$) is ($h_f - H_p$) where H_p = total head delivered by pump. The K value for BC is now due to friction only; the head loss for BC in the table now becomes $h_{L.BC} = (KQ_{BC}^2 - 10 \cdot 0)$ m. Otherwise the iterative procedure is as before.

Solution:

Loop	Pipe	k/D	Q(l/s)	Re(× 10⁵)	λ	K	h_L (m)	$h_{L/Q}$ $\left(\dfrac{m}{m^3/s}\right)$
	AB	0·00024	120·00	5·41	0·0157	664·2	9·56	79·70
1	BE	0·00040	10·00	0·75	0·0208	4526·5	0·45	45·26
	EF	0·00030	−40·00	2·25	0·0175	2711·2	−4·38	108·45
	FA	0·00024	−80·00	3·61	0·0163	413·7	−2·65	33·10

$\Delta Q = -5 \cdot 69$ l/s Σ 3·03   266·51

Loop	Pipe	k/D	Q(l/s)	Re(× 10⁵)	λ	K	h_L (m)	$h_{L/Q}$
	BC	0·00040	50·00	3·76	0·0174	7573·0	8·93	178·66
2	CD	0·00060	10·00	1·13	0·0205	33877·0	3·39	333·77
	DE	0·00040	−20·00	1·50	0·0189	8225·96	−3·29	164·52
	EB	0·00040	−4·31	0·32	0·0242	5266·4	−0·10	22·70

$\Delta Q = -6 \cdot 34$ l/s Σ 8·93   704·65

Loop	Pipe	Q(l/s)	Re(× 10⁵)	λ	K	h_L (m)	$h_{L/Q}$
	AB	114·31	5·15	0·0158	668·4	8·73	76·41
1	BE	10·65	0·80	0·0206	4482·9	0·51	47·74
	EF	−45·69	2·57	0·0173	2680·2	−5·59	122·46
	FA	−85·69	3·66	0·0162	411·2	−3·02	35·24

$\Delta Q = -1 \cdot 11$ l/s Σ 0·63   281·85

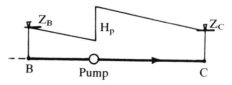

Figure 5.5   Network of example 5.2 with pump

After similar further iterations:

Final values:

Pipe	AB	BE	FE	FA	BC	CD	ED
Q(l/s)	113·21	8·90	46·79	86·79	44·30	4·30	25·70
h_L (m)	8·57	0·37	5·83	3·10	4·95	0·71	5·29

Example 5.4
Determine the discharges in the pipes of the network shown in fig. 5.6 neglecting minor losses.

Pipe	Length (m)	Diameter (mm)
AJ	10 000	450
BJ	2000	350
CJ	3000	300
DJ	3000	250

Roughness size of all pipes = 0·06 mm

The friction factor λ may be obtained from the Moody diagram, or using Barr's equation, using an initially estimated velocity in each pipe. Subsequently λ can be based on the previously calculated discharges. However, unless there is a serious error in the initial velocity estimates, much effort is saved by retaining the initial λ values until perhaps the penultimate or final correction.

Solution:
Estimate Z_J (pressure head elevation at J) = 150·0 m a.o.d. (Note that the elevation of the pipe junction itself does not affect the solution.)
See tables on pages 134 and 135.

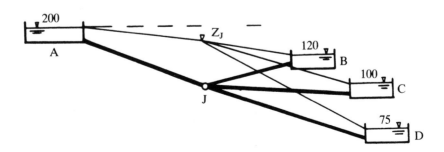

Figure 5.6 Network connecting multi-reservoirs

1st correction

Pipe No. (I–J)	Velocity (estimate) m/s	Re(× 10⁵)	λ	K	$Z_I - Z_J$	Q (m³/s)	(Q/h) × 10⁻³	Q/A (m/s)
AJ	2·0	7·96	0·0145	649	+50	0·2775	5·55	1·75
BJ	2·0	6·20	0·0150	472	−30	−0·2521	8·40	2·62
CJ	2·0	5·31	0·0155	1581	−50	−0·1778	3·56	2·50
DJ	2·0	4·42	0·0165	4188	−75	−0·1338	1·78	2·73
					Σ	−0·2862	0·0193	

$$\text{Correction to } Z_J = \frac{2(-0·2862)}{0·0193} = -29·67: Z_J = 120·33 \text{ m}$$

2nd correction

Pipe No. (I–J)	Velocity (estimate) m/s	Re(× 10⁵)	λ	K	$Z_I - Z_J$	Q (m³/s)	(Q/h) × 10⁻³	Q/A (m/s)
AJ	As	7·96	0·0145	649	79·67	0·3504	4·39	2·20
BJ	initial	6·20	0·0150	472	−0·33	−0·0264	80·12	0·27
CJ	estimate	5·31	0·0155	1581	−20·33	−0·1134	5·58	1·60
DJ		4·42	0·0165	4188	−45·33	−0·1040	2·29	2·20
					Σ	+0·1066	+0·092	

$\Delta Z_J = +2·30$ m; $Z_J = 122·63$ m

Comment: The velocity in BJ has changed significantly but it may oscillate; it is therefore estimated at 1·0 m/s for next correction. Note λ (BJ) altered accordingly.

3rd correction

Pipe No. (I–J)	Velocity (Estimate) m/s	λ	K	$Z_I - Z_J$	Q (m^3/s)	$(Q/h) \times 10^{-3}$	Q/A (m/s)
AJ	2·0	0·0145	649	77·37	0·3452	4·46	2·17
BJ	1·0	0·016	503	−2·63	−0·0723	27·50	0·75
CJ	1·8	0·0155	1581	−22·63	−0·1196	5·29	1·69
DJ	2·3	0·016	4061	47·63	−0·1083	2·27	2·21
				Σ	+0·0450	0·0395	

$\Delta Z_J = 2\cdot 27$ m; $Z_J = 124\cdot 90$ m

Final values:

$Q_{AJ} = 0\cdot 344$ m^3/s; $Q_{JB} = 0\cdot 105$ m^3/s; $Q_{JC} = 0\cdot 127$ m^3/s;

$Q_{JD} = 0\cdot 112$ m^3/s.

Example 5.5

If in the network of Example 5.4 the flow to C is regulated by a valve to 100 l/s, calculate the effect on the flows to the other reservoirs; determine the head loss to be provided by the valve.

The principle of the solution is identical with that of the previous example except that the flow in JC is prescribed and is simply treated as an EXTERNAL OUTFLOW at J. See table on page 134. In this example the flow rates in the pipes have been evaluated directly from equation (5.6).

$$\left(Q = -2A \sqrt{2g\, D\, \frac{h_f}{L}} \log \left[\frac{k}{3\cdot 7\, D} + \frac{2\cdot 51\, v}{D \sqrt{2g\, D\, \dfrac{h_f}{L}}} \right] \right)$$

in which $h_f = Z_I - Z_J$, since there are no minor losses. This approach is ideal for computer analysis; if minor losses are present use the iterative procedure described on page 98.

The method is also suitable for desk analysis using an electronic calculator since for each pipe the only variable is h_f and equation (5.6) can be written:

$$Q = -C_1 \sqrt{h_f} \log \left[C_2 + \frac{C_3}{\sqrt{h_f}} \right]$$

in which C_1, C_2 and C_3 are constant for a particular pipe.

The corresponding velocities and λ values have been evaluated and tabulated; this data may be useful for those who wish to work through the example using the Moody diagram as in Example 5.4.

Note that Q is expressed in l/s; in evaluating $\Sigma Q/h$ the flow is also expressed in l/s so that the units in the correction term: $\Delta Z = 2(\Sigma Q - F)/(\Sigma Q/h)$, are consistent.

Example 5.5 calculations; Estimate Z_J = 150·00 a.o.a.

Pipe	AJ	BJ	DJ
k/D	0·000133	0·000171	0·000240

1st correction

Junction	Pipe (I–J)	$Z_I - Z_J$ (=h) (m)	Q(l/s)	Q/h	V(m/s)	λ
	AJ	50·00	279·32	5·59	1·76	0·0143
J	BJ	−30·00	−255·95	8·53	2·66	0·0146
	DJ	−75·00	−137·90	1·84	2·81	0·0155
	Σ		−114·53	15·96		

$$\text{Correction to } Z_J = \frac{2(\Sigma Q - F)}{\Sigma Q/h} = \frac{2(-144·53 - 100)}{15·96} = -26·89 \text{ m}$$

$$Z_J = 123.11 \text{ m}$$

2nd correction

Junction	Pipe	$Z_I - Z_J$	Q(l/s)	Q/h	V(m/s)	λ
	AJ	76·89	349·70	4·55	2·20	0·0140
J	BJ	−3·11	−77·61	24·96	0·81	0·0164
	DJ	−48·11	−109·50	2·28	2·23	0·0158

ΔZ_J = 3·94 m; Z_J = 127·05 m

3rd correction

Junction	Pipe	$Z_I - Z_J$	Q(l/s)	Q/h	V(m/s)	λ
	AJ	72·95	340·2	4·66	2·14	0·0141
J	BJ	−7·05	−119·94	17·01	1·25	0·0156
	DJ	−52·05	−114·08	2·19	2·32	0·0158

ΔZ_J = 0·52 m; Z_J = 127·57 m

Final values:

$$Z_J = 127·55 \text{ m}$$

Pipe	AJ	JB	JD
Q(l/s)	338·98	124·36	114·65

Head loss due to friction along JC

Diameter = 300 mm; A = 0·0707 m^2; Q = 0·100 m^3/s; V = 1·415 m/s

$$Re = \frac{1 \cdot 415 \times 0 \cdot 3}{1 \cdot 13 \times 10^{-6}} = 3 \cdot 76 \times 10^{5}; \quad k/D = 0 \cdot 0002$$

whence $\lambda = 0 \cdot 016$; $\quad h_f = \dfrac{0 \cdot 016 \times 3000 \times 1 \cdot 415^2}{19 \cdot 62 \times 0 \cdot 3} = 16 \cdot 33$ m

(See fig. 5.7.)

$\therefore$ Head loss at valve $= Z_J - Z_C - h_f$

$$= 127 \cdot 55 - 100 \cdot 00 - 16 \cdot 33 = 11 \cdot 22 \text{ m}$$

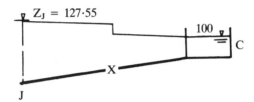

Figure 5.7   Network of example 5.4 with valve losses

Example 5.6
In the network as before, a pump, P, is installed on JB to boost the flow to B. With the flows to C and D uncontrolled and the pump delivering 10 metres head, determine the flows in the pipes. (See fig. 5.8.)

(Note that in the case of rotodynamic pumps the manometric head delivered varies with the discharge (see Chapter 6). Thus it is not strictly possible to specify the head and it is necessary to solve the pump equation $H_p = AQ^2 + BQ + C$ together with the resistance equation for JB. However to illustrate the effect of a pump in this example let us assume that the head does not vary with flow.)

Solution:
The analysis is straightforward, and follows the procedure of Example 5.5.
   The head giving flow along JB is:

$$h_{L,JB} = Z_J - Z_B - H_p$$

The final solution is:

$Z_J = 119 \cdot 66$ m o.d.

Pipe	AJ	JB	JC	JD
Q(l/s)	357·7	141·6	110·8	105·3

Example 5.7
Determine the flows in the network shown in fig. 5.9, neglecting minor losses.

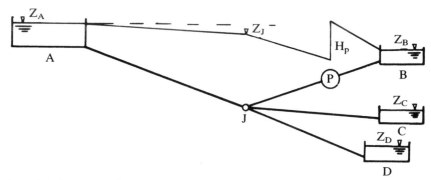

Figure 5.8   Network of example 5.4 with pump

Data:

Pipe	AB	BC	BD	BE	EF	EG
Length (m)	10 000	3000	4000	6000	3000	3000
Diameter (mm)	450	250	250	350	250	200

Roughness of all pipes = 0.03 mm (= k)

Solution:
In this case there are two unknown pressure head elevations which must both therefore be initially estimated and corrected alternately.

Estimate Z_B = 120.0 m o.d.; Z_E = 95.0 m o.d.

1st correction

Junction (J)	Pipe (I–J)	$Z_I - Z_J$ (=h)	Q(l/s)	Q/h	V(m/s)	λ
	AB	30·00	219·77	7·33	1·38	0·0139
B	CB	−20·00	−71·38	3·57	1·45	0·0155
	DB	−40·00	−86·75	4·34	1·77	0·0151
	EB	−25	−135·00	5·40	1·40	0·0145
		Σ −73·35		20·63		

ΔZ_B = +7·11 m; Z_B = 112·89 m

Proceed to junction E noting that the amended value of Z_B is now used:

Junction (J)	Pipe (I–J)	$Z_I - Z_J$	Q(l/s)	Q/h	V(m/s)	λ
	BE	17·89	112·81	6·31	1·17	0·0149
E	FE	−20·00	−71·38	3·57	1·45	0·0155
	GE	−35·00	−53·38	1·53	1·70	0·0159
		Σ 11·95		11·40		

ΔZ_E = −2·1 m; Z_E = 92·9 m

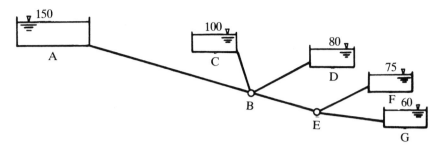

Figure 5.9 Network with multi-reservoirs

2nd correction

Junction	Pipe	$Z_I - Z_J$	Q(l/s)	Q/h	V(m/s)	λ
	AB	37·11	246·21	6·63	1·55	0·0137
B	CB	−12·89	−56·38	4·37	1·15	0·0160
	DB	−32·89	−78·16	6·06	1·59	0·0153
	EB	−19·99	−119·75	5·99	1·25	0·0148
	Σ		−8·07	23·06		

$\Delta Z_B = -0.7$ m; $Z_B = 112.19$ m

Junction	Pipe	$Z_I - Z_J$	Q(l/s)	Q/h	V(m/s)	λ
	BE	92·9	117·48	6·09	1·22	0·0148
E	FE	−17·9	−67·26	3·76	1·37	0·0156
	GE	−32·9	−51·64	1·57	1·64	0·0159
	Σ		−1·43	11·42		

$\Delta Z_E = -0.25$ m; $Z_E = 92.65$ m

Example 5.8
Analyse the network of Example 5.1 by the Linear Theory method.

Solution:
(See fig. 5.3.)

Pipe	AB	BE	EF	FA	BC	CD	DE
Pipe Number	1	2	3	4	5	6	7

Junction Continuity equations (flow rates in m^3/s)

$$- Q_1 - Q_4 = -0.220$$
$$Q_1 - Q_2 - Q_5 = 0.060$$
$$Q_5 - Q_6 = 0.040$$
$$Q_2 + Q_3 - Q_7 = 0.050$$
$$- Q_3 + Q_4 = 0.040$$

The loop energy conservation equations are:

$$K_1'Q_1 + K_2'Q_2 - K_3'Q_3 - K_4'Q_4 = 0$$
$$-K_2'Q_2 + K_5'Q_5 + K_6'Q_6 - K_7'Q_7 = 0$$

K' values are evaluated from

$$K'Q = KQ|Q| = \frac{\lambda L}{2g\,DA^2}\, Q|Q|, \quad (Q \text{ in } m^3/s).$$

λ values have been obtained from Barr's equation (4.12), using the assumed value of 0.030 m³/s in each pipe, in the directions shown in fig. 5.3.

Pipe No.	k/D	Re($\times 10^5$)	λ	K	K'
1	0·00024	1·35	0·0184	931·89	27·95
2	0·00060	3·38	0·0188	31005·0	930·15
3	0·00040	2·25	0·0182	11878·11	356·34
4	0·00030	1·69	0·0182	938·16	28·14
5	0·00040	2·25	0·0182	11878·11	356·34
6	0·00060	3·38	0·0188	31005·00	930·15
7	0·00040	2·25	0·0182	11878·11	356.34

The continuity and energy conservation equations can be written:

$$
\begin{bmatrix}
-1 & & & & -1 & & \\
1 & -1 & & & & -1 & \\
& & & & 1 & -1 & \\
& 1 & 1 & & & & -1 \\
& & -1 & 1 & & & \\
27{\cdot}95 & 930{\cdot}15 & -356{\cdot}34 & -28{\cdot}14 & & & \\
& -930{\cdot}15 & & & 356{\cdot}34 & 930{\cdot}15 & -356{\cdot}34
\end{bmatrix}
\begin{bmatrix}
Q_1 \\ Q_2 \\ Q_3 \\ Q_4 \\ Q_5 \\ Q_6 \\ Q_7
\end{bmatrix}
=
\begin{bmatrix}
-0{\cdot}220 \\ 0{\cdot}060 \\ 0{\cdot}040 \\ 0{\cdot}050 \\ 0{\cdot}040 \\ 0{\cdot}0 \\ 0{\cdot}0
\end{bmatrix}
$$

Solution by Gaussian elimination yields:

$$Q_1 = 0.128; \quad Q_2 = 0.019; \quad Q_3 = 0.052; \quad Q_4 = 0.092;$$
$$Q_5 = 0.049; \quad Q_6 = -0.009; \quad Q_7 = 0.021 \ (\text{in } m^3/s)$$

Revised values of K' are determined using the average of the assumed and revised flow rates in each pipe. After four further corrections the final flows are similar to those given in Example 5.1.

Example 5.9

In the network shown in fig. 5.10 below, a valve in pipe 2-3 is partially closed producing a local head loss of $10V_{2-3}^2/2g$. The head at node1 is 100 m of water. The roughness of all pipes is 0.06 mm. The pipe lengths are in metres and the demand discharges are in l/s.

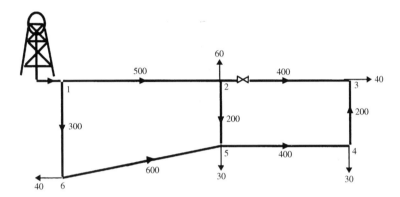

Figure 5.10 Pipe network with valve loss

The pipe diameters are: pipes 1-2 and 1-6, 250 mm; pipe 6-5, 200 mm; pipes 2-3 and 4-5, 150 mm; pipes 2-5 and 3-4, 100 mm. Analyse the network using the Gradient Method.

The iterative process can be summarised in the following steps:

1. Assume initial discharges in each of the network pipes (they can be unbalanced at each node).
2. Solve the system represented by equation (5.17) using a standard method for the solution of simultaneous linear equations.
3. With the calculated $[H_{i+1}]$ (step 2), $[Q_{i+1}]$ is solved by equation (5.18).
4. With the new $[Q_{i+1}]$, equation (5.17) is solved (step 2) to find a new $[H_{i+1}]$.
5. Process continues until

$$[H_{i+1}] \approx [H_i]$$

For all pipes it has been assumed an initial discharge of 100 l/s with the directions as shown in fig. 5.11.

Solution:

All the matrixes and vectors needed for the Gradient are:

  NT = 7
  NN = 5
  NS = 1

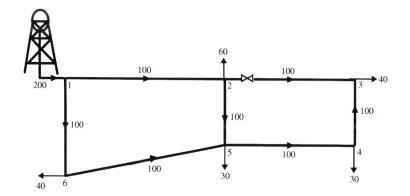

Figure 5.11   Network solution

[A12] = Connectivity matrix; its dimension is (7 × 5).

$$
\begin{vmatrix}
1 & 0 & 0 & 0 & 0 \\
-1 & 1 & 0 & 0 & 0 \\
0 & 1 & -1 & 0 & 0 \\
0 & 0 & 1 & -1 & 0 \\
-1 & 0 & 0 & 1 & 0 \\
0 & 0 & 0 & 1 & -1 \\
0 & 0 & 0 & 0 & 1
\end{vmatrix}
$$

[A21] = Transposed matrix of [A12]

$$
\begin{vmatrix}
1 & -1 & 0 & 0 & -1 & 0 & 0 \\
0 & 1 & 1 & 0 & 0 & 0 & 0 \\
0 & 0 & -1 & 1 & 0 & 0 & 0 \\
0 & 0 & 0 & -1 & 1 & 1 & 0 \\
0 & 0 & 0 & 0 & 0 & -1 & 1
\end{vmatrix}
$$

[A10] = Topologic matrix node to node; dimension (7 × 1).
[Q]   = Discharges vector; dimension (7 × 1).
[H]   = Unknown piezometric head vector; dimension (5 × 1).
[Ho] = Fixed piezometric head vector; dimension (1 × 1).
[q]   = Water demand vector; dimension (5 × 1).

[A10]	[Q] (m³/s)	[H]	[Ho] (m)	[q] (m³/s)
-1	0·10	H_2	| 100 |	0·06
0	0·10	H_3		0·04
0	0·10	H_4		0·03
0	0·10	H_5		0·03
0	0·10	H_6		0·04
0	0·10			
-1	0·10			

[N] = Diagonal matrix; dimension (7 × 7). It has 2 in the diagonal (Darcy-Weisbach head loss equation).

$$
\begin{vmatrix}
2 & 0 & 0 & 0 & 0 & 0 & 0 \\
0 & 2 & 0 & 0 & 0 & 0 & 0 \\
0 & 0 & 2 & 0 & 0 & 0 & 0 \\
0 & 0 & 0 & 2 & 0 & 0 & 0 \\
0 & 0 & 0 & 0 & 2 & 0 & 0 \\
0 & 0 & 0 & 0 & 0 & 2 & 0 \\
0 & 0 & 0 & 0 & 0 & 0 & 2
\end{vmatrix}
$$

[I] = Identity matrix; dimension (7 × 7).

$$
\begin{vmatrix}
1 & 0 & 0 & 0 & 0 & 0 & 0 \\
0 & 1 & 0 & 0 & 0 & 0 & 0 \\
0 & 0 & 1 & 0 & 0 & 0 & 0 \\
0 & 0 & 0 & 1 & 0 & 0 & 0 \\
0 & 0 & 0 & 0 & 1 & 0 & 0 \\
0 & 0 & 0 & 0 & 0 & 1 & 0 \\
0 & 0 & 0 & 0 & 0 & 0 & 1
\end{vmatrix}
$$

First iteration:
The previous matrixes and vectors are valid for all the iterations. The following matrixes change in each iteration.

[A11] = Diagonal matrix; dimension (7 × 7). It has the value $\alpha_i Q_i^{(n_i-1)}$ on the diagonal; coefficients β and γ are zero as no pumps exist in the network.

The following table shows the calculated values for α.

Pipe	Q (m³/s)	λ	V (m/s)	h_f (m)	$h_f + h_m$ (m)	α
1-2	0·10	0·0159	1·974	6·22	6·22	622·28
2-3	0·10	0·0166	5·482	66·89	82·21	8220·77
3-4	0·10	0·0178	12·335	271·02	271·02	27 101·65
5-4	0·10	0·0166	5·482	66·89	66·89	6688·98
2-5	0·10	0·0178	12·335	270·99	270·99	27 098·90
6-5	0·10	0·0161	3·084	23·09	23·09	2308·78
6-1	0·10	0·0159	1·974	3·73	3·73	373·42

Matrix [**A11**]:

$$
\begin{vmatrix}
62\cdot23 & 0 & 0 & 0 & 0 & 0 & 0 \\
0 & 822\cdot08 & 0 & 0 & 0 & 0 & 0 \\
0 & 0 & 2710\cdot16 & 0 & 0 & 0 & 0 \\
0 & 0 & 0 & 668\cdot90 & 0 & 0 & 0 \\
0 & 0 & 0 & 0 & 2709\cdot89 & 0 & 0 \\
0 & 0 & 0 & 0 & 0 & 230\cdot88 & 0 \\
0 & 0 & 0 & 0 & 0 & 0 & 37\cdot34
\end{vmatrix}
$$

[**A11**]′ = Diagonal matrix; dimension (7 × 7). It has the value $\alpha_i Q_i^{(n_i-1)}$ on the diagonal.

For this network [**A11**′] = [**A11**].

$$
\begin{vmatrix}
62\cdot23 & 0 & 0 & 0 & 0 & 0 & 0 \\
0 & 822\cdot08 & 0 & 0 & 0 & 0 & 0 \\
0 & 0 & 2710\cdot16 & 0 & 0 & 0 & 0 \\
0 & 0 & 0 & 668\cdot90 & 0 & 0 & 0 \\
0 & 0 & 0 & 0 & 2709\cdot89 & 0 & 0 \\
0 & 0 & 0 & 0 & 0 & 230\cdot88 & 0 \\
0 & 0 & 0 & 0 & 0 & 0 & 37\cdot34
\end{vmatrix}
$$

To find H_{i+1} by equation (5.17) following a step-by-step analysis, the following matrixes can be found:

$$[N][A11]'$$

$$
\begin{vmatrix}
124\cdot46 & 0 & 0 & 0 & 0 & 0 & 0 \\
0 & 1644\cdot15 & 0 & 0 & 0 & 0 & 0 \\
0 & 0 & 5420\cdot33 & 0 & 0 & 0 & 0 \\
0 & 0 & 0 & 1337\cdot80 & 0 & 0 & 0 \\
0 & 0 & 0 & 0 & 5419\cdot78 & 0 & 0 \\
0 & 0 & 0 & 0 & 0 & 461\cdot76 & 0 \\
0 & 0 & 0 & 0 & 0 & 0 & 74\cdot68
\end{vmatrix}
$$

$$([N][A11]')^{-1}$$

$$
\begin{vmatrix}
0\cdot00804 & 0 & 0 & 0 & 0 & 0 & 0 \\
0 & 0\cdot00061 & 0 & 0 & 0 & 0 & 0 \\
0 & 0 & 0\cdot00018 & 0 & 0 & 0 & 0 \\
0 & 0 & 0 & 0\cdot00075 & 0 & 0 & 0 \\
0 & 0 & 0 & 0 & 0\cdot00018 & 0 & 0 \\
0 & 0 & 0 & 0 & 0 & 0\cdot00219 & 0 \\
0 & 0 & 0 & 0 & 0 & 0 & 0\cdot01339
\end{vmatrix}
$$

$$[A21]([N][A11]')^{-1}$$

$$\begin{vmatrix} 0.00804 & -0.00061 & 0 & 0 & -0.00018 & 0 & 0 \\ 0 & 0.00061 & 0.00018 & 0 & 0 & 0 & 0 \\ 0 & 0 & -0.00018 & 0.00075 & 0 & 0 & 0 \\ 0 & 0 & 0 & -0.00075 & 0.00018 & 0.00219 & 0 \\ 0 & 0 & 0 & 0 & 0 & -0.00219 & 0.01339 \end{vmatrix}$$

$$[A21]([N][A11]')^{-1}[A12]$$

$$\begin{vmatrix} 0.00804 & -0.00061 & 0 & -0.00018 & 0 \\ -0.00061 & 0.00079 & -0.00018 & 0 & 0 \\ 0 & -0.00018 & 0.00093 & -0.00075 & 0 \\ -0.00018 & 0 & -0.00075 & 0.00310 & -0.00217 \\ 0 & 0 & 0 & -0.00217 & 0.01556 \end{vmatrix}$$

$$-([A21]([N][A11]')^{-1}[A12])^{-1}$$

$$\begin{vmatrix} -120.541 & -100.256 & -33.383 & -16.879 & -2.350 \\ -100.256 & -1423.476 & -365.444 & -104.310 & -14.522 \\ -33.383 & -365.444 & -1460.157 & -392.548 & -54.651 \\ -16.879 & -104.310 & -392.548 & -463.689 & -64.555 \\ -2.350 & -14.522 & -54.651 & -64.555 & -73.273 \end{vmatrix}$$

$[A11][Q]$	$[A10][H_o]$	$([A11][Q])+([A10][H_o])$
6.223	−100	−93.777
82.208	0	82.208
271.016	0	271.016
66.890	0	66.890
270.989	0	270.989
23.088	0	23.088
3.734	−100	−96.266

$([A21]([N][A11]')^{-1}$ $([A11][Q][A10][H_o]))$	$[A21][Q]$	$([A21]([N][A11]')^{-1}([A11]$ $[Q]+[A10][H_o])-([A21]$ $[Q]-[q]))$
−0.853	−0.1	−0.6935
0.1	0.2	−0.06
0	0	0.03
0.05	0.1	−0.02
−1.339	0	−1.299

Thus,

$$H_{i+1} = -([A21]([N][A11]')^{-1}[A12])^{-1}([A21]([N][A11]')^{-1}([A11][Q] + [A10][H_o] - ([A21][Q] - [q])))$$

Node		(m)
2		92·000
3		164·922
4	=	80·115
5		99·317
6		97·335

To find Q_{i+1} by equation (5.18) following a step-by-step analysis, the following matrixes can be found:

$[A12][H_{i+1}]$	$[A12][H_{i+1}] + [A10][H_o]$	$([N][A11']^{-1})(([A12][H_{i+1}]) + ([A10][H_o]))$
92·00	−8·00	−0.0643
72·92	72·92	0·0444
84·81	84·81	0·0157
−19·20	−19·20	−0·0144
7·32	7·32	0·0014
1·98	1·98	0·0043
97·33	−2·67	−0·0357

$$([N][A11]')^{-1}[A11]$$

0·5	0	0	0	0	0	0
0	0·5	0	0	0	0	0
0	0	0·5	0	0	0	0
0	0	0	0·5	0	0	0
0	0	0	0	0·5	0	0
0	0	0	0	0	0·5	0
0	0	0	0	0	0	0·5

$([I] - ([N][A11]')^{-1}[A11])$							$([I] - ([N][A11]')^{-1} \times [A11])[Q]$
0·5	0	0	0	0	0	0	0·05
0	0·5	0	0	0	0	0	0·05
0	0	0·5	0	0	0	0	0·05
0	0	0	0·5	0	0	0	0·05
0	0	0	0	0·5	0	0	0·05
0	0	0	0	0	0·5	0	0·05
0	0	0	0	0	0	0·5	0·05

Thus,

$$Q_{i+1} = ([I] - ([N][A11]')^{-1}[A11])[Q] - (([N][A11]')^{-1}([A12][H_{i+1}] + [A10][H_o]))$$

Pipe	(m^3/s)
1–2	0·114
2–3	0·006
3–4	0·034
5–4	0·064
2–5	0·049
6–5	0·046
6–1	0·086

After only five iterations the following are the results:

Heads at each node

Node	(m)
2	92·960
3	81·358
4	81·780
5	89·812
6	96·727

Pipe discharges

Pipe	(m^3/s)
1–2	0·10667
2–3	0·03658
3–4	0·00342
5–4	0·03342
2–5	0·01009
6–5	0·05333

Recommended reading

1. Cornish, R.J. (1939–40) *The Analysis of Flow in Networks of Pipes*. J Inst CE, Vol. 13, p147.
2. Cross, Hardy (1936) *Analysis of Flow in Networks of Conduits or Conductors*. Bulletin No. 286, University of Illinois, Engineering Experimental Station, Urbana, Ill.
3. Featherstone, R.E. and El Jumailly, K.K. (Feb. 1983) *Optimal Diameter Selection for Pipe Networks*. Journal of the Hydraulics Division, ASCE.

4. Martin, D.W. and Peters, G. (1963) *The Application of Newton's Method to Network Analysis by Digital Computer.* Journal of the Institution of Water Engineers, Vol. 17.

5. Wood, D.J. and Charles, C.O.A. (July 1972) *Hydraulic Network Analysis using Linear Theory.* Journal of the Hydraulics Division, ASCE, No. HY7.

6. Salgado, R., Todini, E. and O'Connell, P.E. (Sept. 1987) *Comparison of the gradient method with some traditional methods for the analysis of water supply distribution systems.* International Conference on Computer Applications for Water Supply and Distribution, Leicester University, UK.

7. Saldarriaga, J. (Sept. 1998) *Hidraulica de Tuberias.* Santafe de Bogota. McGraw-Hill Interamericana.

Problems

1.   Calculate the flows in the pipes of the pipe system illustrated in fig. 5.12. Minor losses are given by $C_m V^2/2g$.

Data:

Pipe	Length (m)	Diameter (mm)	Roughness (mm)	Minor loss Coefficients (C_m)
AB	5000	400	0·15	10·0
BC_1	7000	250	0·15	15·0
BC_2	7000	250	0·06	10·0

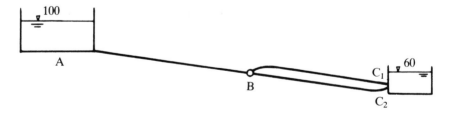

Figure 5.12 Pipes in parallel

(Note: While this problem could be solved by the method of Example 4.1, the method of quantity balance facilitates a convenient method of solution. Note that the pressure head elevations at the ends of C_1 and C_2 are identical.)

2.   In the system shown in problem 1, an axial flow pump producing a total head of 5·0 metres is installed in pipe BC_1 to boost the flow in this branch. Determine the flows in the pipes.

(Note: Although it is not strictly possible to predict the head generated by a rotodynamic pump since this varies with the discharge (see Chapter 6) axial flow pumps often produce a fairly flat head-discharge curve in the mid-discharge range.)

3.   Determine the flows in the network illustrated in fig. 5.13; minor losses are given by $C_m V^2/2g$.

Pipe	Length (m)	Diameter (mm)	k (mm)	C_m
AB	20 000	500	0·3	20
BC	5000	350	0·3	10
BD_1	6000	300	0·3	10
BD_2	6000	250	0·06	10

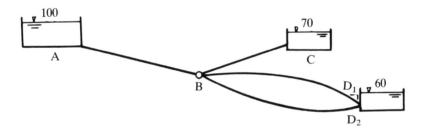

Figure 5.13   Network with reservoirs

4.   In the system illustrated in fig. 5.14 a pump is installed in pipe BC to provide a flow of 40 l/s to reservoir C. Neglecting minor losses calculate the total head to be generated by the pump and the power consumption assuming an overall efficiency of 60 per cent. Determine also the flow rates in the other pipes.

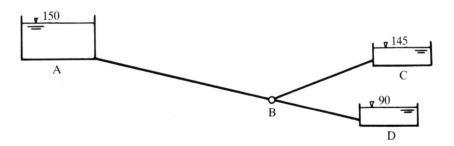

Figure 5.14   Network with reservoirs

Data:

Pipe	Length (m)	Diameter (mm)	Roughness (mm)
AB	10 000	400	0·06
BC	4000	250	0·06
BD	5000	250	0·06

5.   Determine the pressure head elevations at B and D and the discharges in the branches in the system illustrated in fig. 5.15. Neglect minor losses.

Pipe	Length (m)	Diameter (mm)	Roughness (mm)
AB	20 000	600	0·06
BC	2000	250	0·06
BD	2000	450	0·06
DE	2000	300	0·06
DF	2000	250	0·06

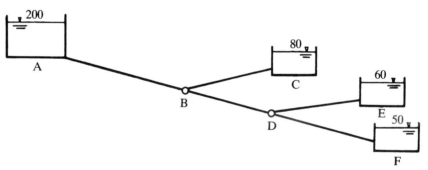

Figure 5.15   Network with reservoirs

6.   Determine the flows in a pipe system similar in configuration to that in Q5. A valve is installed in BC producing a minor loss of $20 \dfrac{V^2}{2g}$; otherwise consider only friction losses.

Pipe	Length (m)	Diameter (mm)	Roughness (mm
AB	20 000	450	0·06
BC	2000	300	0·06
BD	10 000	400	0·06
DE	3000	250	0·06
DF	4000	300	0·06

7.   Determine the flow in the pipes and the pressure head elevations at the junctions of the closed-loop pipe network illustrated, neglecting minor losses. All pipes have the same roughness size of 0·03 mm. The outflows at the junctions are shown in l/s. (See fig. 5.16.)

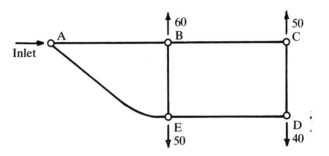

Figure 5.16   Two-looped network

Pipe	AB	BC	CD	DE	EA	BE
Length (m)	500	600	200	600	600	200
Diameter (mm)	200	150	100	150	200	100

Pressure head elevation at A = 60 m a.o.d.

(Note: A more rapid solution is obtained by using the head-balance method. However the network can be analysed by the quantity balance method but in this case FOUR unknown pressure heads, at B, C, D and E are to be corrected. If the quantity balance method is used, set a fixed arbitrary pressure head elevation to A, say 100 m.)

8.   Determine the flow distribution in the pipe system illustrated in fig. 5.17 and the total head loss between A and F. Neglect minor losses. A total discharge of 200 l/s passes through the system.

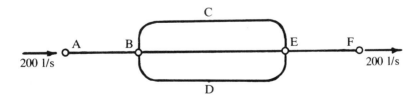

Figure 5.17   Pipes in parallel

Pipe	AB	BCE	BE	BDE	EF
Length (m)	1000	3000	2000	3000	1000
Diameter (mm)	450	300	250	350	450
Roughness (mm)	0·15	0·06	0·15	0·06	0·15

9.   In the system shown in problem 7 (fig. 5.16) a pump is installed in BC to boost the flow to C. Neglecting minor losses determine the flow distribution and head elevations at the junctions if the pump delivers a head of 15·0 m.

10.   Determine the flows in the pipes and the pressure head elevations at the junctions in the network shown in fig. 5.18. Neglect minor losses and take the pressure head elevation at A to be 100 m. The outflows are in l/s. All pipes have a roughness of 0·06 mm.

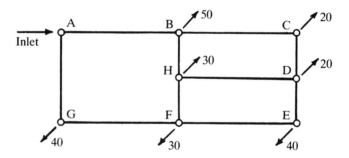

Figure 5.18   Three-looped network

Data:

Pipe	AB	BH	HF	FG	GA
Length (m)	400	150	150	400	300
Diameter (mm)	200	200	150	150	200

Pipe	BC	CD	DH	DE	EF
Length (m)	300	150	300	150	300
Diameter (mm)	150	150	150	150	150

11.   Analyse the flows and pressure heads in the pipe system shown in fig. 5.19. Neglect minor losses.

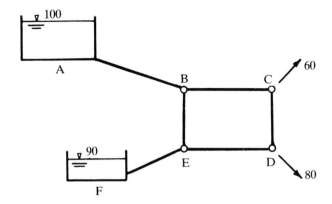

Figure 5.19 Network with reservoirs

Data:

Pipe	AB	BC	CD	DE	EF	EF
Length (m)	1000	400	300	400	800	300
Diameter (mm)	250	200	150	150	250	200
Roughness (mm)	0·06	0·15	0·15	0·15	0·06	0·15

12.   Solve the network in Problem 10 using the gradient method.

13.   Analyse the network of Example 5.1 by the gradient method.

Chapter 6
Pump-pipeline System Analysis and Design

R.E. Featherstone

6.1 Introduction

This section deals with the analysis and design of pipe systems which incorporate rotodynamic pumps. The reader is referred to standard texts (Recommended reading 1 and 4) for details of the construction and performance characteristics of pumps. The civil engineer is mostly concerned with pump selection, in the design of pumping stations, and therefore the design of the shape of pump impellers will not be dealt with here. Rotodynamic pumps can be sub-classified according to the shape of the impellers into three main categories:
(i)   centrifugal (radial flow)
(ii) mixed flow, and
(iii) propeller (axial flow).
For the same power input and efficiency the centrifugal type would generate a relatively large pressure head with a low discharge, the propeller type a relatively large discharge at a low head with the mixed flow having characteristics somewhere between the other two.

Pump types may be more explicitly defined by the parameter called SPECIFIC SPEED (N_s) expressed by

$$N_s = \frac{N\sqrt{Q}}{H^{\frac{3}{4}}}$$

where Q is the discharge, H the total head and N the rotational speed (rev/min). This expression is derived from dynamical similarity considerations and may be interpreted as the speed in rev/min at which a geometrically scaled model would have to operate to deliver unit discharge (e.g. 1 l/s) when generating unit head (e.g. 1 m).

Pump type	N_s range (Q–l/s; H–m)
centrifugal	up to 2600
mixed flow	2600 to 5000
axial flow	5000 to 10 000

The total head generated by a pump is also called the manometric head (H_m) since it is the difference in pressure head recorded by pressure gauges connected to the delivery and inlet pipes on either side of the pump, provided that the pipes are the same diameter.

6.2 Hydraulic gradient in pump-pipeline systems

Figure 6.1 shows a pump delivering a liquid from a lower tank to a higher tank, through a static lift H_{ST} at a discharge Q. It is clear that the pump must generate a total head equal to H_{ST} plus the pipeline head losses.

V_s = velocity in suction pipe

V_d = velocity in delivery pipe

h_{ld} = head loss in delivery pipe (friction, valves, etc.)

h_{ls} = head loss in suction pipe

h_m = local losses

Manometric head is defined as the rise in total head across the pump.

$$H_m = \frac{p_d}{\rho g} + \frac{V_d^2}{2g} - \left(\frac{p_s}{\rho g} + \frac{V_s^2}{2g} \right) \tag{6.1}$$

Now $\dfrac{p_s}{\rho g} = Z_1 - \dfrac{V_s^2}{2g} - h_{ls}$; $\dfrac{p_d}{\rho g} = Z_2 + h_{ld} - \dfrac{V_d^2}{2g}$

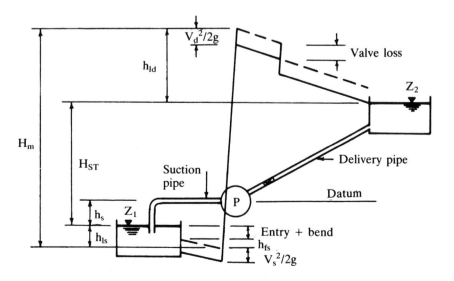

Figure 6.1 Total energy and hydraulic grade lines in pipeline with pump

Thus $H_m = Z_2 - Z_1 + h_{ld} + h_{ls}$

or $H_m = H_{ST} + h_{ld} + h_{ls}$ (6.2)

Note that the energy losses within the pump itself are not included; such losses will affect the efficiency of the pump.

Total head v. discharge and efficiency v. discharge curves (fig. 6.2) for particular pumps are obtained from the manufacturers.

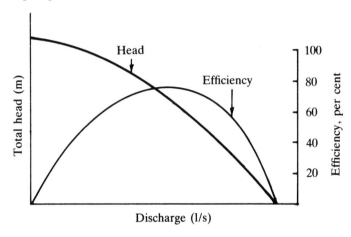

Figure 6.2 Typical performance curve for centrifugal pump

The total head-discharge curves for a centrifugal pump can generally be expressed in the functional form

$$H_m = AQ^2 + BQ + C$$ (6.3)

The coefficients A, B and C can be evaluated by taking three pairs of H_m and Q from a particular curve and solving equation (6.3).

The power consumed by a pump when delivering a discharge Q (m^3/s) at a head H_m (m) with a combined pump/motor efficiency η is

$$P = \frac{\rho g\ QH_m}{\eta} \text{ watts}$$

6.3 Multiple pump systems

(a) Parallel operation
Pumping stations frequently contain several pumps in a 'parallel' arrangement. In this configuration (fig. 6.3) any number of the pumps can be operated simultaneously, the objective being to deliver a range of discharges. This is a common feature of sewage pumping stations where the inflow rate varies during the day. By automatic switching according to the level in the suction well any number of the pumps can be brought into operation.

Manifold

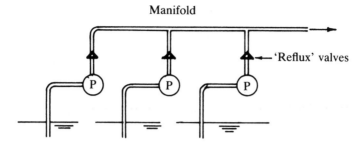

'Reflux' valves

P   P   P

Figure 6.3   Pumps operating in parallel

In predicting the head v. discharge curve for parallel operation it is assumed that the head across each pump is the same. Thus at any arbitrary head the individual pump discharges are added as shown in fig. 6.4.

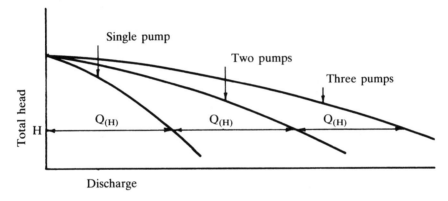

Single pump

Two pumps

Three pumps

Total head

H

$Q_{(H)}$ $Q_{(H)}$ $Q_{(H)}$

Discharge

Figure 6.4   Characteristic curves for identical pumps operating in parallel

(b) Series operation

This configuration is the basis of multistage and borehole pumps; the discharge from the first pump (or stage) is delivered to the inlet of the second pump, and so on. The same discharge passes through each pump receiving a pressure boost in doing so. Figure 6.5 shows the series configuration together with the resulting head v. discharge characteristics which are clearly obtained by adding the individual pump manometric heads at any arbitrary discharges. Note that, of course, all pumps in a series system must be operating simultaneously.

The reader is referred to R.E. Bartlett's[2] and P. Novak *et al.*'s[3] books for practical details of pumping station design and operation.

6.4 Variable speed pump operation

By the use of variable speed motors the discharge of a single pump can be varied to suit the operating requirements of the system.

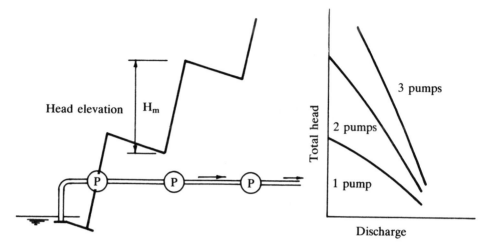

Figure 6.5 Pumps operating in series

Using dimensional analysis and dynamic similarity criteria (see Chapter 9) it can be shown that if the pump delivers a discharge Q_1 at manometric head H_1 when running at speed N_1, the corresponding values when the pump is running at speed N_2 are given by the relationships

$$Q_2 = Q_1 \left(\frac{N_2}{N_1} \right)$$ (6.4)

$$H_2 = H_1 \left(\frac{N_2}{N_1} \right)^2$$ (6.5)

In constructing the characteristic curve for speed N_2, several pairs of values of Q_1, H_1 from the curve for N_1 can be obtained and transformed into homologous points Q_2, H_2 on the N_2 curve. (See fig. 6.6.)

6.5 Suction lift limitations

Cavitation, the phenomenon which consists of local vaporization of a liquid and which occurs when the absolute pressure falls to the vapour pressure of the liquid at the operating temperature, can occur at the inlet to a pump and on the impeller blades, particularly if the pump is mounted above the level in the suction well. Cavitation causes physical damage, reduction in discharge and noise, and to avoid it the pressure head at inlet should not fall below a certain minimum which is influenced by the further reduction in pressure within the pump impeller. (See fig. 6.7.)

If p_s represents the pressure at inlet then $\left(\dfrac{p_s - p_v}{\rho g} \right)$ is the absolute head

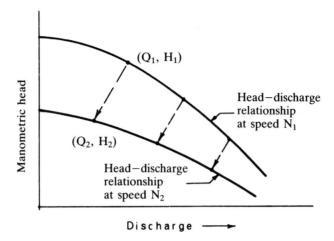

Figure 6.6 Effect of speed change on pump characteristics

at the pump inlet above the vapour pressure (p_v) and is known as the net positive suction head, NPSH.

$$\text{Thus NPSH} = \left(\frac{p_s - p_v}{\rho g}\right) = \frac{p_a}{\rho g} - H_s - \frac{p_v}{\rho g} \qquad (6.6)$$

where p_a = ambient atmospheric pressure

$$H_s = \text{manometric suction head} = h_s + h_{ls} + \frac{V_s^2}{2g} \qquad (6.7)$$

where h_s = suction lift; h_{ls} = total head loss in suction pipe

and V_s = velocity head in suction pipe

ρ = density of liquid.

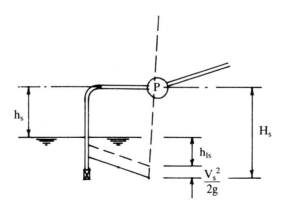

Figure 6.7 Head conditions in suction pipe

Values of NPSH can be obtained from the pump manufacturer and are derived from full-scale or model tests; these values must not be exceeded if cavitation is to be avoided.

Thoma introduced a cavitation number $\sigma \left(= \dfrac{\text{NPSH}}{H_m} \right)$ and from physical tests found this to be strongly related to specific speed.

In recent years electro-submersible pumps in the small to medium size range have been widely used. This type eliminates the need for suction pipes and provided that the pump is immersed to the manufacturer's specified depth, the problems of cavitation and cooling are avoided.

Worked examples

Example 6.1

Tests on a physical model pump indicated a cavitation number of 0·12. A homologous (geometrically and dynamically similar) unit is to be installed where the atmospheric pressure is 950 mb, and the vapour pressure head 0·2 m. The pump will be situated above the suction well, the suction pipe being 200 mm in diameter, of uPVC, 10 m long; it is vertical with a 90° elbow leading into the pump inlet and is fitted with a foot-valve. The foot-valve head loss $(h_v) = 4·5\ V_s^2/2g$; bend loss $(h_b) = 1·0\ V_s^2/2g$. The total head at the operating discharge of 35 l/s is 25 m. Calculate the maximum permissible suction head and suction lift.

Solution:

$$p_a = 950 \text{ mb} = 0·95 \times 10·198 = 9·688 \text{ m of water.}$$

$$\therefore \frac{p_a - p_v}{\rho g} = 9·688 - 0·2 = 9·488 \text{ m of water.}$$

$$\text{NPSH} = \sigma\, H_m = 0·12 \times 25 = 3·0 \text{ m.}$$

From equation (6.6) maximum permissible suction head

$$H_s = (9·488 - 3·0) = 6·488 \text{ m.}$$

Now calculate the losses in the suction pipe.

$$V_s = 1·11 \text{ m/s}; \quad \frac{V_s^2}{2g} = 0·063 \text{ m}; \quad Re = 1·96 \times 10^5;$$

$$k = 0·03 \text{ mm (uPVC)}; \quad k/D = 0·001 \text{ whence } \lambda = 0·0167$$

$$\therefore h_{fs} = 0·053 \text{ m}; \quad h_v = 4·5 \times 0·063 = 0·283 \text{ m}; \quad h_b = 0·063 \text{ m}$$

$$\therefore h_{ls} = 0·4 \text{ m}$$

$$h_s = \text{suction lift} = H_s - h_{ls} - V_s^2/2g = 6·488 - 0·463 = 6·025 \text{ m.}$$

(See also Example 6.7.)

Example 6.2

A centrifugal pump has a 100 mm diameter suction pipe and a 75 mm diameter delivery pipe. When discharging 15 l/s of water, the inlet water-mercury manometer with one limb exposed to the atmosphere recorded a vacuum deflection of 198 mm; the mercury level on the suction side was 100 mm below the pipe centreline. The delivery pressure gauge, 200 mm above the pump inlet, recorded a pressure of 0·95 bar. The measured input power was 3·2 kW. Calculate the pump efficiency. (See fig. 6.8.)

Solution:

Manometric head = rise in total head

$$Hm = \frac{p_2}{\rho g} + \frac{V_2^2}{2g} + z - \left(\frac{p_1}{\rho g} + \frac{V_1^2}{\rho g}\right)$$

1 bar = 10·198 m of water

$$\frac{p_2}{\rho g} = 0·95 \times 10·198 = 9·69 \text{ m of water}$$

$$\frac{p_1}{\rho g} = -0·1 - 0·198 \times 13·6 = -2·793 \text{ m of water}$$

$V_2 = 3·39$ m/s; $V_2^2/2g = 0·588$ m

$V_1 = 1·91$ m/s; $V_1^2/2g = 0·186$ m

Then Hm = 9·69 + 0·588 + 0·2 − (−2·793 + 0·186) = 13·09 m

Efficiency (η) = $\dfrac{\text{output power}}{\text{input power}}$ = $\dfrac{\rho g \, QH_m \text{ (watts)}}{3200 \text{ (watts)}}$

$$\eta = \frac{9·81 \times 0·015 \times 13·09}{3·2} = 0·602 \text{ (60·2 per cent).}$$

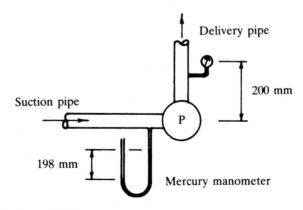

Figure 6.8   Suction and delivery pressures across pump

Example 6.3
Calculate the steady discharge of water between the tanks in the system shown in fig. 6.1, and the power consumption. Pipe diameter ($D_s = D_d$) = 200 mm; Length = 2000 m; k = 0·03 mm (uPVC). Losses in valves, bends plus the velocity head amount to $6·2 \dfrac{V^2}{2g}$. Static lift = 10·0 m.

Pump characteristics:

Discharge (l/s)	0	10	20	30	40	50
Total head (m)	25	23·2	20·8	16·5	12·4	7·3
Efficiency (per cent)	—	45	65	71	65	45

The efficiencies given are the overall efficiencies of the pump and motor combined.

Solution:
The solution to such problems is basically to solve simultaneously the head-discharge relationships for the pump and pipeline:
For the pump, head *delivered* at discharge Q may be expressed by

$$H_m = AQ^2 + BQ + C \tag{i}$$

and for the pipeline, the head required to produce a discharge Q is given by

$$H_m = H_{sT} + \frac{K_m Q^2}{2g A^2} + \frac{\lambda LQ^2}{D 2g A^2} \text{ (from equation (6.2))} \tag{ii}$$

where K_m is the minor loss coefficient.
A graphical solution is the simplest method and also gives the engineer a visual interpretation of the 'matching' of the pump and pipeline.
Equation (ii) when plotted (H v. Q) is called the 'system curve'. Values of H corresponding with a range of Q values will be calculated: k/D = 0·03; values of λ obtained from Moody diagram.

Q (l/s)	10	20	30	40	50
Re($\times 10^5$)	0·56	1·13	1·10	2·25	2·81
λ	0·0210	0·0185	0·0172	0·0165	0·0160
h_f (m)	1·08	3·82	7·99	13·63	20·65
H_m (m)	0·03	0·13	0·29	0·51	0·80
H (m)	11·11	13·95	18·28	24·14	31·45

Alternatively the combined Darcy-Colebrook-White equation can be used,

$$Q = \frac{-2\pi D^2}{4}\sqrt{2g D \frac{h_f}{L}} \log\left(\frac{k}{3·7 D} + \frac{2·51 v}{D\sqrt{2g D \frac{h_f}{L}}}\right)$$

In evaluating pairs of H and Q it is now preferable to take discrete values of h_f, calculate Q explicitly from the above equation and add the static lift and minor head loss.

h_f (m)	2·0	4·0	8·0	16·0
Q (l/s)	14·06	20·57	30·00	43·61
h_m (m)	0·06	0·13	0·29	0·61
H (m)	12·06	14·13	18·29	26·61

The computed system curve data and pump characteristic curve and data are now plotted on fig. 6.9.

The intersection point gives the operating conditions; in this case $H_m = 17\cdot5$ m; Q = 28·0 l/s. The operating efficiency is 71 per cent. Therefore, power consumption

$$P = \frac{1000 \times 9\cdot81 \times 0\cdot028 \times 17\cdot5}{0\cdot71}$$

$$= 6770 \text{ watts } (6\cdot77 \text{ kW}).$$

Example 6.4 (Pipeline selection in pumping system design)
An existing pump, having the tabulated characteristics, is to be used to pump raw sewage to a treatment plant through a static lift of 20 m. A uPVC pipeline

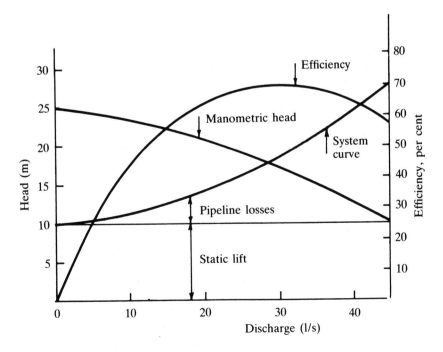

Figure 6.9 Pump and system characteristics

10 km long is to be used. Allowing for minor losses totalling $10\ V^2/2g$ and taking an effective roughness of 0·15 mm because of sliming, select a suitable commercially available pipe size to achieve a discharge of 60 l/s. Calculate the power consumption.

Discharge (l/s)	0	10	20	30	40	50	60	70
Total head (m)	45	44·7	43·7	42·5	40·6	38·0	35·0	31·0
Overall efficiency (per cent)	—	35	50	57	60	60	53	40

Solution:
At 60 l/s, total head = 35·0 m, therefore the sum of the static lift and pipeline losses must not exceed 35·0 m.

Try 300 mm diameter: $A = 0.0707$ m^2; $V = 0.85$ m/s;

$Re = 2.25 \times 10^5$; $k/D = 0.0005$; $\lambda = 0.019$

$$\text{Friction head loss} = \frac{0.019 \times 10\,000 \times 0.85^2}{0.3 \times 19.62} = 23.32 \text{ m}$$

$H_s + h_f = 43.32\ (> 35)$ − pipe diameter too small.

Try 350 mm diameter: $A = 0.0962$ m^2; $V = 0.624$ m/s;

$Re = 1.93 \times 10^5$; $k/D = 0.00043$; $\lambda = 0.0185$

$h_f = 10.48$ m; $h_m = \dfrac{10 \times 0.624^2}{19.612} = 0.2$ m

$H_s + h_f + h_m = 30.68\ (< 35$ m) − O.K.

The pump would deliver approximately 70 l/s through the 350 mm pipe and to regulate the flow to 60 l/s an additional head loss of 4·32 m by valve closure would be required.

$$\text{Power consumption P} = \frac{1000 \times 9.81 \times 0.06 \times 35}{0.55 \times 1000} = 38.85 \text{ kW}.$$

Example 6.5 (Pumps in parallel and series)
Two identical pumps having the tabulated characteristics are to be installed in a pumping station to deliver sewage to a settling tank through a 200 mm uPVC pipeline 2·5 km long. The static lift is 15 m. Allowing for minor head losses of $10.0\ V^2/2g$ and assuming an effective roughness of 0·15 mm calculate the discharge and power consumption if the pumps were to be connected: (a) in parallel, and (b) in series.

Pump characteristics

Discharge (l/s)	0	10	20	30	40
Total head (m)	30	27·5	23·5	17·0	7·5
Overall efficiency (per cent)	—	44	58	50	18

Solution:
The 'system curve' is computed as in the previous examples; this is, of course, independent of the pump characteristics. Calculated system heads (H) are tabulated below for discrete discharges (Q)

$$H = H_{sT} + h_f + h_m$$

Q (l/s)	10	20	30	40
H (m)	16·53	20·80	27·37	36·48

(a) Parallel operation
The predicted head v. discharge curve for dual pump operation in parallel mode is obtained as described in section 6.3 (a), i.e. by doubling the discharge over the range of heads (since the pumps are identical in this case). The system and efficiency curves are added as shown in fig. 6.10. From the intersection of the characteristic and system curves the following results are obtained:

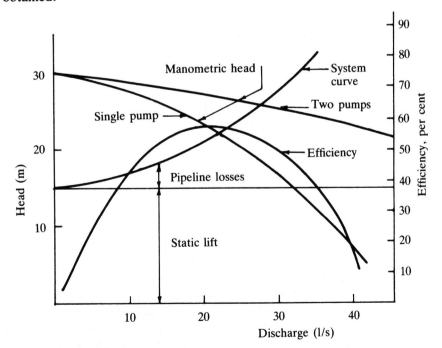

Figure 6.10 Parallel operation

Single pump operation, Q = 22·5 l/s; H_m = 24 m; η = 0·58

Power consumption = 9·13 kW

Parallel operation, Q = 28·5 l/s; H_m = 26 m; η = 0·51

(corresponding with 14·25 l/s per pump)

Power input = 14·11 kW.

(b) Series operation
Using the method described in section 6.3 (b) and plotting the dual-pump
characteristic curve, intersection with the system curve yields (see fig. 6.11):

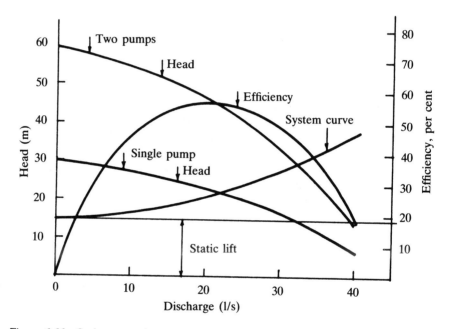

Figure 6.11   Series operation

Q = 32·5 l/s; H_m = 28 m; η = 0·41

Power input = 21·77 kW.

Note that for this particular pipe system, comparing the relative power
consumptions the parallel operation is more efficient in producing an increase
in discharge than the series operation.

Example 6.6 (Pump operation at different speeds)
A variable speed pump having the tabulated characteristics, at 1450 rev/min, is
installed in a pumping station to handle variable inflows. Static lift = 15 m;
diameter of pipeline = 250 mm; length 2000 m, k = 0·06 mm. Minor loss =
10·0 $V^2/2g$.

Determine the total head of the pump and discharge in the pipeline at pump speeds of 1000 rev/min and 500 rev/min.
Pump characteristics at 1450 rev/min.

Discharge (l/s)	0	10	20	30	40	50	60	70
Total head (m)	45·0	44·0	42·5	39·5	35·0	29·0	20·0	6·0

Solution:
The characteristic curve for speed N_2 using that for speed N_1 can be constructed using equations (6.4) and (6.5) (section 6.4) i.e.

$$H_2 = H_1 \left(\frac{N_2}{N_1}\right)^2 \tag{i}$$

$$Q_2 = Q_1 \left(\frac{N_2}{N_1}\right) \tag{ii}$$

where (H_1, Q_1) are pairs of values taken from the N_1 curve and (H_2, Q_2) are the corresponding points on the N_2 curve.
The system curve is computed giving the following data:

Discharge l/s	20	40	60	80
System head (m)	16·40	20·08	26·12	34·23

Construct the pump head-discharge curves for speeds of 1000 rev/min and 500 rev/min using equations (i) and (ii). Q_1, H_1 values (at 1450 rev/min) can be taken from the tabulated data (or from the plotted curve in fig. 6.12). For example, taking $Q_1 = 20$ l/s, $H_1 = 42·5$ m at 1450 rev/min, the corresponding values at 1000 rev/min are:

$$Q_2 = 20 \times \left(\frac{1000}{1450}\right) = 13·79 \text{ l/s}; \quad H_2 = 42·5 \left(\frac{1000}{1450}\right)^2 = 20·2 \text{ m}$$

Taking three other pairs of values the following table can be constructed.

N (rev/min)					
1450	Q_1	0	20	40	60
	H_1	45	42·5	35	20
1000	Q_2	0	13·79	27·59	41·38
	H_2	21·4	20·2	16·65	9·50
500	Q_2	0	6·9	13·90	20·70
	H_2	5·35	5·05	4·16	2·40

The computed values are now plotted together with the system curve (see fig. 6.12).
Operating conditions:

at N = 1450 rev/min: Q = 55 l/s; H_m = 25·0 m

at N = 1000 rev/min: Q = 33 l/s; H_m = 18·5 m

at N = 500 rev/min: no discharge produced.

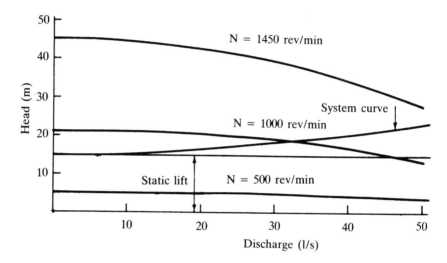

Figure 6.12   Pump operation with variable speed

Example 6.7
A laboratory test on a pump revealed that the onset of cavitation occurred, at a discharge of 35 l/s, when the total head at inlet was reduced to 2·5 m and the total head across the pump was 32 m. Barometric pressure was 760 mm Hg and the vapour pressure 17 mm Hg. Calculate the Thoma cavitation number. The pump is to be installed in a situation where the atmospheric pressure is 650 mm Hg and water temperature 10°C (vapour pressure 9·22 mm Hg) to give the same total head and discharge. The losses and velocity head in the suction pipe are estimated to be 0·55 m of water. What is the maximum height of the suction lift?

Solution:

$$NPSH = \left(\frac{p_a}{\rho g} - H_s\right) - \frac{p_v}{\rho g} \quad \text{(equation (6.6))} \tag{i}$$

where H_s = manometric suction head.

$\dfrac{p_a}{\rho g}$ = 10·3 m of water; $\dfrac{p_v}{\rho g}$ = 0·23 m of water

$$\left(\frac{p_a}{\rho g} - H_s \right) = 2 \cdot 5 \text{ m}$$

$$\therefore \text{NPSH} = 2 \cdot 5 - 0 \cdot 23 = 2 \cdot 27 \text{ m}$$

Cavitation number

$$\sigma = \frac{\text{NPSH}}{H} = \frac{2 \cdot 27}{32} = 0 \cdot 071$$

Installed conditions: $\dfrac{p_a}{\rho g} = 8 \cdot 84$ m; $\dfrac{p_v}{\rho g} = 0 \cdot 1254$ m (at 10°C).

$$\text{NPSH} = \frac{p_a}{\rho g} - H_s - \frac{p_v}{\rho g} \quad \text{(equation (i))}$$

$$\therefore 2 \cdot 27 = 8 \cdot 84 - H_s - 0 \cdot 1254$$

whence $H_s = 6 \cdot 44$ m

$$H_s = h_s + h_{ls} + V_s / 2g$$

whence $h_s = 6 \cdot 44 - 0 \cdot 55 = 5 \cdot 89$ m

where h_s = suction lift.

Example 6.8

An impounding reservoir at elevation 200 m delivers water to a service reservoir at elevation 80 m through a 20 km long 500 mm diameter coated C.I. pipeline (k = 0·03 mm). Minor losses amount to $20 \dfrac{V^2}{2g}$. Determine the steady discharge. (410·9 l/s.)

A booster pump having the tabulated characteristics is to be installed on the pipeline. Determine the improved discharge and the power consumption. (See fig. 6.13).

Q (l/s)	0	100	200	300	400	500	600
H$_m$ (m)	60·0	58·0	54·0	47·0	38·4	26·0	8·0
Overall efficiency (per cent)	—	33·0	53·0	62·0	62·0	54·0	28·0

The effective gross head is now $H_e = H + H_m$ where H_m is a function of the total discharge passing through the pump.

Thus $H_e = f_1 (Q^2)$ (i)

The head H_e is overcome by the pipeline losses

$H_L = f_2 (Q^2)$ (ii)

The discharge in the system is therefore evaluated by equating (i) and (ii); this can be done graphically as in the previous examples.

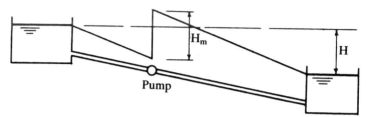

Figure 6.13   Pipe delivery with booster pump

Compute the head loss discharge curve ($f_2(Q^2)$) for the pipeline using one of the methods described in Chapter 4. The relationship is (equation (ii)):

Total pipeline head loss (m)	120	130	140	150	160
Discharge (l/s)	410·9	428·7	445·75	462·25	478·25

Using a common head datum of 120 m equations (i) and (ii) are now plotted (see fig. 6.14):

The point of intersection yields: $Q = 465$ l/s; $H_m = 32$ m; $\eta = 58$ per cent

$$\text{Power consumption} = \frac{9·81 \times 0·465 \times 32·0}{0·58}$$

$$= 251·67 \text{ kW.}$$

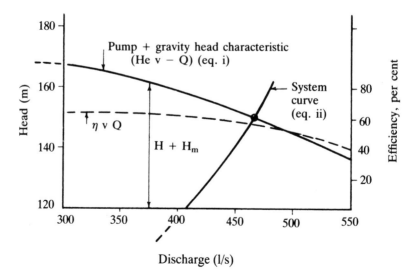

Figure 6.14   Pump and gravity and system characteristics

Example 6.9 (Pipe network with pump, using head discharge curve)
Neglecting minor losses, determine the discharges in the pipes of the network
illustrated in fig. 6.15 (a) with the pump in BC absent, (b) with the pump,
which boosts the flow to C in operation, and calculate the power consumption.

Pipe	Length (m)	Diameter (mm)	Roughness (mm)
AB	5000	300	0·06
BC	2000	200	0·06
BD	3000	150	0·06

Pump characteristics

Discharge (l/s)	0	20	40	60	80	100
Total head (m)	40.0	38·8	35·4	29·5	21·0	10·0
Efficiency (per cent)	—	50·0	70·0	73·0	58·0	22·0

Solution:
Plot the pump head v. discharge curve (see fig. 6.16).

The analysis is carried out using the quantity balance method, noting that
with the pump in operation in BC the head producing flow is: $H_{BC} = Z_C -$
$Z_B - H_p$ where H_p is the head delivered by the pump at the discharge in the
pipeline. Initially H_p can be obtained using an estimated pipe velocity but
subsequently the computed discharges can be used.
(a)

Pipe	AB	BC	BD
k/D	0·0002	0·0003	0·0004

λ values obtained from the Moody diagram using initially assumed values
of pipe velocity.

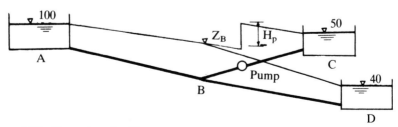

Figure 6.15　Pipe network with pump

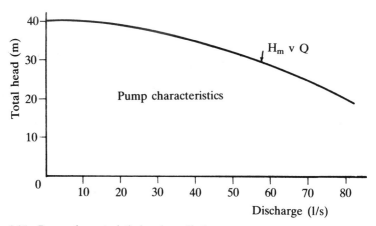

Figure 6.16 Pump characteristic head vs. discharge

With the pump not installed:

$Z_B = 80.25$ m

$Q_{AB} = 84.87$ l/s; $Q_{BC} = 58.97$ l/s; $Q_{BD} = 25.9$ l/s

(b) Estimate $Z_B = 80$ m.

Units in table: V = m/s; H_p (pump total head) = m;   Q = m³/s.

'Head' = head loss in pipeline.

A = area of pipe (m²). H_p is initially obtained from an estimated flow in BC of 60 l/s (29 m).

Pipe	V(est) (m/s)	Re × 10⁵	λ	H_p (m)	Head (m)	Q (m³/s)	Q/h × 10⁻³	Q/A (m/s)
AB	2·0	5·3	0·0155	—	20·0	0·0871	4·36	1·23
BC	2·0	5·3	0·0170	29·0	−59·0	−0·0820	1·39	2·6
BD	2·0	5·3	0·0178	—	−40·0	−0·0262	0·65	1·5
					Σ	−0·0211	6·40	

$\Delta Z_B = -6.6$ m; $Z_B = 73.41$ m.

Note that H_p at each step is obtained from the H v. Q curve corresponding with the value of Q_{BC} at the previous step.

Pipe	V(est)	Re × 10⁵	λ	H_p	Head	Q	Q/h × 10⁻³	Q/A
AB	1·5	3·98	0·016	—	26·59	0·0989	3·72	1·40
BC	2·5	4·42	0·017	22·0	−45·41	−0·0720	1·58	2·30
BD	1·5	1·99	0·018	—	−33·41	−0·0238	0·71	1·35
					Σ	0·0031	6·01	

$\Delta Z_B = 1.03$ m; $Z_B = 74.44$ m.

Pipe	V(est)	Re × 10⁵	λ	H_p	Head	Q	Q/h × 10⁻³	Q/A
AB	1·4	3·7	0·016	—	25·56	0·0969	3·79	1·37
BC	2·3	4·07	0·0165	24·8	−49·24	−0·0760	1·54	2·4
BD	1·35	1·79	0·0185	—	−34·44	−0·0239	0·69	1·35
					Σ	−0·003	6·02	

$\Delta Z_B = -0.99$ m; $Z_B = 73.45$ m.

Pipe	V(est)	Re × 10⁵	λ	H_p	Head	Q	Q/h × 10⁻³	Q/A
AB	1·4	3·7	0·016	—	26·55	0·0988	3·72	1·4
BC	2·4	4·2	0·0165	23·0	−46·45	−0·0738	1·59	2·35
BD	1·35	1·79	0·0185	—	−33·45	−0·0236	0·70	1·33
					Σ	0·0014	6·01	

$\Delta Z_B = 0.47$ m; $Z_B = 73.92$ m.

Final values:

$$Z_B = 73.79 \text{ m}$$

$$Q_{AB} = 98.1 \text{ l/s}; \quad Q_{BC} = 74.5 \text{ l/s}; \quad Q_{BD} = 23.6 \text{ l/s}$$

$$H_p = 23.5 \text{ m}$$

At the pump discharge of 74·5 l/s, efficiency = 64 per cent,

$$\text{whence Power} = \frac{9.81 \times 0.0745 \times 23.5}{0.64}$$

$$= 26.83 \text{ kW.}$$

Recommended reading

1. Anderson, H.H. (1981) 'Liquid Pumps', in *Kempe's Engineers Year Book*, Vol. 1, F9, London: Morgan-Grampian.
2. Bartlett, R.E. (1974) *Pumping Stations for Water and Sewage.* Barking: Elsevier Applied Science Publishers.
3. Novak, P., Moffat, A.I.B., Nalluri, C. and Narayanan, R. (1997) *Hydraulic Structures*, 2nd edn. London: E&FN Spon.
4. Wislicenus, G.F. (1965) *Fluid Mechanics of Turbo-machinery.* Vols 1 and 2. London: Dover Publications.

Problems

1.   A rotodynamic pump having the characteristics tabulated below delivers water from a river at elevation 52·0 m o.d. to a reservoir with a water

level of 85 m o.d., through a 350 mm diameter coated cast iron pipeline, 2000 m long (k = 0·15 mm). Allowing $10 \frac{V^2}{2g}$ for losses in valves, etc. calculate the discharge in the pipeline and the power consumption.

Q l/s	0	50	100	150	200
H_m (m)	60	58	52	41	25
$\eta\%$	—	44	65	64	48

2.   If in the system described in problem 1, the discharge is to be increased to 175 l/s by the installation of a second identical pump,

(a) determine the unregulated discharges produced by connecting the pumps, (i) in parallel, and (ii) in series.
(b) Calculate the power demand when the discharge is regulated (by valve control) to 175 l/s in the case of (i) parallel operation, and (ii) series operation.

3.   A pump is required to discharge 250 l/s against a calculated system head of 6·0 m. Assuming that the pump will run at 960 rev/min, what type of pump would be most suitable?

4.   The performance characteristics of a variable speed pump when running at 1450 rev/min are tabulated below, together with the calculated system head losses. The static lift is 8·0 m. Determine the discharge in the pipeline when the pump runs at 1450, 1200 and 1000 rev/min.

Q l/s	0	10	20	30	40
H_m (m)	20·0	19·2	17·0	13·7	8·7
System head loss (m)	—	0·7	2·3	4·8	9·0

5.   A pump has the characteristics tabulated when operating at 960 rev/min. Calculate the specific speed and state what type of pump this is. What discharge will be produced when the pump is operating at a speed of 700 rev/min in a pipeline having the system characteristics given in the table. Static lift is 2·0 m. What power would be consumed by the pump itself?

Q (l/s)	0	50	100	150	200	250	300
H_m (m)	7·0	6·3	5·5	5·0	4·6	4·1	3·5
Pump efficiency %	—	20·0	40·0	56·0	71·0	81·0	82·0
System head loss (m)	—	0·10	0·35	0·80	1·40	2·10	3·40

6. (a) Tests on a rotodynamic pump revealed that cavitation started when the manometric suction head, H_S, was 5 m, the discharge 60 l/s and the total head 40 m. Barometric pressure was 986 mb, and the vapour pressure 23·4 mb. Calculate the NPSH and the Thoma cavitation number.

   (b) Determine the maximum suction lift if the same pump is to operate at a discharge of 65 l/s and total head 35 m under field conditions where the barometric pressure is 950 mb and vapour pressure 12·5 mb. The sum of the suction pipe losses and velocity head are estimated to be 0·6 m.

7.   The characteristics of a variable speed rotodynamic pump when operating at 1200 rev/min are as follows:

Q (l/s)	0	10	20	30	40	50	60
H_m (m)	47·0	46·0	42·5	38·4	34·0	27·2	20.0

The pump is required to be used to deliver water through a static lift of 10 m in a 300 mm diameter pipeline 5000 m long and roughness size 0·15 mm, at a rate of 70 l/s. At what speed will the pump have to operate?

8.   The steady level, below ground level, in an abstraction well in a confined aquifer is calculated from the equation

$$z_w = 2·0 + \frac{Q}{2\pi Kb} \log_e \frac{R_o}{r_w} \text{ (m)}$$

where Q is the abstraction rate (m³/day), K the coefficient of permeability of the aquifer (m³/day/m²), b the aquifer thickness (m), R_o the radius of influence of the well, and r_w the radius of the well.
K = 50 m³/day/m²;   b = 20 m; r_w = 0·15 m.
   During a pumping test the observed z_w was 5·0 m when an abstraction rate of 30 l/s was applied.
   Under operating conditions the submersible borehole pump delivers the groundwater to the surface from where an in-line booster pump delivers the water to a reservoir the level in which is 20 m above the ground level at the well site. The pipeline is 500 m long, 200 mm in diameter and of roughness size 0·3 mm. Minor losses total $10 \frac{V^2}{2g}$.

Pump Characteristics

(a) Borehole pump						(b) Booster pump				
Discharge (l/s)	0	10	20	30	40	0	10	20	30	40
Total head (m)	10·0	9·6	8·7	7·4	5·6	22·0	21·5	20·4	19·0	17·4

Assuming that the radius of influence of the well is linearly related to the abstraction rate determine the maximum discharge which the combined pumps would deliver to the reservoir.

9.   The discharge in a pipeline delivering water under gravity between two reservoirs at elevations 150 m and 60 m is to be boosted by the installation of a rotodynamic pump, the characteristics of which are shown tabulated.
     The pipeline is 15 km long, 350 mm in diameter and has a roughness value of 0·3 mm. Determine the discharges (a) under gravity flow conditions, and (b) with the pump installed. Assume a minor loss of 20 $V^2/2g$ in both cases.

Pump characteristics

Discharge (l/s)	0	50	100	150	200	250
Manometric head (m)	50·0	49·0	46·5	42·0	36·0	28·2

10.   Reservoir A delivers water to service reservoirs C and D through pipelines AB, BC and BD. A pump is installed in pipeline BD to boost the flow to D.
     Elevations of water in reservoirs: $Z_A = 100$ m; $Z_C = 60$ m; $Z_D = 50$ m.

Pipe	Length (m)	Diameter (mm)	Roughness (mm)
AB	10 000	350	0·15
BC	4000	200	0·15
BD	5000	150	0·15

Pump characteristics

Discharge (l/s)	0	20	40	60	80
Total Head (m)	30·0	27·5	23·0	17·0	9·0
Efficiency (per cent)	—	44·0	68·0	66·0	44·0

Neglecting minor losses calculate the flows to the service reservoirs:
(a) with the pump not installed;
(b) with the pump operating and determine the power consumption.

Chapter 7
Boundary Layers on Flat Plates and in Ducts

R.E. Featherstone

7.1 Introduction

A boundary layer will develop along either side of a flat plate placed edgewise into a fluid stream (fig. 7.1). Initially the flow in the layer may be laminar with a parabolic velocity distribution. The boundary layer increases in thickness with distance along the plate and at a Reynolds number $\dfrac{U_0\,x}{\nu} \simeq 500\,000$ turbulence develops in the boundary layer. The frictional force due to the turbulent portion of the boundary layer may be considered as that which would be found if the entire length were turbulent *minus* that corresponding to the hypothetical turbulent layer up to the critical point.

For rough plates Schlichting gave an expression for the maximum height of roughness elements such that the surface may be considered hydraulically smooth:

$$k \leqslant \frac{100\ \nu}{U_0} \tag{7.1}$$

The theory of boundary layers on smooth flat plates is to be found in standard texts.

7.2 The laminar boundary layer

In 1908 Blasius developed analytical equations for the flow in a laminar boundary layer formed on a flat plate for the case of zero pressure gradient along the plate. Taking the outer limit of the boundary layer as the position where $v = 0.99\,U_0$ the boundary layer thickness δ_x at distance x from the leading edge was given as:

$$\delta_x = \frac{5x}{Re_x^{1/2}} \tag{7.2}$$

$$\text{where } Re_x = \frac{U_0\,x}{\nu}$$

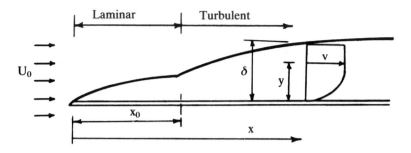

Figure 7.1   Boundary layer formation on flat plate

Local boundary shear stress $\tau_o = 0.332\ \mu\ \dfrac{U_0}{x}\ Re_x^{1/2}$ (7.3)

Drag along one side of a plate of length L and width B;

$$F_s = \int_o^L \tau_o\ B\ dx = 0.664\ B\mu\ U_0\ Re_L^{1/2}$$ (7.4)

or $F_s = C_f\ BL\ \rho\ \dfrac{U_0^2}{2}$ (7.5)

where $C_f = \dfrac{1.33}{Re_L^{1/2}}$ (7.6)

7.3 The turbulent boundary layer

Studies have shown that the velocity profile in the turbulent boundary layer is approximated closely over a wide range of Reynolds numbers by the equation:

$$\frac{v}{U_0} = \left(\frac{y}{\delta}\right)^{1/7}$$ (7.7)

If the turbulent boundary layer is assumed to develop at the leading edge of the plate

$$\delta_x = 0.37\ x\ \left(\frac{v}{U_0\ x}\right)^{1/5} = \frac{0.37\ x}{Re_x^{1/5}}$$ (7.8)

Boundary shear stress: $\tau_o = 0.0225\ \rho\ U_0^2\ \left(\dfrac{v}{U_0\ \delta_x}\right)^{1/4}$ (7.9)

Drag on one side of plate: $F_s = C_f\ BL\ \rho\ \dfrac{U_0^2}{2}$ (7.10)

where $C_f = \dfrac{0.074}{Re_L^{1/5}}$ (7.11)

7.4 Combined drag due to both laminar and turbulent boundary layers

Where the plate is sufficiently short that the laminar boundary layer forms over a significant proportion of the length the total drag may be calculated by considering the turbulent boundary layer to start at the leading edge, deducting the drag in the turbulent boundary layer up to x_o and adding the drag in the laminar boundary layer.

$$\text{Hence } F_s = \left(\frac{1 \cdot 33 \, x_o}{\text{Re}_{x,o}^{1/2}} + \frac{0 \cdot 074 \, L}{\text{Re}_L^{1/2}} - \frac{0 \cdot 074 \, x_o}{\text{Re}_{x,o}^{1/5}} \right) B \, \rho \, \frac{U_0^2}{2} \tag{7.12}$$

7.5 The displacement thickness

Due to the reduction in velocity in the boundary layer the discharge past a point on the surface is reduced by an amount

$$\delta q = \int_o^\delta (U_0 - v) \, dy. \tag{7.13}$$

The displacement thickness is the distance, δ^* by which the surface would have to be moved in order to reduce the discharge of an ideal fluid at velocity U_0 by the same amount.

$$\text{Then } U_0 \, \delta^* = \int_o^\delta (U_0 - v) \, dy \tag{7.14}$$

Assuming that the velocity distribution is:

$$\frac{v}{U_0} = \left(\frac{y}{\delta} \right)^{1/7} \quad (\text{equation (7.7)})$$

$$\delta^* = \frac{1}{U_0} \int_o^\delta \left(U_0 - U_0 \left(\frac{y}{\delta} \right)^{1/7} \right) dy \tag{7.15}$$

$$\delta^* = \left[y - \frac{7}{8} \left(\frac{1}{\delta} \right)^{1/7} y^{8/7} \right]_o^\delta$$

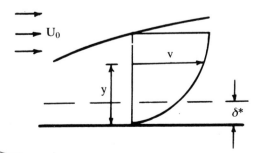

Figure 7.2 Displacement thickness

whence $\delta^* = \dfrac{\delta}{8}$ (7.16)

7.6 Boundary layers in turbulent pipe flow

In Chapters 4 and 5 the analysis and design of pipelines was demonstrated using the Darcy-Weisbach and Colebrook-White equations. The latter equation

$$\frac{1}{\sqrt{\lambda}} = -2 \log \left[\frac{k}{3 \cdot 7 \, D} + \frac{2 \cdot 51}{Re \sqrt{\lambda}} \right]$$

owes its origin to the theoretical hypothesis of Professor L. Prandtl for the general form of the velocity distribution in full pipe flow supported by the experimental work of J. Nikuradse. Using the concept of the exchange of momentum due to the transverse velocity components in turbulent flow Prandtl developed his 'mixing theory' expressing the shear stress τ, in terms of the velocity gradient $\dfrac{dv}{dy}$ and 'mixing length' ℓ in the form:

$$\tau = \rho \, \ell^2 \left(\frac{dv}{dy} \right)^2$$ (7.17)

Prandtl further assumed the mixing length to be directly proportional to the distance from the boundary and that the shear stress was constant and hence equal to the boundary shear stress τ_o

Hence $\tau_o = \rho \, (\kappa y)^2 \left(\dfrac{dv}{dy} \right)^2$ (7.18)

whence $\dfrac{dv}{dy} = \dfrac{1}{\kappa} \sqrt{\dfrac{\tau_o}{\rho}} \dfrac{1}{y}$ (7.19)

The term $\sqrt{\dfrac{\tau_o}{\rho}}$ has the dimensions of velocity and is called the 'shear velocity' and represented by the symbol v^*.

From equation (7.19) $\dfrac{v}{v^*} = \dfrac{1}{\kappa} \log_e y + C$ (7.20)

or $\dfrac{v}{v^*} = \dfrac{1}{\kappa} \log_e \dfrac{y}{y'}$ (7.21)

where y' is the 'boundary condition', i.e. the distance from the boundary at which the velocity becomes zero. (See fig. 7.3.)

Observations of friction head loss and velocity distribution on smooth and artificially roughened pipes, conveying water, by J. Nikuradse revealed that the forms of the velocity distribution in the smooth and rough turbulent

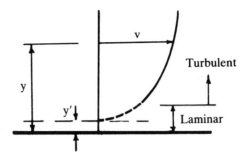

Figure 7.3 Velocity distribution – log law

zones were different. For the smooth turbulent zone he assumed y' to be proportional to $\dfrac{v}{v^*}$ and equation (7.21) therefore becomes

$$\frac{v}{v^*} = A \log \left(\frac{v^* y}{v}\right) + B \qquad (7.22)$$

where A and B are constants.

From the experimental data Nikuradse plotted $\dfrac{v}{v^*}$ against $\log \left(\dfrac{v^* y}{v}\right)$ and a straight line fit having the equation:

$$\frac{v}{v^*} = 5.75 \log \left(\frac{v^* y}{v}\right) + 5.5 \qquad (7.23)$$

verified the hypotheses. The mean velocity is obtained by integration:

$$V = \frac{1}{\dfrac{\pi D^2}{4}} \int_0^{\frac{D}{2}} 2 \pi \, rv \, dr$$

and substitution of v from equation (7.23) yields:

$$\frac{V}{v^*} = 5.75 \log \frac{v^* \, D}{2 \, v} + 1.75 \qquad (7.24)$$

The equation for the Darcy friction factor can consequently be obtained:

Since $\tau_0 = \rho g \, R \, \dfrac{h_f}{L} = \rho g \, \dfrac{D}{4} \dfrac{h_f}{L}$ (see Example 7.5)

and $\dfrac{h_f}{L} = \dfrac{\lambda \, V^2}{2g \, D};\ \sqrt{\dfrac{\tau_0}{\rho}} = V \sqrt{\dfrac{\lambda}{8}} \qquad (7.25)$

substitution into (7.25) yields:

$$\frac{1}{\sqrt{\lambda}} = 2 \log \left(\frac{VD}{v} \frac{\sqrt{\lambda}}{2.51}\right) \text{ or } \frac{1}{\sqrt{\lambda}} = 2 \log \frac{Re \sqrt{\lambda}}{2.51} \qquad (7.26)$$

Similarly for the rough turbulent zone Nikuradse found y' to be proportional to k and obtained the velocity distribution in the form:

$$\frac{v}{v^*} = 5 \cdot 75 \log \frac{y}{k} + 8 \cdot 5 \tag{7.27}$$

$$\text{and} \quad \frac{V}{v^*} = 5 \cdot 75 \log \frac{D}{2k} + 4 \cdot 75 \tag{7.28}$$

$$\text{whence} \quad \frac{1}{\sqrt{\lambda}} = 2 \log \frac{D}{2k} + 1 \cdot 74 \tag{7.29}$$

$$\text{or} \quad \frac{1}{\sqrt{\lambda}} = 2 \log \frac{3 \cdot 7 \, D}{k} \tag{7.30}$$

Equations (7.26) and (7.30) are referred to as the Kármán-Prandtl equations. While equation (7.30) was obtained for artificially roughened pipes Colebrook and White (1939) found that the addition of the Kármán-Prandtl equations produced a 'universal' function which fitted data collected on commercial pipes covering a wide range of Reynolds numbers and relative roughness values. The Colebrook-White equation was expressed as

$$\frac{1}{\sqrt{\lambda}} = -2 \log \left(\frac{k}{3 \cdot 7 \, D} + \frac{2 \cdot 51}{Re \sqrt{\lambda}} \right) \tag{7.31}$$

k now becomes the 'effective roughness size' equivalent to Nikuradse's uniform roughness elements.

7.7 The laminar sub-layer

The λ v. Re curves for artificially roughened pipes are shown in fig. 7.4.

At the lower range of Reynolds numbers the curves for the rough pipes merge into the single smooth pipe curve. Thus, rough pipes can behave like smooth ones at low Reynolds numbers. This phenomenon is explained by the presence of a sub-layer, formed adjacent to the boundary, in which the flow is laminar. The presence of such a layer was also confirmed by the velocity distributions obtained by Nikuradse. At low Reynolds numbers the sub-layer is sufficiently thick to cover the boundary roughness elements so that the turbulent boundary layer is, in effect, flowing over a smooth boundary, fig. 7.5a.

The sub-layer thickness decreases with increasing Reynolds number and at high Reynolds numbers the roughness elements are fully exposed to the turbulent boundary layer. At intermediate Reynolds numbers, in the transitional turbulent zone, the friction factor is influenced by both the relative roughness and Reynolds number.

Figure 7.6 shows a sub-layer formed on a smooth boundary beneath a turbulent boundary layer.

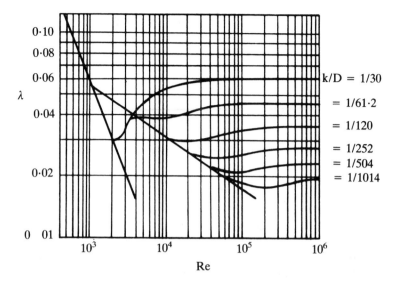

Figure 7.4 Variation of λ with Re for artificially roughened pipes

In the laminar sub-layer, $\tau = \mu \dfrac{dv}{dy}$

whence $\dfrac{dv}{dy} = \dfrac{\tau}{\rho\,v}$ and $v = \dfrac{\tau y}{\rho\,v}$ $\qquad\qquad$ (7.32)

At the upper boundary of the sub-layer, $y = \delta'$, the velocities given by equations (7.23) and (7.32) are identical and $\tau = \tau_0$.

Whence $v^* \left(5 \cdot 75 \log \dfrac{v^*\,\delta'}{v} + 5 \cdot 5 \right) = \dfrac{(v^*)^2\,\delta'}{v}$ the solution to

which is: $\delta' = \dfrac{11 \cdot 6\,v}{v^*}$ $\qquad\qquad$ (7.33)

Substituting $v^* = V \sqrt{\dfrac{\lambda}{8}}$ and introducing D on both sides yields

$\dfrac{\delta'}{D} = \dfrac{32 \cdot 8}{Re\sqrt{\lambda}}$ $\qquad\qquad$ (7.34)

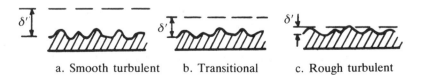

a. Smooth turbulent $\qquad$ b. Transitional $\qquad$ c. Rough turbulent

Figure 7.5 Variation in thickness of laminar sub-layer with Reynolds number

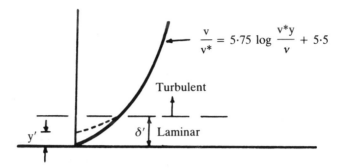

Figure 7.6   Laminar sub-layer on smooth boundary

Thus the thickness of the laminar sub-layer decreases with Reynolds number.
   For rough surfaces the zone of turbulent flow is clearly related to the ratio
of the magnitudes of δ' and k. Since $\dfrac{\delta'}{D}$ is inversely proportional to $\mathrm{Re}\sqrt{\lambda}$
the quantity

$$\frac{\mathrm{Re}\sqrt{\lambda}}{\dfrac{D}{2k}} \text{ will be proportional to } \frac{k}{\delta'}.$$

Each of the curves in fig. 7.4 should therefore deviate from the smooth pipe
law at the same value of $\dfrac{k}{\delta'}$. Thus superimposing all curves by plotting

$$\frac{1}{\sqrt{\lambda}} - 2 \log \frac{D}{2k} \text{ on a base of}$$

$$\frac{\mathrm{Re}\sqrt{\lambda}}{\dfrac{D}{2k}} \quad \text{(fig. 7.7)}$$

shows that the transition from the smooth law begins when δ' is approximately
4 k and ends when k is approximately 6 δ'. (See fig. 7.7.)

Worked examples

Example 7.1
A thin smooth plate 2 m long and 3 m wide is towed edgewise through
water, at 20°C, at a speed of 1 m/s.
(a) Calculate the total drag and the thickness of the boundary layer at the
    trailing edge.
(b) If the plate were towed with the 3 m side in the direction of flow, what
    would be the drag?

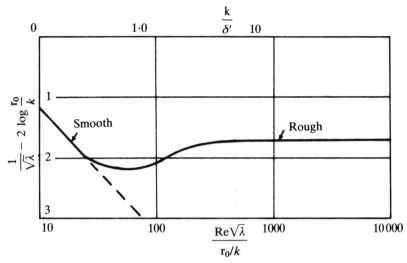

Figure 7.7 The transition zone for artificially roughened pipes

Solution (a):

$$\nu_{20°C} = 1 \times 10^{-6} \text{ m}^2/\text{s}; \quad \rho = 1000 \text{ kg/m}^3$$

Assuming the laminar layer to become unstable at $Re_x = 500\,000$, the length of the layer, $x_o = 0.5$ m

Using equation (7.11) for the combined drag in the laminar and turbulent layers,

$$F_s = 2 \left(\frac{1.33 \, x_o}{Re_{x,o}^{1/2}} + \frac{0.074 \, L}{Re_L^{1/5}} - \frac{0.074 \, x_o}{Re_{x,o}^{1/5}} \right) B \rho \frac{U_0^2}{2}$$

$$F_s = 2 \left(\frac{1.33 \times 0.5}{707.1} + \frac{0.074 \times 2}{18.2} - \frac{0.074 \times 0.5}{13.8} \right) \frac{3 \times 1000 \times 1}{2}$$

$$F_s = 19.17 \text{ N}$$

Boundary layer thickness: experiments have shown that when the turbulent boundary layer develops downstream of the laminar layer, the characteristics of the turbulent layer are those of one which develops at the leading edge.

Thus $\delta_L = \dfrac{0.37 \, L}{Re_L^{1/5}} = 0.041$ m

If the drag were assumed to be due entirely to the turbulent boundary layer:

$$F_s = 2 \, C_f \, BL \, \rho \, \frac{U_0^2}{2}$$

$$C_f = \frac{0.074}{Re_L^{1/5}} = \frac{0.074}{18.2} = 0.0041$$

$$F_s = 2 \times 0.0041 \times 2 \times 3 \times 1000 \times \frac{1}{2} = 24.6 \text{ N}$$

Solution (b):

   B = 2·0;   L = 3·0

From equation (7.11)

$$F_s = 2 \times 2 \left(\frac{1 \cdot 33 \times 0 \cdot 5}{707 \cdot 1} + \frac{0 \cdot 074 \times 3}{19 \cdot 74} - \frac{0 \cdot 074 \times 0 \cdot 5}{13 \cdot 8} \right) 1000 \times \frac{1}{2}$$

   $= 19 \cdot 01$ N.

Example 7.2
A 5 m long smooth model of a ship is towed in fresh water of kinematic viscosity 1×10^{-6} m²/s at 3·5 m/s. The wetted hull area is 1·4 m². What is the skin friction drag?

Solution:
Assuming that a turbulent boundary layer develops at the leading edge:

$$\text{Drag} = C_f \, A \, \rho \, \frac{U_0^2}{2}$$

$$C_f = \frac{0 \cdot 074}{Re_L^{1/5}}; \quad Re_L = \frac{3 \cdot 5 \times 5}{1 \times 10^{-6}} = 17 \cdot 5 \times 10^6$$

whence $C_f = 0 \cdot 00263$

$$\text{And Drag} = 0 \cdot 00263 \times 1 \cdot 4 \times 1000 \times \frac{3 \cdot 5^2}{2}$$

$$= 22 \cdot 55 \text{ N}.$$

Example 7.3
Water enters a 300 mm diameter test section of a water tunnel at a uniform velocity of 15 m/s. Assuming that the boundary layer starts 0·5 m upstream of the test section estimate the increase in axial velocity at the end of a 3 m test section due to the growth of the boundary layer. Take $v = 1 \times 10^{-6}$ m²/s.

Solution:
Length of boundary layer $= 3 \cdot 5$ m.

$$Re_L = 15 \times 3 \cdot 5 \times 10^6 = 52 \cdot 5 \times 10^6$$

$$Re_L^{1/5} = 35 \cdot 0$$

$$\delta_L = \frac{0 \cdot 37 \, L}{Re_L^{1/5}} = \frac{0 \cdot 37 \times 3 \cdot 5}{35} = 0 \cdot 037 \text{ m}$$

From section 7.5, assuming that the velocity distribution in the boundary layer is of the form

$$\frac{v}{U_0} = \left(\frac{y}{\delta} \right)^{1/7} \quad \text{(equation (7.15))}$$

the displacement thickness is expressed by

$$\delta^* = \frac{\delta}{8} \quad (\text{equation } (7.16))$$

or $\delta^* = 0.004625$ m

Effective duct diameter $= 300 - 9.25 = 290.75$ mm

Then $U_0 A_0 = U_L A_L$

whence $U_L = 15 \left(\frac{300}{290.75}\right)^2 = 15.97$ m/s

Change in pressure due to boundary layer formation: apply Bernoulli equation assuming ideal fluid flow:

$$\frac{p_0}{\rho g} + \frac{U_0^2}{2g} = \frac{p_L}{\rho g} + \frac{U_L^2}{2g}$$

$$\frac{p_0 - p_L}{\rho g} = \Delta h = \frac{U_L^2 - U_0^2}{2g}$$

$$\Delta h = \frac{15.97^2 - 15^2}{2g} = 1.53 \text{ m.}$$

Example 7.4
Water at 20°C enters the 250 mm square working section of a water tunnel at 20 m/s with a turbulent boundary layer of thickness equal to that from a starting point 0.45 m upstream. Estimate the length of the sides of the divergent duct for constant pressure core flow at 1 m, 2 m and 3 m downstream from the duct entrance.

Solution:

From equation (7.16) displacement thickness: $\delta_x^* = \frac{\delta_x}{8}$

From equation (7.7) $\delta_x = \frac{0.37 \text{ x}}{\text{Re}_x^{1/5}}$

At 1 m from the entrance of the working section the length of the boundary layer $= 1.45$ m.

$$\text{Re}_x = \frac{20 \times 1.45}{1 \times 10^{-6}} = 29 \times 10^6$$

whence $\delta = 0.01726$ m

and $\delta^* = 0.00216$ m $= 2.16$ mm

Thus section size $= 254.32$ mm square.

At 2 m from entrance, boundary layer is 2.45 m long

$$\delta^* = 3\cdot29 \text{ mm}$$

$$\therefore \text{ section size } = 256\cdot58 \text{ mm}$$

At 3 m from entrance

$$\delta^* = 4\cdot32 \text{ m}$$

$$\therefore \text{ section size } = 258\cdot64 \text{ mm}.$$

Example 7.5
The velocity distribution in the rough turbulent zone is expressed by:

$$\frac{v}{\sqrt{\dfrac{\tau_o}{\rho}}} = 5\cdot75 \log \frac{y}{k} + 8\cdot5 \text{ (equation (7.27))} \qquad \text{(i)}$$

Local axial velocities measured at 25 mm and 75 mm across a radius from the inner wall of a 150 mm diameter pipe conveying water at 15°C were 0·815 m/s and 0·96 m/s respectively. Calculate the effective roughness size, the hydraulic gradient and the discharge.

Writing $v^* = \sqrt{\dfrac{\tau_o}{\rho}}$ and inserting the pairs of values of v and y in the velocity distribution equation:

$$\frac{0\cdot96}{v^*} = 5\cdot75 \log \frac{75}{k} + 8\cdot5 \qquad \text{(ii)}$$

$$\frac{0\cdot815}{v^*} = 5\cdot75 \log \frac{25}{k} + 8\cdot5 \qquad \text{(iii)}$$

(Note that y can be expressed in millimetres since $\dfrac{y}{k}$ is dimensionless; the calculated value of k will then be in millimetres.)

Hence $\dfrac{0\cdot96 - 0\cdot815}{v^*} = 5\cdot75 \log \dfrac{75}{25}$

whence $v^* = 0\cdot0528$

and $\tau_o = 2\cdot79 \text{ N/m}^2$

Substituting for v^* in (ii), $\dfrac{0\cdot96}{0\cdot0528} = 5\cdot75 \log \dfrac{75}{k} + 8\cdot5$

whence $k = 1\cdot55 \text{ mm}$

The hydraulic gradient $\left(S_f = \dfrac{h_f}{L}\right)$ can be related to the boundary shear stress thus:
Consider an element of flow in a pipeline (see fig. 7.8):

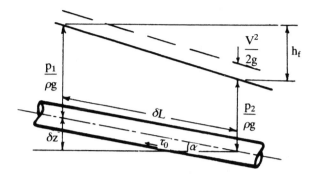

Figure 7.8   Flow analysis in pipeline

For steady, uniform flow, equate the forces on the element of length δL.

$$(p_1 - p_2) A + \rho g A \, \delta L \sin \alpha = \tau_o P \, \delta L$$

i.e. $\dfrac{p_1 - p_2}{\rho g} + \delta z = \dfrac{\tau_o P \, \delta L}{\rho g A}$

or $h_f = \dfrac{\tau_o \, \delta L}{\rho g R}$ where $R = A/P = D/4$

$$\frac{h_f}{L} = S_f = \frac{0 \cdot 0528^2}{9 \cdot 806 \times 0 \cdot 0375} = 0 \cdot 00758$$

The discharge can now be evaluated using the Darcy-Colebrook-White combination:

$$Q = -2A \sqrt{2g D S_f} \log \left[\frac{k}{3 \cdot 7 D} + \frac{2 \cdot 51 \, v}{D \sqrt{2g D S_f}} \right]$$

$Q = 13 \cdot 37$ l/s

Mean velocity $V = 0 \cdot 757$ m/s

$$Re = 1 \cdot 004 \times 10^5$$

$$\frac{k}{D} = \frac{1 \cdot 55}{150} = 0 \cdot 0103$$

Referring to the Moody chart this is in the rough turbulent zone.

Alternatively, as shown previously (section 7.6) the mean velocity V can be expressed thus:

$$\frac{V}{\sqrt{\tau_o / \rho}} = 5 \cdot 75 \log \frac{D}{2k} + 4 \cdot 75$$

whence $V = 0 \cdot 762$ m/s   and $Q = 13 \cdot 45$ l/s

Location of local velocity equal to mean velocity.

$$\frac{v - V}{\sqrt{\tau_o/\rho}} = 5{\cdot}75 \log \frac{2y}{D} + 3{\cdot}75 \quad \text{(from equations (7.28) and (7.29))}$$

i.e. $\log \dfrac{2y}{D} = -0{\cdot}6522$

$$\frac{2y}{D} = 0{\cdot}22275$$

$$y = 16{\cdot}7 \text{ mm.}$$

Example 7.6

The velocity distribution in the smooth turbulent zone is given by:

$$\frac{v}{v^*} = 5{\cdot}75 \log \frac{v^* \, y}{v} + 5{\cdot}5 \quad \text{(equation (7·23))}$$

The axial velocity at 100 mm from the wall across a radius in a 200 mm perspex pipeline conveying water at 20°C was found to be 1·2 m/s. Calculate the hydraulic gradient and discharge. $v = 1{\cdot}0 \times 10^{-6}$ m²/s

$$\frac{1{\cdot}2}{v^*} = 5{\cdot}75 \log \frac{v^* \times 0{\cdot}1}{1 \times 10^{-6}} + 5{\cdot}5$$

(Note that y must be expressed here in metres.)
Solve by trial or graphical interpolation:

Solution:

$$v^* = 0{\cdot}045 \text{ m/s} \left(= \sqrt{\frac{\tau_o}{\rho}} \right)$$

Now $S_f = \dfrac{\tau_o}{\rho g \, R}$ (see previous Example)

$$\therefore \; S_f = \frac{0{\cdot}045^2}{9{\cdot}806 \times 0{\cdot}05} = 0{\cdot}00413$$

Now if the flow is assumed to be in the smooth turbulent zone (k/D = 0) the discharge may be calculated from the Darcy-Colebrook-White equation in the form:

$$Q = -\, 2A \sqrt{2g \, D \, S_f} \, \log \left(\frac{2{\cdot}51 \, v}{D \sqrt{2g \, D \, S_f}} \right)$$

$$= 32{\cdot}03 \text{ l/s}$$

$$V = 1{\cdot}02 \text{ m/s}; \quad Re = 2{\cdot}04 \times 10^5$$

Example 7.7

Sand grains 0·5 mm in diameter are glued to the inside of a 200 mm, diameter pipeline. At what velocity of flow of water at 15°C will the surface roughness (a) cause the flow to depart from the smooth pipe, and (b) enter the rough pipe curve.

Solution (a):

The transition from the smooth law begins when $\delta' = 4\,k$

whence $\delta' = 4 \times 0.5 = 2.0$ mm

Also $\quad \delta' = \dfrac{32 \cdot 8\ D}{\mathrm{Re}\sqrt{\lambda}}$ (equation (7.34))

$$2\cdot 0 = \frac{32\cdot 8 \times 200}{\mathrm{Re}\sqrt{\lambda}}$$

whence $\mathrm{Re}\sqrt{\lambda} = 3280 \cdot 0$ $\hspace{4cm}$ (i)

The smooth turbulent zone is represented by:

$$\frac{1}{\sqrt{\lambda}} = 2\,\log\frac{\mathrm{Re}\sqrt{\lambda}}{2\cdot 51}$$

$$\frac{1}{\sqrt{\lambda}} = 2\,\log\frac{52\,480}{2\cdot 51} = 6\cdot 232$$

Then from (i) $\mathrm{Re} = 20\,441$ $\left(= \dfrac{VD}{\nu} \right)$

whence $\hspace{2cm} V = 0\cdot 102$ m/s.

Solution (b):

At $k = 6\delta'$ the flow enters the rough turbulent zone.

$$\delta' = \frac{32\cdot 8\ D}{\mathrm{Re}\sqrt{\lambda}} = 0\cdot 0833$$

$\mathrm{Re}\sqrt{\lambda} = 78751\cdot 0$ $\hspace{4cm}$ (ii)

The rough turbulent zone is represented by

$$\frac{1}{\sqrt{\lambda}} = 2\,\log\frac{3\cdot 7\ D}{k} = 2\,\log\frac{3\cdot 7 \times 200}{0\cdot 5}$$

$$= 6\cdot 341$$

Substituting in (ii): $\mathrm{Re} = 499372\cdot 0$

whence $V = 2\cdot 5$ m/s.

Recommended reading

1. Douglas, J.F., Gasiorek, J.M. and Swaffield, J.A. (1995) *Fluid Mechanics*, 3rd edn. London: Pitman.
2. Nikuradse, J. (1950) *Laws of Flow in Rough Pipes* (Translation of *Stromungsgesetze in rauhen Rohren VDI* – (Forshungshaft 361 (1933)) Washington: NACA Tech Mem 1292.
3. Rouse, H. (1946) *Elementary Mechanics of Fluids* (Chapter VII). Chichester: Wiley.

Problems

1.   A pontoon 15 m long 4 m wide with vertical sides floats to a depth of 0·5 m. The pontoon is towed in sea water at 10°C (ρ_s = 1024 kg/m³, μ = 1·31 × 10⁻³ Ns/m²) at a speed of 2 m/s. (a) Determine the viscous resistance and the thickness of the boundary layer at the downstream end. (b) What is the shear stress at the mid-length?

2.   Sealed hollow pipes 2 m in external diameter and 6 m long, fitted with rounded nose-pieces to reduce wave-making drag, are towed in a river to a construction site; the pipes float to a depth of 1·5 m. The towing speed is 5 m/s in the water which has a density of 1000 kg/m³ and dynamic viscosity 1·2 × 10⁻³ Ns/m². Determine the viscous drag of each pipe assuming (a) that a turbulent boundary layer exists over the entire length, and (b) that the drag is a combination of that due to the laminar and turbulent boundary layers.

3.   Air of density 1·3 kg/m³ and dynamic viscosity 1·8 × 10⁻⁵ Ns/m² enters the test section 1 m wide × 0·5 m deep of a wind tunnel at a velocity of 20 m/s. Determine the increase in axial velocity 10 m downstream from the entrance to the test section due to the development of the boundary layer assuming that this forms at the entrance to the test section.

4.   Wind velocities over flat grassland were observed to be 3·1 m/s and 3·3 m/s at heights of 3 m and 6 m above the ground respectively. Determine the effective roughness of the surface and estimate the wind velocity at a height of 25 m.

5.   The centre line velocity in a 100 mm diameter brass pipeline (k = 0·0) conveying water at 20°C was found to be 3·5 m/s. Determine the boundary shear stress, the hydraulic gradient, the discharge and the thickness of the laminar sub-layer.

6.   The velocities at 50 mm and 150 mm from the pipe wall of a 300 mm diameter pipeline conveying water at 15°C were 1·423 m/s and 1·674 m/s

respectively. Determine the effective roughness size, the hydraulic gradient, the discharge and the Darcy friction factor and using equation (7.35) et seq. verify that the flow is in the rough turbulent zone.

7.   Show that the equation

$$\frac{v \, max \, - \, v}{\sqrt{\tau_o/\rho}} = 5{\cdot}75 \, \log \frac{D}{2y}$$

applies to full-bore flow in circular pipes both in the rough and smooth turbulent zones.

Chapter 8
Steady Flow in Open Channels

R.E. Featherstone and C. Nalluri

8.1 Introduction

Open channel flow, for example, flow in rivers, canals and sewers not flowing full, is characterised by the presence of the interface between the liquid surface and the atmosphere. Therefore, unlike full pipe flow, where the pressure is normally above atmosphere pressure, but sometimes below it, the pressure on the surface of the liquid in open channel flow is always that of the ambient atmosphere.

The energy per unit weight of liquid flowing in a channel at a section where the depth of flow is y and the mean velocity is V is

$$H = z + y \cos \theta + \frac{\alpha V^2}{2g} \qquad \text{(see fig. 8.1)} \qquad (8.1)$$

where z is the position energy (or head), θ is the bed slope and α the Coriolis coefficient (see also Eqn. 3.11)

$$\left(\alpha = \frac{1}{AV^3} \int_o^A v^3 \, dA \right)$$

The motivating force establishing flow is predominantly the gravity force component acting parallel with the bed slope but net pressure forces and inertia forces may also be present. Flow in channels may be unsteady, resulting from changes in inflow such as floods or changes in depth caused

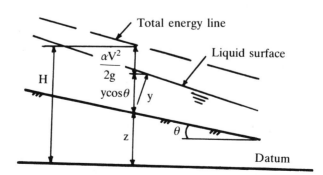

Figure 8.1 **Energy components**

by control gate operation, etc. Steady flow can either be uniform or varied depending upon whether or not the mean velocity is constant with distance. In gradually varied flow there is a gentle change in depth with distance; a common example is the back-water curve, fig. 8.2 (a). Rapidly varied surface profiles are created by changes in channel geometry, for example flow through a venturi flume, fig. 8.2 (b).

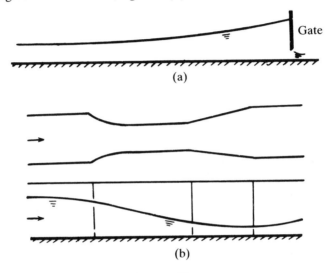

(a)

(b)

Figure 8.2   Steady, varied flow surface profiles

Steady uniform flow occurs when the motivating forces and drag forces are exactly balanced over the reach under consideration. This type of flow is analogous with steady pressurised flow in a pipeline of constant diameter. Thus the area of flow in the channel must remain constant with distance, a condition requiring the bed slope and channel geometry to remain constant. The liquid surface is parallel with the bed.

8.2 Uniform flow resistance

The nature of the boundary resistance is identical with that of full pipe flow (Chapter 4) and the Darcy-Weisbach and Colebrook-White equations for non-circular sections may be applied. (See section 4.2.)

Noting that the energy gradient S_f is equal to the bed slope S_o in uniform flow:

$$\text{Darcy-Weisbach} \quad \frac{h_f}{L} = S_f = S_o = \frac{\lambda \, V^2}{8g \, R} \tag{8.2}$$

$$\text{Colebrook-White} \quad \frac{1}{\sqrt{\lambda}} = -2 \log \left[\frac{k}{14 \cdot 8 \, R} + \frac{2 \cdot 51 \, v}{4R \, V \, \sqrt{\lambda}} \right] \tag{8.3}$$

Eliminating λ from (8.2) and (8.3) yields

$$V = - \sqrt{32g\ R\ S_o}\ \log \left[\frac{k}{14\cdot 8\ R} + \frac{1\cdot 255\ v}{R\sqrt{32g\ R\ S_o}} \right] \qquad (8.4a)$$

where $R = \dfrac{A}{P} = \dfrac{\text{Area of flow}}{\text{Wetted perimeter}}$

In addition to the Darcy-Weisbach equation the Manning equation is widely used in open channel water flow computations. This was derived from the Chezy equation $V = C\ \sqrt{R\ S_o}$ by writing $C = \dfrac{R^{1/6}}{n}$ resulting in

$$V = \frac{R^{2/3}}{n}\ S_o^{1/2} \quad \text{(S.I. units)} \qquad (8.4b)$$

n is called the Manning roughness factor and its value is related to the type of boundary surface. If the value of n is taken to be constant regardless of depth then, unlike the Darcy friction factor it does not account for changes in relative roughness, nor does it include the effects of viscosity. See table 8.1; the Manning equation is in general applicable to shallow flows in rough boundaries and for a boundary of roughness k, the Manning n may be written as (Strickler's equation):

$$n = \frac{k^{1/6}}{26} \qquad (8.4c)$$

where k is in metres.

Table 8.1 Typical values of Manning's n

Type of Surface	n
Concrete	
Culvert, straight and free of debris	0·011
Culvert, with bends, connections and some debris	0·013
Cast on steel forms	0·013
Cast on smooth wood forms	0·014
Unfinished, rough wood form	0·017
Excavated or dredged channels	
Earth, after weathering, straight and uniform	0·022
Earth, winding, clean	0·025
Earth bottom, rubble sides	0·030

For a comprehensive list see *Open-channel Hydraulics* by Ven Te Chow.[5]

8.3 Channels of composite roughness

In applying the Manning formula to channels having different n values for the bed and sides it is necessary to compute an equivalent n value to be used for the whole section. The water area is 'divided' into N parts having wetted perimeters P_1, $P_2 . \_ . P_N$ with associated roughness coefficients n_1, $n_2 . \_ . n_N$. Horton and Einstein assumed that each sub-area has a velocity equal to the mean velocity.

Thus
$$n = \left[\frac{\sum\limits_{i=1}^{N} P_i n_i^{3/2}}{P} \right]^{2/3} \tag{8.5}$$

where P = total wetted perimeter.

Pavlovskij and others equated the sum of the component resisting forces to the total resisting force and thus found

$$n = \left[\frac{\sum\limits_{i=1}^{N} (P_i n_i^2)}{P} \right]^{1/2} \tag{8.6}$$

Lotter applied the Manning equation to sub-areas and equated the sum of the individual discharge equations to the total discharge. Thus the equivalent roughness coefficient is:

$$n = \frac{PR^{5/3}}{\sum\limits_{i=1}^{N} \left(\frac{P_i R_i^{5/3}}{n_i} \right)} \tag{8.7}$$

8.4 Channels of compound section

A typical example of a compound section (two-stage channel) is a river channel with flood plains. The roughness of the side channels will be different (generally rougher) than that of the main channel and the method of analysis is to consider the total discharge to be the sum of component discharges computed by the Manning equation.

Thus in the channel shown in fig. 8.3, assuming that the bed slope is the same for the three sub-areas:

$$Q = \left(\frac{A_1}{n_1} R_1^{2/3} + \frac{A_2}{n_2} R_2^{2/3} + \frac{A_3}{n_3} R_3^{2/3} \right) S_o^{1/2}$$

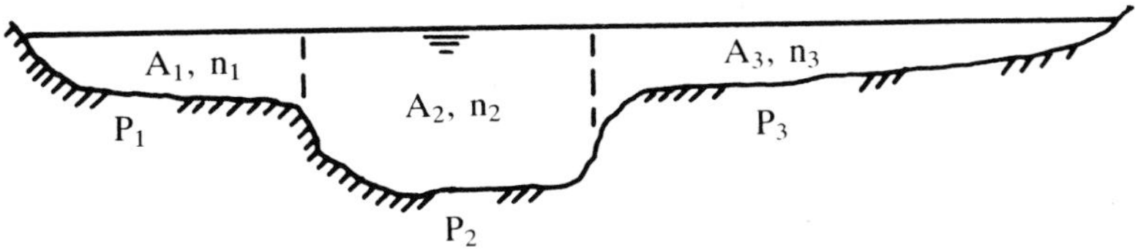

Figure 8.3 Compound channel section

The above assumption leads to large discrepancies between computed and measured discharges under flood flow (above bank-full stages) conditions. The interaction between the slower moving berm flows and the fast moving main channel flow significantly increases head losses. As a result, the discharge computed by this conventional method will overestimate the flow. Utilising the recent research data from the Flood Channel Facility at Wallingford, Ackers[1,2] has shown that the discrepancy between the conventional calculations and the measured flow is dependent on flood flow levels. He formulated appropriate correction factors for each region of flow; a detailed exposure of the analysis of the research is beyond the scope of the book.

8.5 Channel design

The design of open channels involves the selection of suitable sectional dimensions such that the maximum discharge will be conveyed within the section. The bed slope is sometimes constrained by the topography of the land in which the channel is to be constructed.

In the design of an open channel a resistance equation, the Darcy or Chezy or Manning, may be used. However at least one other equation is required to define the relationship between width and depth. This second series of equations incorporates the design criteria; for example in rigid boundary (non-erodible) channels the designer will wish to minimise the construction cost resulting in what is commonly termed 'the most economic section'. In addition there may be a constraint on the maximum velocity to prevent erosion or on the minimum velocity to prevent settlement of sediment.

In the case of erodible (unlined channels excavated in natural ground e.g. clay, silts, etc.) the design criterion will be that the boundary shear stress exerted by the moving liquid will not exceed the 'critical tractive force' of the bed and side material.

(a) Rigid boundary channels — economic section.
Using the Darcy-type resistance equation,

$$Q = A \sqrt{\frac{8g}{\lambda} \frac{A}{P} S_o} = \frac{KA^{3/2}}{P^{1/2}}$$

$$A = f(y); \qquad P = f(y).$$

Q max. is achieved when $\dfrac{dQ}{dy} = 0$ i.e. $\dfrac{d}{dy}\left(\dfrac{A^3}{P}\right) = 0$

$$\frac{3A^2}{P}\frac{dA}{dy} - \frac{A^3}{P^2}\frac{dP}{dy} = 0$$

$$\text{whence } 3P\frac{dA}{dy} - A\frac{dP}{dy} = 0$$

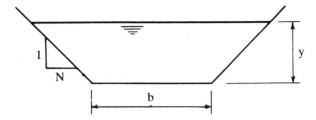

Figure 8.4 Trapezoidal channel

For a given area $\dfrac{dA}{dy} = 0$; then for Q max., $\dfrac{dP}{dy} = 0$, i.e. the wetted perimeter is a minimum. For a trapezoidal channel (see fig. 8.4):

$$A = (b + Ny)y$$

$$P = b + 2y \sqrt{1 + N^2}$$

For a given area A,

$$P = \frac{A}{y} - Ny + 2y \sqrt{1 + N^2}$$

For Q max. $\dfrac{dP}{dy} = -\dfrac{A}{y^2} - N + 2 \sqrt{1 + N^2} = 0$

i.e. $\dfrac{dP}{dy} = -(b + Ny) - Ny + 2y \sqrt{1 + N^2} = 0$

or $\quad b + 2Ny = 2y \sqrt{1 + N^2}$ $\hspace{3cm}$ (8.8)

It can be shown that if a semicircle of radius y is drawn with its centre in the liquid surface it will be tangential to the sides and bed. Thus the most economic section approximates as closely as possible to a circular section which is known to have the least perimeter for a given area.

For a rectangular section (N = 0) and b = 2y.

(b) Mobile boundary channels (erodible)

The 'critical tractive force' theory and the 'maximum permissible velocity' concept are commonly used in the design of erodible channels for stability.

(i) Critical tractive force theory

The force exerted by the water on the wetted area of a channel is called the 'tractive force'. The average 'unit tractive force' is the average shear stress given by $\bar{\tau}_o = \rho g\, R\, S_o$. Boundary shear stress is not, however, uniformly distributed; the distribution varies somewhat with channel shape but not with size. For trapezoidal sections the maximum shear stress on the bed may be taken as $\rho g\, y\, S_o$ and on the sides as $0.76\, \rho g\, y\, S_o$ (See fig. 8.5); however, the shear distribution depends on the channel aspect ratio, b/y (see Table 8.2).

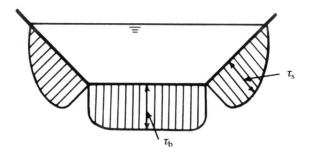

Figure 8.5 Distribution of shear stress on channel boundary

Table 8.2 Maximum bed/side shear stress

Aspect ratio, b/y	$\tau_{b\ max}/\rho gyS_0$	$\tau_{s\ max}/\rho gyS_0$
2	0·890	0·735
4	0·970	0·750
>8	0·985	0·780

If the shear stresses can be kept below that which will cause the material of the channel boundary to move, the channel will be stable. The critical tractive force of a particular material is the unit tractive force which will not cause erosion of the material on a horizontal surface. Material on the sides of the channel is subjected, in addition to the shear force due to the flowing water, to a gravity force down the slope. It can be shown (see e.g. Chow[5]) that if τ_{cb} is the critical tractive force the maximum critical shear stress due to the water flow on the sides is

$$\tau_{cs} = \tau_{cb} \sqrt{1 - \frac{\sin^2 \theta}{\sin^2 \phi}} \tag{8.9}$$

where θ is the slope of the sides to the horizontal, and ϕ is the angle of repose of the material.

Table 8.3 gives some typical values of critical tractive force and permissible velocity.

(ii) Maximum permissible mean velocity concept.

This appears to be a rather uncertain concept since the depth of flow has a significant effect on the boundary shear stress. Fortier and Scobey[6] published the values in Table 8.3 for well-seasoned channels of small bed slope and depths below 1 m.

8.6 Uniform flow in part-full circular pipes

Circular pipes are widely used for underground storm sewers and waste-water sewers. Storm sewers are usually designed to have sufficient capacity

Table 8.3 Critical tractive force and mean velocity for different bed materials

Material	Size mm	Critical tractive force N/m²	Approximate mean velocity m/sec	Manning's coefficient of roughness
Sandy loam (non-colloidal)		2·0	0·50	0·020
Silt loam (non-colloidal)		2·5	0·60	0·020
Alluvial silt (non-colloidal)		2·5	0·60	0·020
Ordinary firm loam		3·7	0·75	0·020
Volcanic ash		3·7	0·75	0·020
Stiff clay (very colloidal)		1·22	1·15	0·025
Alluvial silts (colloidal)		12·2	1·15	0·025
Shales and hard-pans		31·8	1·85	0·025
Fine sand (non-colloidal)	0·062–0·25	1·2	0·45	0·020
Medium sand (non-colloidal)	0·25–0·5	1·7	0·50	0·020
Coarse sand (non-colloidal)	0·5–2·0	2·5	0·60	0·020
Fine gravel	4–8	3·7	0·75	0·020
Coarse gravel	8–64	14·7	1·25	0·025
Cobbles and shingles	64–256	44·0	1·55	0·035
Graded loam and cobbles (non-colloidal)	0·004–64	19·6	1·15	0·30
Graded silts to cobbles (colloidal)	0–64	22·0	1·25	0·30

so that they do not run full when conveying the computed surface run off resulting from a storm of a specified average return period. Under these conditions 'open channel flow' conditions exist. However more intense storms may result in the capacity of the pipe, when running full at a hydraulic gradient equal to the pipe slope, being exceeded and pressurised pipe conditions will follow. Waste-water sewers, on the other hand, generally carry very small discharges and the design criterion in this case is that the mean velocity under the design flow conditions should exceed a 'self-cleansing velocity' of 0·61 m/s so that sediment will not be permanently deposited.

Although the flow in sewers is rarely steady (and hence non-uniform), some commonly used design methods adopt the assumption of uniform flow at design flow conditions; the Rational Method for storm sewer design is an example and foul sewers are invariably designed under assumed uniform flow conditions. Mathematical simulation and design models of flow in storm sewer networks take account of the unsteady flows using the dynamic equations of flow but such models often incorporate the steady uniform flow relations in storage-discharge relationships.

Geometrical properties; flow equations (see fig. 8.6).

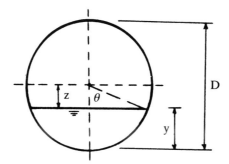

Figure 8.6   Circular channel section

$$z = \frac{D}{2} - y$$

$$\theta = \cos^{-1}\left(\frac{2z}{D}\right)$$

$$A = r^2\left(\theta - \frac{\sin 2\theta}{2}\right)$$

$$P = D\,\theta$$

$$R = \frac{A}{P}$$

$$V = -\sqrt{32g\,R\,S_o}\,\log\left[\frac{k}{14\cdot8\,R} + \frac{1\cdot255\,v}{R\sqrt{32g\,R\,S_o}}\right] \qquad (8.10)$$

$$Q = -A \sqrt{32g\ R\ S_o}\ \log \left[\frac{k}{14\cdot8\ R} + \frac{1\cdot255\ v}{R\sqrt{32\ g\ R\ S_o}} \right] \tag{8.11}$$

8.7 Steady, rapidly varied channel flow-energy principles

The computation of non-uniform surface profiles caused by changes of channel section, etc., requires the application of energy and momentum principles.

The energy per unit weight of liquid at a section of a channel above some horizontal datum is

$$H = z + y \cos \theta + \alpha\ V^2/2g \quad \text{(see fig. 8.1)}$$

For mild slopes $\cos \theta = 1\cdot0$

$$H = z + y + \alpha\ V^2/2g \tag{8.12}$$

Specific energy is measured relative to the bed:

$$E_s = y + \alpha\ V^2/2g$$

$$\text{or } E_s = y + \alpha\ Q^2/2g\ A^2$$

For a steady fixed inflow into a channel the specific energy at a particular section can be varied by changing the depth by means of a structure such as sluice gate with an adjustable opening. Plotting E_s v. y for a section where the relationship $A = f(y)$ is known and Q is fixed results in the 'specific energy curve'; fig. 8.7 shows that at a given energy level two alternative depths are possible.

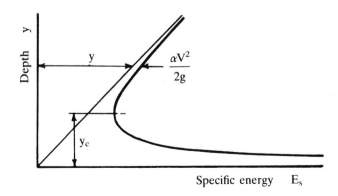

Figure 8.7 Specific energy curve

Specific energy becomes a minimum at a certain depth called the critical depth (y_c). For this condition

$$\frac{dE_s}{dy} = 1 - \frac{\alpha Q^2}{A^3 g} \frac{dA}{dy} = 0$$

Now $dA/dy = B$, the surface width

$$\text{whence } \frac{\alpha Q^2 B}{A^3 g} = 1 \text{ or } \frac{Q}{\sqrt{g/\alpha}} = A\sqrt{\frac{A}{B}} \tag{8.13}$$

For the special case of a rectangular channel and where $\alpha = 1\cdot0$.

$$\frac{V^2}{gy} = 1; \text{ i.e. the Froude number is unity and } y_c = \sqrt[3]{\frac{Q^2}{gb^2}} = \sqrt[3]{\frac{q^2}{g}} \tag{8.14}$$

where q = discharge/unit width.
At the critical depth,

$$E_{s,min} = y_c + \frac{V_c^2}{2g} \text{ and } V_c = \sqrt{gy_c}$$

$$E_{s,min} = y_c + \frac{y_c}{2}$$

In a rectangular channel the depth of flow at the critical flow is $2/3 \times$ the specific energy at critical flow.

The velocity corresponding with the critical depth is called the critical velocity.

At depths below the critical the flow is called supercritical and at depths above the critical the flow is subcritical or 'tranquil'.

From the specific energy equation we can write

$$Q = A\sqrt{\frac{2g}{\alpha}(E_s - y)}$$

which shows that $Q = f(y)$ for a constant E_s. This exhibits a maximum discharge occurring at a depth equal to critical depth (see fig. 8.21).

8.8 The momentum equation and the hydraulic jump

The hydraulic jump is a stationary surge and occurs in the transition from a supercritical to subcritical flow (fig. 8.8).

A smooth transition is not possible; if this were to occur the energy would vary according to the route ABC on the E_s curve. At B the energy would be less than that at C, corresponding with the downstream depth, y_s. Therefore a rapid depth change occurs corresponding with the route AC on the E_s curve. The depth at which the jump starts is called the 'initial depth', y_i and the downstream depth the 'sequent depth', y_s. For a given channel and

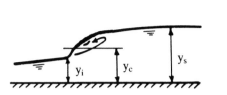
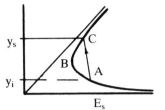

Figure 8.8 The hydraulic jump

discharge there is a unique relationship between y_i and y_s which requires application of the momentum equation.
Steady state momentum equation. (See fig. 8.9.)
Assuming hydrostatic pressure distribution:

$$\rho g A_1 \bar{y}_1 + \rho Q (\beta_1 V_1 - \beta_2 V_2) - \rho g A_2 \bar{y}_2 + \rho g \bar{A} S_o - \rho g \bar{A} S_f = 0$$

$\bar{y}$ = depth of centre of area of cross-section

and $\bar{A} = (A_1 + A_2)/2$

i.e. $A_1 \bar{y}_1 + \dfrac{Q}{g} (\beta_1 V_1 - \beta_2 V_2) - A_2 \bar{y}_2 + \bar{A} (S_o - S_f) = 0$ (8.15)

Equation 8.15 may be rewritten as $M_1 = M_2$, where

$$M = A \bar{y} + \frac{Q}{g} \beta V = f(y)$$

The M function (specific force) exhibits a minimum value at critical depth. When applied to the analysis of the hydraulic jump the term $(S_o - S_f)$ may be neglected. In the special case of a rectangular channel the above equation

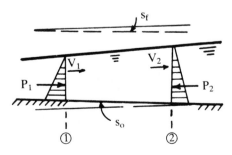

Figure 8.9 Reference diagram for momentum equation

reduces to a quadratic which can be solved either for y_i or y_s. Taking $\beta_1 = \beta_2$.

$$\left(\beta = \text{Boussinesq coefficient} = \frac{1}{AV^2} \int_o^A v^2 \, dA \right)$$

$$y_i = \frac{y_s}{2} \left(\sqrt{1 + 8\beta F_s^2} - 1 \right) \tag{8.16}$$

$$\text{or } y_s = \frac{y_i}{2} \left(\sqrt{1 + 8\beta F_i^2} - 1 \right) \tag{8.17}$$

$$\text{where } F_s = \frac{V_s}{\sqrt{gy_s}} \text{ and } F_i = \frac{V_i}{\sqrt{g_i y_i}} \quad \text{(Froude numbers)}$$

The energy loss through the jump ($= E_1 - E_2$) can also be shown as

$$E_L = \frac{(y_2 - y_1)^3}{4y_1 y_2} \tag{8.18}$$

The length of the jump is a function of the approach flow Froude number and for $F_{r1} > 9$, the length is approximately equal to $7y_2$. The excess kinetic energy of the downstream flow over a control structure (such as a spillway) is often destroyed by the formation of a hydraulic jump (energy dissipator) over a confined solid structure known as a stilling basin (see Chow[5]).

8.9 Steady gradually varied open channel flow

This condition occurs when the motivating and drag forces are not balanced with the result that the depth varies gradually along the length of the channel (fig. 8.10).

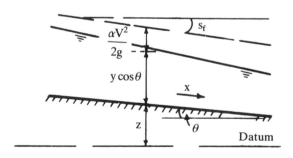

Figure 8.10 Varied flow in channel

The dynamic equation of gradually varied flow is obtained by differentiating the energy equation: $H = z + y \cos \theta + \alpha V^2/2g$ with respect to distance along the channel bed (x-direction)

$$\frac{dH}{dx} = \frac{dz}{dx} + \cos \theta \frac{dy}{dx} + \frac{d}{dx}\left(\alpha \frac{V^2}{2g}\right)$$

$$\text{Now } S_f = -\frac{dH}{dx}; S_o = \sin \theta = -\frac{dz}{dx}$$

$$\frac{dy}{dx} = \frac{S_o - S_f}{\cos \theta + \frac{d}{dy}\left(\alpha \frac{V^2}{2g}\right)}$$

$$\frac{d}{dy}\left(\frac{V^2}{2g}\right) = \frac{d}{dy}\left(\frac{Q^2}{2g\,A^2}\right) = -\frac{2Q^2}{2g\,A^3}\frac{dA}{dy}$$

$$\text{and } \frac{dA}{dy} = T, \text{ the width at the liquid surface.}$$

$$\frac{d}{dy}\left(\frac{V^2}{2g}\right) = -\frac{Q^2 T}{A^3 g}$$

Since channel slopes are usually small $\cos \theta = 1\cdot0$

$$\text{whence } \frac{dy}{dx} = \frac{S_o - S_f}{1 - \alpha \dfrac{Q^2 T}{A^3 g}} \tag{8.19}$$

which gives the slope of the water surface relative to the bed. Space does not permit a full discussion of the various types of surface profile which can occur, nor on the 'classification' of these surfaces according to the channel slope. Treatment of these topics is to be found for example in *Open-channel Hydraulics* by Ven Te Chow.[5]; see example 8.24 on water surface profiles.

8.10 Computations of gradually varied flow

Equation (8.19) has analytical solutions (by direct integration) which require tables of varied flow functions to facilitate the computation e.g. Bresse (wide channels) and Bakhmeteff.[3] The graphical integration method is widely applicable. However, with the advent of electronic calculators and especially digital computers it is often more convenient to use numerical methods; such methods are applicable to general cases of non-prismatic channels of varying slope.

Computations of gradually varied surface profiles should proceed upstream from the control section in subcritical flow and downstream from the control section in supercritical flow.

8.11 Graphical and numerical integration methods

Consider two channel sections at distances x_2 and x_1 at which the depths are y_2 and y_1.

$$x_2 - x_1 = \int_{x_1}^{x_2} dx = \int_{y_1}^{y_2} \frac{dx}{dy} \, dy = \text{area under the } \frac{dx}{dy} \text{ v. y curve (fig. 8.11).}$$

Therefore, take a series of y values, compute $\dfrac{dx}{dy}$ from equation (8.19) and

find the area, ΔA, between successive y values on the $\dfrac{dx}{dy}$ v. y curve to find

the distance between them. Alternatively if dy is small the curve of $\dfrac{dx}{dy}$ v. y

may be approximated by straight lines between y_2 and y_1 (Δy)

i.e. $\Delta x = \dfrac{\overline{dx}}{dy} \Delta y$

whence a numerical method may be exclusively used. (See fig. 8.11.)

8.12 The direct step method

The direct step method is a simple method applicable to prismatic channels. As in the graphical integration method depths of flow are specified and the distances between successive depths calculated.

Consider an element of the flow (fig. 8.12).

Equating total heads at 1 and 2

$$S_o \, \Delta x + y_1 + \alpha_1 \frac{V_1^2}{2g} = y_2 + \alpha_2 \frac{V_2^2}{2g} + S_f \, \Delta x$$

i.e. $\Delta x = \dfrac{E_2 - E_1}{S_o - S_f}$ \hfill (8.20)

where E is the specific energy.

In the computations S_f is calculated for depths y_1 and y_2 and the average taken denoted by $\overline{S_f}$.

8.13 The standard step method

The standard step method is applicable to non-prismatic channels and therefore to natural rivers. The station positions are predetermined and the objective is to calculate the surface elevations, and hence the depths, at the stations. A trial and error method is employed. (See fig. 8.13.)

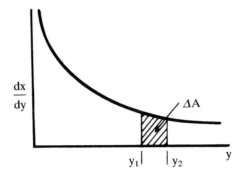

Figure 8.11 Graphical integration method

$$y_1 + \frac{\alpha V_1^2}{2g} + h_f = S_o \Delta x + y_2 + \frac{\alpha V_2^2}{2g}$$

$$Z_1 = y_1; \quad Z_2 = S_o \Delta x + y_2 \quad \text{and assuming } \alpha_1 = \alpha_2 = \alpha$$

$$Z_1 + \frac{\alpha V_1^2}{2g} + h_f = Z_2 + \frac{\alpha V_2^2}{2g} \tag{8.21}$$

$$\text{writing } H_1 = Z_1 + \frac{\alpha V_1^2}{2g}; \quad H_2 = Z_2 + \frac{\alpha V_2^2}{2g},$$

equation (8.21) becomes $H_1 + h_f = H_2$

Proceeding upstream (in subcritical flow), for example, H_1 is known and Δx is predetermined. Z_2 is estimated, for example by adding a small amount to Z_1; y_2 is obtained from: $y_2 = Z_2 - z_2$. The area and wetted perimeter, and hence hydraulic radius corresponding with y_2 are obtained from the known geometry of the section.

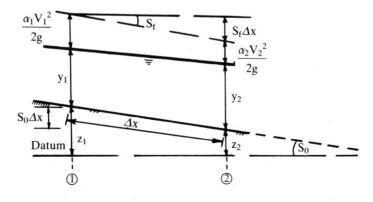

Figure 8.12 The direct step method

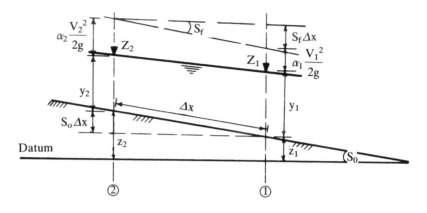

Figure 8.13 The standard step method

Calculate $S_{f,2} = \dfrac{n^2 V_2^2}{R_2^{4/3}}$ and $\bar{S}_f = \dfrac{S_{f,2} + S_{f,1}}{2}$

Calculate $\dfrac{\alpha V_2^2}{2g}$ and $H_{(2)} = Z_2 + \dfrac{\alpha V_2^2}{2g}$

Calculate $H_{(1)} = H_1 + \bar{S}_f \Delta x$

Compare $H_{(2)}$ and $H_{(1)}$; if the difference is not within prescribed limits (e.g. 0·001 m), re-estimate Z_2 and repeat until agreement is reached; Z_2, y_2 and $H_{(2)} = H_2$ are then recorded and Z_2 and H_2 become Z_1 and H_1 for the succeeding station.

8.14 Canal delivery problems

When a channel is connected to two reservoirs its discharge capacity depends upon inlet (upstream) and outlet (downstream) conditions imposed by the water levels in the reservoirs. The reservoir−canal−reservoir interaction depends upon the channel characteristics such as its boundary roughness, slope, length between reservoirs and the state of water levels in the reservoirs.

Case 1: upstream reservoir water level is constant
For a given boundary characteristic the discharge rate in the channel depends on whether it is a long or short channel with a mild or critical or steep bed slope. A long channel delivers water with no interference from the downstream water levels, i.e. with no downstream control; only the inlet controls the flow rate. This suggests that any water surface profiles likely to develop with the available downstream water levels will not be long enough to reach the inlet, thus allowing the inlet to discharge freely. On the other hand if the channel is short, the water surface profiles could submerge the inlet and this

submergence affects the flow rate. The types of profile and their appropriate lengths would depend on the channel slope.

(a) Long channel with mild slope (see fig. 8.20)
The flow on entering the inlet establishes uniform conditions from a short distance downstream of the entry. Thus at entry we have two simultaneous equations to compute the flow depth, y_0 and discharge, Q_0 or velocity:

energy equation: upstream water level above inlet, $H = y_0' + \alpha V^2/2g + K\alpha V^2/2g$
uniform flow resistance equation (say Manning's) $V = (1/n)R^{2/3}S^{1/2}$

See worked example 8.7 for the solution of these two equations.

(b) Long channel with critical slope
The flow now establishes with its normal flow depth equal to the critical depth from the inlet, thus allowing maximum possible discharge through the channel. We now have at the channel entry two equations enabling the computations of Q and y:

energy equation at inlet: $H = y_c + \alpha V_c^2/2g + \alpha K V_c^2/2g$
either the critical depth criterion: $\alpha Q^2 B/gA^3 = 1$

or the appropriate uniform flow resistance equation.

(c) Long channel with steep slope
The flow depth at entry is critical, the channel delivering maximum possible discharge, and if the channel is sufficiently long uniform flow will establish further downstream of the entry. The flow up to this point will be nonuniform with the development of an S_2 profile which asymptotically merges with the uniform flow depth (see worked example 8.8).

The problems are much more complicated if the channels are short, i.e. any downstream control or disturbance (e.g. downstream water level variations) extends its influence right up to the entry, thus submerging the entry and changing the delivery capacities of the channel. Such problems are solved iteratively by computer.

Case 2: downstream water level is constant and upstream level varies
Long channel with mild slope: here the discharge gradually increases with increasing upstream level (y_1) with the formation of M_1 profiles and attains uniform flow conditions ($Q = Q_0$) when $y_1 = y_2$, the downstream level. Further increases in y_1 produce M_2 profiles, ultimately delivering a maximum discharge whose critical depth is equal to y_2. Any further rise in y_1 would develop an M_2 profile terminating with its corresponding critical depth, now greater than y_2; this necessitates a corresponding increase in y_2.

Case 3: both water levels varying (mild slope)
For a constant Q the levels y_1 and y_2 are fluctuating, thus leading to a number of possible surface profiles. With $y_1 = y_2$, uniform flow is established.

However, for water levels above uniform flow depth M_1 profiles develop with the upper limit occurring when $y_2 = y_1 + S_0L$, L being the length of the channel between reservoirs. For water levels below uniform depth M_2, profiles develop with the minimum depth of flow occurring when $y_2 = y_c$, the critical depth corresponding to the given discharge.

8.15 Culvert flow

Highway cross-drainage is normally provided with culverts, bridges and dips. Culverts are submerged structures buried under a high level embankment (see fig. 8.14). The culvert consists of a pipe barrel (conveyance part, i.e. the channel) with protection works at its entrance and exit. It creates a backwater effect to the approach flow, causing a pondage of water above the culvert entrance. The hydraulic design of the culvert is based upon the characteristics of the barrel flow (free surface flow, orifice flow or pipe flow) conditions which depend on its length, roughness, gradient and upstream and downstream water levels.[10]

Free entrance conditions:
(1) $H/D < 1\cdot2$; $y_0 > y_c < y_2 < D$; any length; mild slope: open channel subcritical flow.
(2) $H/D < 1\cdot2$; $y_0 > y_c > y_2 < D$; any length; mild slope: open channel subcritical flow.
(3) $H/D < 1\cdot2$; $y_0 < y_c > y_2 < D$; any length; steep slope: open channel supercritical flow; critical depth at inlet.
(4) $H/D < 1\cdot2$; $y_0 < y_c < y_2 < D$; any length; steep slope: open channel supercritical flow; formation of hydraulic jump in barrel.

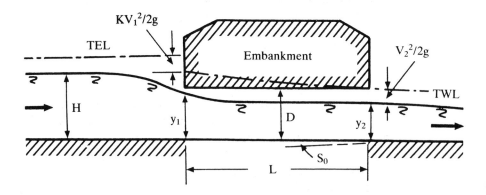

(L = length of culvert; D = height of culvert; S_0 = bed slope; y_1 = depth at entrance; y_2 = depth at exit; y_0 = uniform flow depth; y_c = critical depth; K = entry loss coefficient; TEL: total energy line; TWL: tail water level.)

Figure 8.14 Culvert flow

Submerged entrance conditions:

(5) $H/D > 1·2$; $y_2 < D$; short; any slope: orifice flow.
(6) $H/D > 1·2$; $y_2 < D$; long; any slope: pipe flow.
(7) $H/D > 1·2$; $y_2 > D$; any length; any slope: pipe flow.

See worked example 8.27 for a complete analysis of culvert flow.

8.16 Spatially varied flow in open channels

Spatially varied flow (SVF) is represented by the discharge variation along the length of the channel due to lateral inflow (side spillway channel) or outflow (side weir or bottom racks).

(i) Increasing flow (q_*, inflow rate per unit length)
In this case there exists a considerable amount of turbulence due to the addition of the incoming flow, and the energy equation is not of much use. With the usual assumptions introduced in the development of nonuniform flow equations, and assuming the lateral inflow has no x-momentum added to the channel flow, we can deduce an equation for the surface slope as:

$$\frac{dy}{dx} = \frac{S_0 - S_f - \dfrac{2\beta Q q_*}{gA^2}}{1 - \dfrac{\beta Q^2 T}{gA^3}} \tag{8.22}$$

In the case of subcritical flow all along the channel, the control (critical depth) of the profile is at the downstream end of the channel. For all other flow situations the establishment of the control point is essential to initiate the computational procedures.

In a rectangular channel ($T = B$, the bed width) the location of the control point, x_c, may be written approximately (see Henderson[8]) as:

$$x_c = \frac{8q_*^2\beta^3}{gB^2\left(S_0 - \dfrac{gP}{C^2B}\right)^3} \tag{8.23}$$

where C is the Chezy coefficient and P is the wetted perimeter of the channel. The control point in a channel will exist only if the channel length $L > x_c$. In a given length of the channel, for a control section to exist its slope, S_0, should have a minimum value given by:

$$S_0 > \frac{gP}{C^2B} + \frac{2\beta}{B}\left(\frac{q_*^2B}{gL}\right)^{1/3} \tag{8.24}$$

and the flow upstream of the control is subcritical.

(ii) Decreasing flow (q_*, outflow rate per unit length) – side weir

Assuming the energy loss (due to diversion of the water) in the parent channel is zero, the water surface slope equation can be deduced as:

$$\frac{dy}{dx} = \frac{S_0 - S_f - \dfrac{\alpha Q q_*}{gA^2}}{1 - \dfrac{\alpha Q^2 T}{gA^3}}$$ (8.25)

Equation 8.25 can be extended to be applicable to a side weir of short length with $S_0 = S_f = 0$ and $\alpha = 1$. In the case of a rectangular channel, equation 8.25 is rewritten as:

$$\frac{dy}{dx} = \frac{Qy\left(-\dfrac{dQ}{dx}\right)}{gB^2 y^3 - Q^2}$$ (8.26)

The outflow per unit length, $q_* = (-dQ/dx)$ is given by the weir equation:

$$-\frac{dQ}{dx} = \frac{2}{3} C_M \sqrt{2g} \, (y - s)^{3/2}$$ (8.27)

in which C_M is De Marchi's discharge coefficient, s is sill height and y is flow depth in the channel. If the specific energy in the channel, E, is assumed constant, the discharge in the channel at any section is given by

$$Q = B \, y \, \sqrt{2g \, (E - y)}$$ (8.28)

Combining equations 8.26, 8.27 and 8.28 and integrating we obtain

$$x = \frac{3B}{2C_M} \phi_M(y, E, s) + \text{const.}$$ (8.29)

in which

$$\phi_M (y, E, S) = \frac{2E - 3s}{E - s} \sqrt{\frac{E - y}{y - s}} - 3 \sin^{-1} \sqrt{\frac{E - y}{y - s}}$$ (8.30)

The weir length, L, between two sections is then given by

$$L = x_2 - x_1 = \frac{3}{2} \frac{B}{C_M} (\phi_{M2} - \phi_{M1})$$ (8.31)

The De Marchi coefficient, C_M, for a rectangular sharp crested side weir is given by

$$C_M = 0.81 - 0.60 F_{r1}$$ (8.32)

for both subcritical and supercritical approach flows, F_{r1}, being the flow Froude number. For a broad crested side weir, the discharge coefficient is given by

$$C_M = (0.81 - 0.60 F_{r1}) \, K$$ (8.33)

where K is a parameter depending on the crest length, W, and for a 90°
branch channel is given by

$$K = 1 \cdot 0 \text{ for } \frac{y_1 - s}{W} > 2 \cdot 0 \tag{8.34}$$

and

$$K = 0 \cdot 80 + 0 \cdot 10 \left(\frac{y_1 - s}{W} \right) \text{ for } \frac{y_1 - s}{W} < 2 \cdot 0 \tag{8.35}$$

(iii) Decreasing flow (bottom racks)
The flow over bottom racks (e.g. kerb openings) is spatially varied with the
surface slope given by

$$\frac{dy}{dx} = \frac{2 \varepsilon C \sqrt{y (E - y)}}{3y - 2E} \tag{8.36}$$

in which ε is the void ratio (opening area to total rack area), E is the specific
energy (constant) and C is a coefficient of discharge dependent on the
configuration of openings. Further treatment of these topics can be found
for example in *Open Channel Hydraulics* by French[7].

Worked examples

Example 8.1
Measurements carried out on the uniform flow of water in a long rectangular
channel 3·0 m wide and of bed slope 0·001, revealed that at a depth of flow
of 0·8 m the discharge of water at 15°C was 3·6 m³/s. Estimate the discharge
of water at 15°C when the depth is 1·5 m using (a) the Manning equation,
and (b) the Darcy equation and state any assumptions made.

Solution:
From the flow measurement the value of n and the effective roughness size
(k) can be found

(a) $\quad Q = \dfrac{A}{n} R^{2/3} S_o^{1/2}$ (equation (8.4)) $\tag{i}$

$\quad\quad n = \dfrac{A}{Q} R^{2/3} S_o^{1/2}$

$\quad\quad\quad = 0 \cdot 0137$

when $y = 1 \cdot 5$ m; $\quad Q = 8 \cdot 60$ m³/s.

(b) Using equation (8.11)

$$Q = - A \frac{\sqrt{32g \ R \ S_o}}{2 \cdot 303} \log_e \left[\frac{k}{14.8 \ R} + \frac{1 \cdot 255 \ v}{R \sqrt{32g \ R \ S_o}} \right] \tag{ii}$$

(noting the conversion to $\log_e$)

$$\text{whence} \left(\frac{k}{14 \cdot 8 \ R} + \frac{1 \cdot 255 \ v}{R \sqrt{32g \ R \ S_o}} \right) = \text{EXP} \left(- \frac{Q \times 2 \cdot 303}{A \sqrt{32g \ R \ S_o}} \right)$$

$$k = \left[\text{EXP} \left(- \frac{Q \times 2 \cdot 303}{A \sqrt{32g \ R \ S_o}} \right) - \frac{1 \cdot 255 \ v}{R \sqrt{32g \ R \ S_o}} \right] \times 14 \cdot 8 \times R = 0 \cdot 00146 \ m$$

Substitution in equation (ii) with $y = 1 \cdot 5$ m and $k = 0 \cdot 00146$ m yields: $Q = 8 \cdot 44 \ m^3/s$.

Example 8.2
A concrete-lined trapezoidal channel has a bed width of 3·5 m, side slopes at 45° to the horizontal, a bed slope 1 in 1000 and Manning roughness coefficient of 0·015. Calculate the depth of uniform flow when the discharge is 20 m³/s. (See fig. 8.15.)

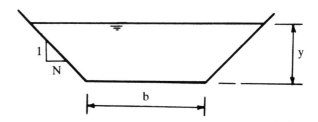

Figure 8.15   Flow through trapezoidal channel

Solution:

$$A = (b + Ny) \ y = (3 \cdot 5 + y) \ y$$

$$P = b + 2y \sqrt{1 + N^2} = 3 \cdot 5 + 2 \sqrt{2} \ y$$

$$R = \frac{A}{P} = \frac{(3 \cdot 5 + y) \ y}{(3 \cdot 5 + 2 \sqrt{2y})}$$

Manning equation: $Q = \dfrac{A}{n} R^{2/3} S^{1/2}$

$$\text{i.e.} \ Q = \frac{(3 \cdot 5 + y) \ y}{0 \cdot 015} \left(\frac{(3 \cdot 5 + y) \ y}{(3 \cdot 5 + 2 \sqrt{2y})} \right)^{2/3} (0 \cdot 001)^{1/2} \tag{i}$$

Setting $Q = 20 \ m^3/s$ equation (i) may be solved for y by trial or by graphical interpolation from a plot of discharge against depth for a range of y values substituted into (i). (See fig. 8.16.) (See table below.)

At 20 m³/s depth of uniform flow $= 1 \cdot 73$ m.

Of course the graph (fig. 8.16) will enable the depth at any other discharge to be determined.

Depth (y) (m)	A (m²)	P (m)	R (m)	Q (m³/s)
1·0	4·50	6·33	0·711	7·56
1·2	5·64	6·89	0·818	10·40
1·4	6·86	7·46	0·920	13·67
1·6	8·16	8·02	1·017	17·39
1·8	9·54	8·59	1·110	21·57
2·0	11·00	9·16	1·200	26·21

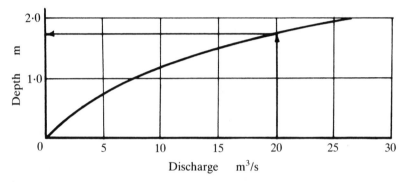

Figure 8.16 Normal depth vs. discharge

Example 8.3

Assuming that the flow in a river is in the rough turbulent zone, show that in a wide river a velocity measurement taken at 0·6 of the depth of flow will approximate closely to the mean velocity in the vertical. (See fig. 8.17.)

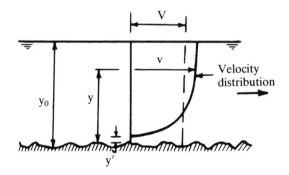

Figure 8.17 Velocity profile in rough turbulent zone

Solution:

In Chapter 7 it was shown that the velocity distribution in the turbulent boundary layer formed in the fluid flow past a rough surface is:

$$\frac{v}{\sqrt{\dfrac{\tau_o}{\rho}}} = 5{\cdot}75 \log \frac{y}{k} + 8{\cdot}5 \tag{i}$$

$$\text{or } \frac{v}{\sqrt{\dfrac{\tau_o}{\rho}}} = 5{\cdot}75 \log \frac{30\,y}{k} \tag{ii}$$

Noting that the local velocity given by (i) is reduced to zero at $y' = \dfrac{k}{30}$ from the boundary, the mean velocity in the vertical is obtained from:

$$V = \frac{1}{y_o} \int_{y'}^{y_o} v\,dy$$

$$\text{whence } V = \frac{5{\cdot}75 \sqrt{\dfrac{\tau_o}{\rho}}}{2{\cdot}303\,y_o} \int_{y'}^{y_o} \ln \frac{30\,y}{k}\,dy$$

$$= \frac{5{\cdot}75 \sqrt{\dfrac{\tau_o}{\rho}}}{2{\cdot}303\,y_o} \left[y_o \ln \frac{30}{k} + y_o \ln y_o - y_o - \left(y' \ln \frac{30}{k} + y' \ln y' - y' \right) \right]$$

$$= \frac{5{\cdot}75 \sqrt{\dfrac{\tau_o}{\rho}}}{2{\cdot}303\,y_o} \left(y_o \ln \frac{30\,y_o}{k} - y_o + y' \right)$$

$$\text{i.e. } V = 5{\cdot}75 \sqrt{\frac{\tau_o}{\rho}} \left(\log \frac{30\,y_o}{k} - \frac{1}{2{\cdot}303} \right) \tag{iii}$$

ignoring the single y' term.

The distance above the bed at which the mean velocity coincides with the local velocity is obtained by equating (ii) and (iii)

$$\log \frac{30\,y}{k} = \log \frac{30\,y_o}{k} - 0{\cdot}434$$

$$\text{or } y = 0{\cdot}37\,y_o \simeq 0{\cdot}4\,y_o$$

This verifies the field practice of taking current meter measurements at 0·6 of depth to obtain a close approximation to the mean velocity in the depth.

Example 8.4
Derive Chezy's resistance equation for uniform flow in open channels and show that the Chezy coefficient, C, is a function of the flow Reynolds number and the channel relative roughness and is given by

$$C = 5 \cdot 75 \sqrt{g} \log \left[\frac{12R}{k + \dfrac{\delta'}{3 \cdot 5}} \right]$$

where δ' is the sublayer thickness given by equation 7.33.

Solution:
Balancing the gravity and resisting forces along a reach length, L, of the channel (for uniform flow) we obtain

$$\rho \, g \, A \, L \, S_0 = \tau \, P \, L \tag{i}$$

giving the uniform boundary shear stress,

$$\tau = \rho \, g \, R \, S_0 \tag{8.37}$$

In turbulent flows $\tau \propto V^2$ (V is the mean velocity of flow) and we can hence write

$$V = C \sqrt{R \, S_0} \tag{8.38}$$

Comparing equation 8.38 (Chezy's) with the Darcy-Weisbach equation (equation 8.2) we obtain

$$C = \sqrt{\frac{8g}{\lambda}} \tag{8.39}$$

In pipe flow $\lambda = f \, (R_e, \, k/D)$ and extending this to open channel flow $(D_e = 4R)$

$$\lambda = f \, (4RV/v, \, k/4R) \tag{8.40}$$

Equation 8.40 is represented by the same Moody diagram constructed for pipe flow.

Evaluation of the Chezy coefficient:
The velocity distributions (two-dimensional flows) for smooth and rough boundaries given by equations 7.23 and 7.27 can be written as:

$$u/u_* = 5 \cdot 75 \log (9u_* y/v) \tag{ii}$$

$$\text{and} \quad u/u_* = 5 \cdot 75 \log (30y/k) \tag{iii}$$

In turbulent two-dimensional flows $u = V$ at $y = 0 \cdot 4 y_0$ (see worked example 8.3), y_0 being the flow depth. Also, equation 7.33 suggests that $u_* = 11 \cdot 6 \, v/\delta'$. Combining these with (ii) and (iii) and replacing y_0 by R yields

$$V = 5 \cdot 75 \, u_* \log \left[\frac{12R}{k + \dfrac{\delta'}{3 \cdot 5}} \right] \tag{iv}$$

By writing $u_* = (g \, R \, S_0)^{1/2}$ in equation (iv) and comparing with the Chezy equation (equation 8.38) we can deduce

$$C = 5 \cdot 75 \sqrt{g} \log \left[\frac{12R}{k + \dfrac{\delta'}{3 \cdot 5}} \right] \tag{8.41}$$

Example 8.5
A trapezoidal channel with side slopes $1:1$ and bed slope $1:1000$ has a 3 m wide bed composed of sand ($n = 0 \cdot 02$) and sides of concrete ($n = 0 \cdot 014$). Estimate the discharge when the depth of flow is 2·0 m.

Solution (see fig. 8.18):

$P_1 (= P_3) = 2 \cdot 828$ m; $P_2 = 3 \cdot 0$ m; $P = 8 \cdot 656$ m (on solid surface only)

$A_1 (= A_3) = 2 \cdot 0$ m^2; $A_2 = 6 \cdot 0$ m^2; $A = 10 \cdot 0$ m^2

$R_1 (= R_3) = 0 \cdot 7072$ m; $R_2 = 2 \cdot 0$ m; $R = 1 \cdot 155$ m

Evaluate composite roughness:

Horton and Einstein: $n = \left[\dfrac{\Sigma P_i n_i^{1 \cdot 5}}{P} \right]^{2/3}$

$$n = \left[\frac{2(2 \cdot 828) \times 0 \cdot 014^{1 \cdot 5} + 3 \times 0 \cdot 02^{1 \cdot 5}}{8 \cdot 656} \right]^{2/3} = 0 \cdot 0162$$

Pavlovskij: $n = \left[\dfrac{\Sigma P_i n_i^2}{P} \right]^{1/2}$

$$n = \left[\frac{2(2 \cdot 828) \times 0 \cdot 014^2 + 3 \times 0 \cdot 02^2}{8 \cdot 656} \right]^{1/2} = 0 \cdot 0163$$

Lotter: $n = \dfrac{PR^{5/3}}{\displaystyle\sum_{i=1}^{N} \left(\dfrac{P_i R_i^{5/3}}{n_i} \right)} = 0 \cdot 0157$

With $y = 2 \cdot 0$ m;

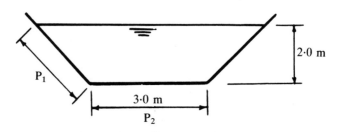

3·0 m

P_2

2·0 m

P_1

Figure 8.18   Equivalent Manning's n in channel of variable boundary roughness

Discharge (Horton n) $= \dfrac{A}{n} R^{2/3} S_o^{1/2} = 21\cdot49$ m^3/s

Discharge (Pavlovskij n) $= 21\cdot36$ m^3/s

Discharge (Lotter n) $= 22\cdot17$ m^3/s.

Example 8.6
The cross-section of the flow in a river during a flood was as shown in fig. 8.19. Assuming the roughness coefficients for the side channel and main channel to be 0·04 and 0·03 respectively, estimate the discharge.
Bed slope = 0·005
Area of main channel (bank full) = 280 m^2
Wetted perimeter of main channel = 54 m
Area of flow in side channel = 152·25 m^2

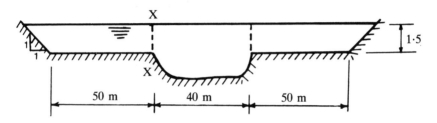

Figure 8.19   Two-stage (compound) channel

Wetted perimeter of side channel around the solid boundary only (excluding the interfaces X − X between the main and side channel flows) = 104·24 m
Area of main channel component = 280 + 40 × 1·5 = 340 m^2

$$\text{Discharge} = \frac{340}{0\cdot03}\left(\frac{340}{54}\right)^{2/3}\sqrt{0\cdot005} + \frac{152\cdot25}{0\cdot04}\left(\frac{152\cdot25}{104\cdot24}\right)^{2/3}\sqrt{0\cdot005}$$

$$= 3079 \text{ m}^3/\text{s}.$$

Note that the treatment of this problem by the equivalent roughness methods of Horton and Pavlovskij will produce large errors in the computed discharge due to the inherent assumptions. However the Lotter method should produce a similar result to that computed above since it uses basically the same method.

$$\text{Lotter equivalent roughness } n = \frac{PR^{5/3}}{\displaystyle\sum_{i=1}^{N}\left(\frac{P_i R_i^{5/3}}{n_i}\right)}$$

$n = 0\cdot0241$

$$\text{and } Q = \frac{492\cdot25}{0\cdot024}\left(\frac{492\cdot25}{158\cdot242}\right)^{2/3}\sqrt{0\cdot005}$$

$$= 3077 \text{ m}^3/\text{s}.$$

Example 8.7

A long rectangular concrete-lined channel (k = 0·3 mm) 4·0 m wide, bed slope 1:500 is fed by a reservoir via an uncontrolled inlet. Assuming that uniform flow is established a short distance from the inlet and that entry losses = 0·5 $V^2/2g$ determine the discharge and depth of uniform flow in the channel when the level in the reservoir is 2·5 m above the bed of the channel at inlet.

Figure 8.20 is an example of natural channel control; the discharge is affected both by the resistance of the channel and the energy available at the inlet.

Two simultaneous equations therefore need to be solved:

(i) Apply the energy equation to sections 1 and 2

$$2 \cdot 5 = y + \frac{V^2}{2g} + h_L = y + \frac{V^2}{2g} + \frac{0 \cdot 5\ V^2}{2g} = y + \frac{Q^2}{(by)^2\ 2g}\ (1 + 0 \cdot 5) \ \text{(i)}$$

$$\text{or } Q_2 = by \sqrt{2g \frac{(2 \cdot 5 - y)}{1 \cdot 5}} \tag{ii}$$

(ii) Resistance equation applied downstream of section 2:

$$Q_3 = -A \sqrt{32g\ R\ S_o}\ \log \left[\frac{k}{14 \cdot 8\ R} + \frac{1 \cdot 255\ v}{R \sqrt{32g\ R\ S_o}} \right] \tag{iii}$$

Solution:

Equation (ii) could be incorporated in (iii) to yield an implicit equation by y which could then be found iteratively. However a graphical solution can be obtained by generating curves for Q v. y from (ii) and (iii).

y (m)	0·4	0·8	1·2	1·6	2·0	2·4	2·5
Q_2 (m³/s)	8·38	15·09	19·79	21·95	20·45	10·98	0
Q_3 (m³/s)	2·14	5·94	10·50	15·51	20·80	26·30	27·70

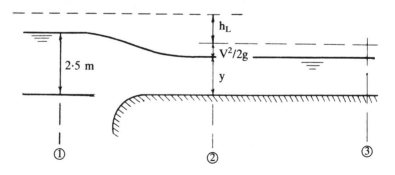

Figure 8.20   Channel inlet

Q_2 and Q_3 are plotted against y in fig. 8.21 whence discharge = 20·5 m³/s at a uniform flow depth of 1·98 m, given by the point of intersection of the two curves.

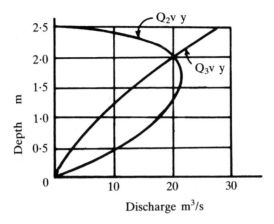

Figure 8.21 Depth at inlet vs. discharge

Note: Care must be taken in treating this method of solution as a universal case. For example, if the channel slope is steep the flow may be supercritical and the plots of equations (ii) and (iii) would appear thus (see fig. 8.22).

The solution is not now the point of intersection of the two curves. The depth passes through the critical depth at inlet and this condition controls the discharge, given by Q_c. Channel resistance no longer controls the flow and the depth of uniform flow corresponds with Q_c on the curve of equation (iii).

Example 8.8
Using the data of Example 8.7 but with a channel bed slope of 1 : 300 calculate the discharge and depth of uniform flow.

Solution:
The discharge v. depth curve using the inlet energy relationship, equation (ii) of Example 8.7 is unaffected by the bed slope. Q_3 is recomputed from equation (iii) of Example 8.7 with $S_o = \dfrac{1}{300}$.

y (m)	0·4	0·8	1·2	1·6	2·0	2·4	2·5
Q_2 (m³/s)	8·38	15·06	19·79	21·95	20·45	10·98	0
Q_3 (m³/s)	3·94	10·91	19·27	28·44	38·14	48·20	50·77

The plotted curves of Q_2 v. y and Q_3 v. y appear as in fig. 8.22. The flow at the channel inlet is critical thus controlling the discharge ($= Q_c$). Downstream, the uniform depth of flow is supercritical.

Summary: Discharge $= 21 \cdot 95$ m³/s;   depth in channel $= 1 \cdot 31$ m.

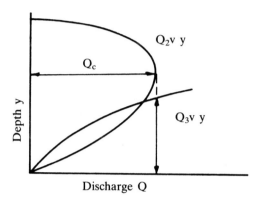

Figure 8.22 Q–y curves in steep sloped channel

Example 8.9

Determine the dimensions of a trapezoidal channel, lined with concrete ($k = 0 \cdot 15$ mm) with side slopes at 45° to the horizontal and bed slope 1 : 1000 to discharge 20 m³/s of water at 15°C under uniform flow conditions such that the section is the most economic.

Solution (see fig. 8.4):

$$N = 1 \cdot 0$$

$$b + 2Ny = 2y \sqrt{1 + N^2}$$

$$b + 2y = 2 \sqrt{2} \, y$$

$$\text{or} \quad b = 0 \cdot 828 \, y$$

Then $A = 1 \cdot 828 \, y^2$; $P = 3 \cdot 656 \, y$

$$Q = - A \sqrt{32g \, R \, S_o} \, \log \left[\frac{k}{14 \cdot 8 \, R} + \frac{1 \cdot 255 \, v}{R \sqrt{32g \, R \, S_o}} \right]$$

This must be solved by trial, calculating Q for a series of y.

y (m)	0·5	1·0	1·5	1·8	1·9	2·0
Q (m³/s)	0·54	3·3	9·5	15·26	17·56	20·06

Adopt $y = 2 \cdot 0$ m

Then $b = 1 \cdot 414$ m.

Example 8.10
A trapezoidal irrigation channel excavated in silty sand having a critical tractive force on the horizontal of $2\cdot4$ N/m^2 and angle of friction $30°$ is to be designed to convey a discharge of 10 m^3/s on a bed slope of $1:10\,000$. The side slopes will be 1 (vert):2 (hor). n $= 0\cdot02$.

Solution:
The channel bed is almost horizontal and the critical tractive force on the bed may therefore be taken as $2\cdot4$ N/m^2.
    The limiting tractive force on the sides is:

$$\tau_{cs} = \tau_{cb} \sqrt{1 - \frac{\sin^2 \theta}{\sin^2 \phi}} \quad \text{(see section 8.5 (b))}$$

$$= 2\cdot4 \sqrt{1 - \frac{\sin^2 (26\cdot565°)}{\sin^2 (30°)}}$$

$$= 2\cdot4 \sqrt{1 - 0\cdot8} = 1\cdot073 \text{ N/m}^2$$

$$\therefore \ 0\cdot76 \ \rho g \ y \ S_o \ngtr 1\cdot073 \text{ N/m}^2$$

$$\therefore \ y \ngtr \frac{1\cdot073}{0\cdot76 \times 1000 \times 9\cdot81 \times 0\cdot0001}$$

$$y \ngtr 1\cdot44 \text{ m.}$$

Now $Q = \dfrac{A}{n} \cdot R^{2/3} S_o^{1/2}$

$$10 = \frac{(b + 2y) \ y}{n} \left(\frac{(b + 2y) \ y}{b + 2y \sqrt5} \right)^{2/3} S_o^{1/2}$$

$$10 = \frac{(b + 2\cdot88) \times 1\cdot44}{0\cdot02} \left(\frac{(b + 2\cdot88) \times 1\cdot44}{(b + 6\cdot44)} \right)^{2/3} \times \sqrt{0\cdot0001}$$

Solving by trial (graphical interpolation) for series of values of b,

b	1·0	2·0	4·0	8·0	10·0	12·0
R.H.S.	2·31	3·11	4·78	8·27	10·05	11·84

required b $= 9\cdot95$ m

$$V = 0\cdot54 \text{ m/s}$$

which agrees reasonably with the maximum mean velocity criterion (table 8.3).

Example 8.11 (Maximum mean velocity criterion)
Using the data of the previous Example determine the channel dimensions such that the mean velocity does not exceed $0\cdot6$ m/s when conveying the discharge of 10 m^3/s.

$$Q = A \, V; \quad 10 = A \times 0{\cdot}6$$

whence $A = 16{\cdot}67 \text{ m}^2$

and $A = (b + 2y) \, y$

whence $b = 16{\cdot}67/y - 2y$

$$Q = \frac{A}{n} R^{2/3} S_o^{1/2} = \frac{A^{5/3}}{n \, P^{2/3}} S^{1/2}$$

i.e. $10 = \dfrac{16{\cdot}67^{5/3} \times \sqrt{0{\cdot}0001}}{0{\cdot}02 \times P^{2/3}}$

whence $P = 12{\cdot}865 \text{ m}$

and $P = b + 2y \sqrt{5} = \dfrac{16{\cdot}67}{y} + 2y \, (\sqrt{5} - 1)$

i.e. $0{\cdot}472 \, y^2 - 12{\cdot}865 \, y + 16{\cdot}67 = 0$

whence $y = 1{\cdot}365 \text{ m}$

and $b = 9{\cdot}48 \text{ m}$.

Example 8.12

Check the proposed design of a branch of a wastewater sewerage system receiving the flow from 250 houses. The pipe is 150 mm diameter of vitrified clay with a proposed bed gradient of 1 in 100. Take the per capita daily water supply to be 200 1/day (= 1 dry-weather flow, (dwf)) and assume that the population density is 3·5 persons per house.

Notes: Sewers conveying crude sewage develop a coating of slime on the inner boundary. The Hydraulics Research Station[6] recommend an effective roughness of 3·0 mm for slimed vitrified clay pipes and 6·0 mm for similar concrete pipes. Due to variations in rate of water consumption during the day, in general a flow of at least twice the average (2 dwf) is achieved each day around midday. This flow value is generally used to check the minimum velocity criterion, and a figure of 6 dwf used to check the pipe capacity.

Solution:

Equations (8.10) and (8.11) can be used to generate curves of velocity and discharge with variation in depth (see fig. 8.23).

$$\text{2 dwf discharge} = \frac{2 \times 250 \times 3{\cdot}5 \times 200}{24 \times 3600} = 4{\cdot}05 \text{ l/s}$$

At this discharge, from the graph (fig. 8.23), depth of flow = 0·056 m and at this depth $V = 0{\cdot}66 \text{ m/s}$.

The design is satisfactory since a velocity of 0·61 m/s is exceeded.

Note: If the self-cleansing velocity had not been attained the slope of the pipe would have to be increased, not the diameter.

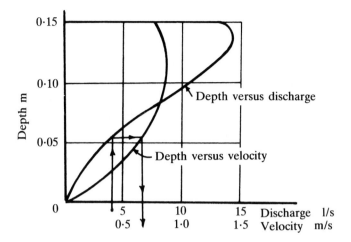

Figure 8.23 Depth vs. discharge and velocity in pipe channel

6 dwf discharge = 12·15 l/s.

The pipe will convey 14·8 l/s at 0·95 of depth and 13·6 l/s when full.
This example illustrates the basis of the tables of proportional velocity,

$$\frac{V \text{ (part-full)}}{V \text{ (full)}} \text{ and proportional discharge } \frac{Q \text{ (part-full)}}{Q \text{ (full)}}$$

with proportional depth (y/D) which appear in *Tables for the hydraulic design of pipes* (Recommended reading 9).

Example 8.13
Design a branch within a storm sewer network which has a length of 100 m, a bed slope of 1 in 150 and roughness size of 0·15 mm and which receives the storm run off from 3·5 hectares of impermeable surface using the Rational (Lloyd-Davies) Method. In designing the upstream pipes the maximum 'time of concentration' at the head of the pipe has been found to be 6·2 minutes. The relationship between rainfall intensity and average storm duration is tabulated below.

storm duration (min)	2·0	3·0	4·0	5·0	6·0	7·0	8·0
average rainfall intensity (mm/hr)	94·0	82·7	73·8	70·0	61·4	57·2	53·5

Notes: The (Rational) Lloyd-Davies Method gives the peak discharge (Q_p) from an urbanised catchment in the form:

$$Q_p = \frac{1}{360} A_p i \ (m^3/s)$$

where A_p = impermeable area (hectares)

 i = average rainfall intensity (mm/hr) during the storm.

Since the average rainfall intensity of storms of a given average return period decreases with increase in storm duration the critical design storm is that which has a duration equal to the 'time of concentration of the catchment' T_c. T_c is the longest time of travel of a liquid element to the point in question in the catchment and includes the times of overland and pipeflow; the time of pipe flow is based on full-bore velocities.

Solution:
The selection of the appropriate pipe diameter is by trial. Noting that the increment in the pipe size of sewer pipes is 75 mm from 150 mm, try D = 375 mm.
Full-bore conditions; D = 4 R whence, using equation (8.10)

$$V_F = -2 \sqrt{19\cdot62 \times 0\cdot375 \times 0\cdot0067} \ \log \left[\frac{0\cdot15}{3\cdot7 \times 0\cdot375} \right.$$

$$\left. + \frac{2\cdot51 \times 1\cdot13 \times 10^{-6}}{0\cdot375 \times \sqrt{19\cdot62 \times 0\cdot375 \times 0\cdot0067}} \right]$$

(noting that $S_o = 0\cdot0067$),

 whence $V_F = 1\cdot703$ m/s and $Q_F = 0\cdot188$ m³/s (from equation (8·11))

$$\text{Travel time along pipe} = \frac{100}{1\cdot70 \times 60} = 0\cdot98 \text{ min}$$

Thus $T_c = 6\cdot2 + 0\cdot98 = 7\cdot18$ min; whence i = 56·5 mm/hr

$$\therefore \text{ inflow} = \frac{3\cdot5 \times 56\cdot5}{360} = 0\cdot55 \text{ m}^3/\text{s}.$$

This is greater than the full-bore discharge (Q_F) of the 375 mm pipe which is therefore too small. Try 600 mm pipe.

$V_F = 2\cdot28$ m/s

$Q_F = 0\cdot645$ m³/s

$$\text{Travel time along pipe} = \frac{100}{2\cdot28 \times 60} = 0\cdot73 \text{ min}$$

$T_c = 6\cdot93$ min; whence i = 57·5 mm/hr

$$\text{Inflow} = \frac{3\cdot5 \times 57\cdot5}{360} = 0\cdot56 \text{ m}^3/\text{s}.$$

A 600 mm diameter pipe is required.
(Note that a 525 mm pipe has a full-bore capacity of 0·454 m³/s.)

Example 8.14
A rectangular channel 5 m wide laid to a mild bed slope conveys a discharge
of 8 m³/s at a uniform flow depth of 1·25 m.
(a) Determine the critical depth.
(b) Neglecting the energy loss, show how the height of a streamlined sill
    constructed on the bed affects the depth upstream of the sill and the
    depth at the crest of the sill.
(c) Show that if the flow at the crest becomes critical the structure can
    be used as a flow measuring device using only an upstream depth
    measurement.

Solution:

(a) $y_c = \sqrt[3]{\dfrac{q^2}{g}}$; $\quad q = \dfrac{8}{5} = 1\cdot6$ m³/s.m

$\quad\quad y_c = \sqrt[3]{\dfrac{1\cdot6^2}{9\cdot81}} = 0\cdot639$ m.

(b)

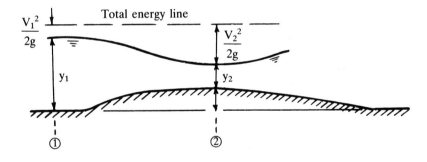

Figure 8.24 Flow over hump

Neglecting losses between 1 and 2 (see fig. 8.24)

$\quad\quad E_1 = E_2$

$\quad\quad E_{s,1} = E_{s,2} + z$

In the case of uniform rectangular channel the specific energy curve is the
same for any section and if this is drawn for the specified discharge it can be
used to show the variation of y_1 and y_2.

y (m)	$V\left(=\dfrac{Q}{by}\right)$ (m/s)	$V^2/2g$ (m)	$E_s = y + V^2/2g$ (m)
0·2	8·0	3·262	3·462
0·3	5·33	1·45	1·75
0·4	4·0	0·815	1·215
0·6	2·67	0·362	0·962
0·8	2·00	0·204	1·004
1·0	1·60	0·130	1·130
1·2	1·33	0·091	1·291
1·4	1·14	0·067	1·467
1·6	1·00	0·051	1·651

For small values of z (crest height) and assuming that the upstream depth is the uniform flow depth (y_n), (see fig. 8.25) the equation:

$$E_{s,1} = E_{s,2} + z$$

can be evaluated (for y_2) by entering the diagram with y_1 ($= y_n$) moving horizontally to meet the E_s curve (at x) setting off z to the left to meet the E_s curve again at W, which corresponds with the depth at the crest y_2.

This procedure can be repeated for all values of z up to z_c at which height the flow at the crest will just become 'critical'. Within this range of crest heights the upstream depth (the uniform flow depth, y_n) remains unaltered.

Of course the solution can also be obtained from the equation:

$$y_1 + \frac{V_1^2}{2g} = y_2 + \frac{V_2^2}{2g} + z$$

but it is important to realise that if z exceeds z_c then y_1 will not remain equal to y_n.

If z exceeds z_c the E_s curve can still be used to predict the surface profile;

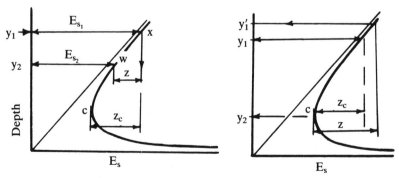

Figure 8.25   E–y curves with hump

y_2 remains equal to y_c. Set off z to the right from C (right-hand diagram of fig. 8.25).

Note that y_1 has increased (to y_1') to give the increased energy to convey the discharge over the crest.

The solution, using a numerical method of solution of the energy equation for greater accuracy is tabulated: the graphical method described gives similar values. (See fig. 8.26.) (See table below.)

z (m)	0·1	0·2	0·3	0·4	0·5	0·6	0·7	0·8	0·9
y_1 (m)	1·25	1·25	1·25	1·28	1·39	1·50	1·61	1·715	1·82
y_2 (m)	1·13	1·00	0·85	0·639	0·639	0·639	0·639	0·639	0·639

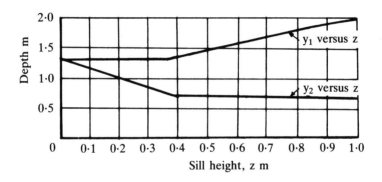

Figure 8.26 Sill height vs. depths upstream and over hump

(c) (See fig. 8.27.)

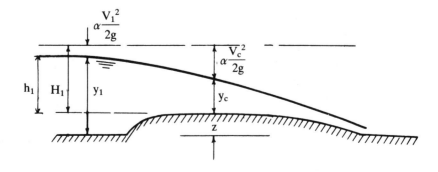

Figure 8.27 Hump as flow measuring structure

$$H_1 = y_c + \frac{V_c^2}{2g}$$

now $y_c = 2/3 \ H_1$ (see section 8.7)

$$H_1 = \frac{2H_1}{3} + \frac{V_c^2}{2g}$$

$$V_c = \sqrt{\frac{2g}{3}} \ H_1^{1/2}$$

and $Q = b \ y_c \ V_c = \frac{2}{3} \ b \ H_1 \ \sqrt{\frac{2g}{3}} \ H_1^{1/2}$

i.e. $Q = \frac{2b}{3} \ \sqrt{\frac{2g}{3}} \ H_1^{3/2}$

Note that H_1 is the upstream energy measured relative to the crest of the sill. In practice the upstream depth above the crest (h_1) would be measured and the velocity head $\dfrac{\alpha \ V_1^2}{2g}$ is allowed for by a coefficient C_v and energy losses by C_d

$$Q = \frac{2}{3} \ b \ \sqrt{\frac{2g}{3}} \ C_v \ C_d \ h_1^{3/2}$$

(see BS 3680 Part 4A).
See also Example 8.18 which illustrates the effect of downstream conditions on the existence of critical flow over the sill.

Example 8.15
Venturi Flume: A rectangular channel 2·0 m wide is contracted to a width of 1·2 m. The uniform flow depth at a discharge of 3 m³/s is 0·8 m. (a) Calculate the surface profile through the contraction, assuming that the profile is unaffected by downstream conditions. (b) Determine the maximum throat width such that critical flow in the throat will be created. (See fig. 8.28.)

Solution:
In principle the system, and method of solution is similar to that of Example 8.14. However the specific energy diagrams for any of the contracted sections are no longer identical with that for section 1.

$$E_1 = E_2 = \left(y_2 + \frac{V_2^2}{2g} \right)$$

Entering with $y_1 \ (= y_n)$ to meet the E_s curve for section 1 (width B) at X and moving vertically to meet the curve for width b' yields y_2', the depth in the throat. (See fig. 8.29.)

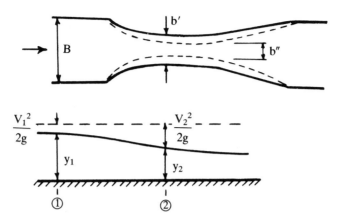

Figure 8.28 Channel contraction – venturi flume

If the throat is further contracted to b″ the vertical through X no longer intercepts the E_s curve for width b″; this means that the energy at 1 is insufficient and must rise to meet the specific energy at 2 corresponding with the critical depth at 2 ($E_{s,2min}$).

For a given discharge a minimum degree of contraction is required to establish critical flow at the throat; this is b_c corresponding with the specific energy curve which is just tangential to the vertical through X. Provided it is recognised that if $b < b_c$, y_1 will be greater than y_n the problem can be solved numerically.

$$\text{At uniform flow } E_1 = y_n + \frac{Q^2}{2g\ B^2\ y_n^2}$$

$$= 0{\cdot}8 + \frac{3^2}{19{\cdot}62 \times 2^2 \times 0{\cdot}8^2}$$

$$= 0{\cdot}979 \text{ m}$$

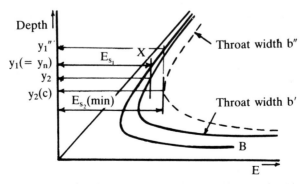

Figure 8.29 E–y curves at throat

If flow at throat were critical, the minimum specific energy would be $y_c + y_c/2 = 1.5 y_c$

and $y_c = \sqrt[3]{\dfrac{q^2}{g}} = \sqrt[3]{\dfrac{Q^2}{g\, b^2}}$

$$y_c = \sqrt[3]{\dfrac{3^2}{9 \cdot 81 \times 1 \cdot 2^2}} = 0 \cdot 86 \text{ m}$$

and $E_{s,min} = 1 \cdot 291$ m.

This is greater than the energy upstream at uniform flow: thus the flow in the throat will be critical.

Therefore $y_2 = 0 \cdot 86$ m and y_1 is obtained from:

$$1 \cdot 291 = y_1 + \dfrac{Q^2}{2g\, B^2\, y_1^2}$$

$$1 \cdot 291 = y_1 + \dfrac{3^2}{19 \cdot 62 \times 2^2 \times y_1^2}$$

whence $y_1 = 1 \cdot 215$ m.

(b) Let b_c = the throat width to create critical flow at the specified discharge. Critical flow can just be achieved with the upstream energy of $0 \cdot 979$ m.

$$\therefore 0 \cdot 979 = y_c + \dfrac{Q^2}{2g\, b_c^2\, y_c^2} \tag{i}$$

Since $0 \cdot 979$ m is also the energy corresponding with critical flow at the throat. $y_c = 2/3 \times 0 \cdot 979 = 0 \cdot 653$ m

whence from (i), $0 \cdot 326 = \dfrac{3^2}{19 \cdot 62 \times b_c^2 \times 0 \cdot 653^2}$

The solution to which is $b_c = 1 \cdot 816$ m.

Notes: In a similar manner to that of Example 8.14 (c), it can be shown that if the flow in the throat is critical the discharge can be calculated from the theoretical equation

$$Q = \frac{2}{3} \sqrt{\frac{2g}{3}}\, b\, H_1^{3/2} \left(\text{practical form: } Q = \frac{2}{3} \sqrt{\frac{2g}{3}}\, b\, C_v\, C_d\, h_1^{3/2} \right)$$

where H_1 is the upstream energy $(h_1 + V_1^2/2g)$ and h_1 the upstream depth. In practice the throat would be made narrower than that calculated in the Example above in order to create SUPERCRITICAL flow conditions in the expanding section downstream of the throat followed by a hydraulic jump in the downstream channel. The reader is referred to BS 3680 Part 4C and also to Example 8.19.

Example 8.16

A vertical sluice gate with an opening of 0·67 m produces a downstream jet depth of 0·40 m when installed in a long rectangular channel 5·0 m wide conveying a steady discharge of 20·0 m³/s. Assuming that the flow downstream of the gate eventually returns to the uniform flow depth of 2·5 m.

(a) Verify that a hydraulic jump occurs. Assume $\alpha = \beta = 1\cdot0$.
(b) Calculate the head loss in the jump.
(c) If the head loss through the gate is 0·05 $V_J^2/2g$ calculate the depth upstream of the gate and the force on the gate.
(d) If the downstream depth is increased to 3·0 m analyse the flow conditions at the gate.

Solution:
(a) (See fig. 8.30.)

If a hydraulic jump is to form the required initial depth (y_i) must be greater than the jet depth.

$$y_i = \frac{y_s}{2} (\sqrt{1 + 8 F_s^2} - 1); \quad F_s = \frac{V_s}{\sqrt{g\, y_s}} = \frac{20}{5\cdot0 \times 2\cdot5 \sqrt{9\cdot8 \times 2\cdot5}}$$

i.e. $F_s = 0\cdot323$ and hence $y_i = 0\cdot443$ m (equation (8·16))
Therefore a jump will form.

(b) Head loss at jump

$$= \left(y_i + \frac{V_i^2}{2g}\right) - \left(y_s + \frac{V_s^2}{2g}\right)$$

$$h_L = 0\cdot443 - 2\cdot5 + \frac{1}{2g}\left[\left(\frac{4}{0\cdot443}\right)^2 - \left(\frac{4}{2\cdot5}\right)^2\right]$$

$$= 1\cdot97 \text{ m}.$$

(c) (i) Apply the energy equation to 1 and 2:

$$y_1 + \frac{V_1^2}{2g} = y_2 + \frac{V_2^2}{2g} + 0\cdot05\, \frac{V_2^2}{2g}$$

$$V_2 = 4/0\cdot4 = 10 \text{ m/s}; \quad \frac{V_2^2}{2g} = 5\cdot099 \text{ m}$$

$$y_1 + \frac{q^2}{2g\, y_1^2} = 5\cdot752 \text{ m}$$

whence $y_1 = 5\cdot73$ m.

(ii) F_x = gate reaction per unit width
Apply momentum equation to element of water between 1 and 2

$$\frac{\rho g\, y_1^2}{2} + \rho q\, (V_1 - V_2) - \frac{\rho g\, y_2^2}{2} - F_x = 0$$

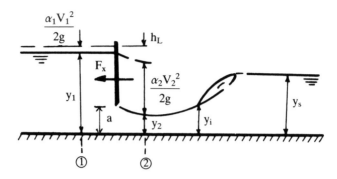

Figure 8.30 Sluice gate control and hydraulic jump

(Note that the force due to the friction head loss through the gate is implicitly included in the above equation since this affects the value of y_1.)

$$1000 \left[\frac{9\cdot806}{2} (5\cdot73^2 - 0\cdot4^2) + 4 (0\cdot693 - 10) \right] - F_x = 0$$

whence $F_x = 123$ kN/m width.

(d) With a sequent depth of $3\cdot0$ m the initial depth required to sustain a jump is $0\cdot327$ m (following the procedure of (a)). Therefore the jump will be submerged (see fig. 8.31), since the depth at the vena contracta is $0\cdot4$ m.

Apply the momentum equation to 2 and 3 neglecting friction and gravity forces.

$$\frac{\rho g \, y_G^2}{2} + \rho q \, (V_2 - V_s) - \frac{\rho g \, y_s^2}{2} = 0$$

$$y_G^2 - y_s^2 + \frac{2q^2}{g} \left(\frac{1}{y_2} - \frac{1}{y_s} \right) = 0$$

whence $y_G = y_s \sqrt{1 + 2F_s^2 \left(1 - \frac{y_s}{y_2} \right)}$; where $F_s = \frac{V_s}{\sqrt{g \, y_s}}$

$y_s = 3\cdot0$ m; $y_2 = 0\cdot4$ m; whence $y_G = 1\cdot39$ m

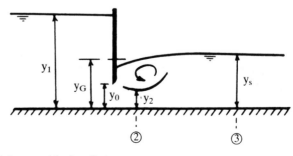

Figure 8.31 Submerged hydraulic jump

Applying the energy equation to 1 and 2:

$$y_1 + \frac{V_1^2}{2g} = y_G + \frac{V_2^2}{2g} + 0 \cdot 05 \frac{V_2^2}{2g}$$

$$y_1 + \frac{V_1^2}{2g} = 6 \cdot 74 \text{ m}.$$

whence the upstream depth, y_1 is now 6·73 m.

Example 8.17
A sluice gate is discharging water freely (modular flow) under a head of 5 m (upstream of the gate) with a gate opening of 1·5 m. Compute the discharge rate per unit width of the gate. If the water depth immediately downstream of the gate is 2 m (drowned or nonmodular flow), determine the discharge rate.

Solution:
Referring to fig. 8.30, for the modular/free flow case

$$q = C_d \, a \, \sqrt{2gy_1} \tag{8.42}$$

where

$$C_d = \frac{C_c}{\sqrt{1 + C_c \dfrac{a}{y_1}}} \tag{8.43}$$

The contraction coefficient C_c (= y_2/a) is a function of a/y_1; a reasonable constant value of $C_c = 0 \cdot 60$ may be assumed for most conditions. Equation 8.42 can also be written as

$$q = C'_d \, a \, \sqrt{2g(y_1 - C_c a)} \tag{8.44}$$

where

$$C'_d = \frac{C_c}{\sqrt{1 - \left(C_c \dfrac{a}{y_1}\right)^2}} \tag{8.45}$$

For the submerged (nonmodular) flow condition, the discharge

$$q_s = C'_{ds} \, a \, \sqrt{2g(y_1 - y_G)} \tag{8.46}$$

where $C'_{ds} = C'_d$ with $C_c = 0 \cdot 60$. The depth of submergence, y_G, downstream of the gate (see fig. 8.31) is computed by momentum equation (see Example 8.16) as

$$\frac{y_G}{y_s} = \sqrt{1 + 2F_s^2 \left(1 - \frac{y_s}{C_c a}\right)} \tag{8.47}$$

(i) Modular flow:

$$a/y_1 = 0.3; \therefore C_d = 0.552 \text{ (equation 8.43)}$$

hence, $q = 8.2 \text{ m}^3/\text{s/m}$ (equation 8.42)

(ii) Nonmodular flow:

$$C'_{ds} = C'_d = 0.610 \text{ (equation 8.45)}$$

$$\therefore q = 0.61 \times 1.5 \times \sqrt{[2g(5 - 2)]} = 7.08 \text{ m}^3/\text{s/m}$$

Note: if we assume the flow condition immediately downstream of the gate remains unaffected by submergence, we can obtain V_1 by the energy and continuity equations

$$5 + V_1^2/2g = 2 + V_2^2/2g; \; 5V_1 = 0.6 \times 1.5V_2$$

as $V_1 = 1.4 \text{ m/s}$; hence $q = y_1 \times V_1 = 7.0 \text{ m}^3/\text{s/m}$

Example 8.18
A broad-crested weir is to be constructed in a long rectangular channel of mild bed slope for discharge monitoring by single upstream depth measurement.
    Bed width = 4.0. Discharge measurement range from 3.0 m³/s to 20.0 m³/s. Depth-discharge (uniform flow) rating curve for channel:

Depth (m)	0.5	1.0	1.5	2.0	2.5
Dicharge m³/s	3.00	8.15	14.22	20.8	27.7

Select a suitable crest height for the weir. (See fig. 8.32.)

Solution:
Ideally the design criterion is that a hydraulic jump should form downstream of the sill. From the table the depth of uniform flow at 20 m³/s is 1.95 m

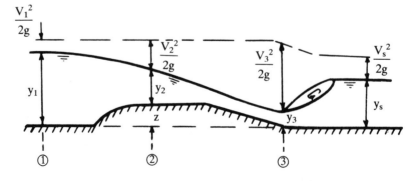

Figure 8.32 Design of broad crested weir

(y_n). Required initial depth to sustain a hydraulic jump of sequent depth 1·95 m, (y_n) is

$$y_i = \frac{y_s}{2} (\sqrt{1 + 8F_s^2} - 1)$$

$y_i = 0.913$ m; $(= y_3)$

$$E_{s,3} (= E_3) = y_3 + \frac{Q^2}{2g \, (by_3)^2} = 0.913 + 1.529 = 2.442 \text{ m}$$

For critical flow conditions at the crest of the sill,

$$y_2 = y_c = \sqrt[3]{\frac{q^2}{g}} = 1.37 \text{ m}$$

and $\dfrac{V_2^2}{2g} = \dfrac{V_c^2}{2g} = \dfrac{Q^2}{2g \, (by_c)^2} = 0.679$ m

Thus the specific energy at critical flow at the crest of the sill

$$E_{s,2} \text{ (crit)} = y_c + \frac{V_c^2}{2g} = 1.37 + 0.679 = 2.049 \text{ m}$$

To find the minimum sill height equate $E_2 = E_3$

i.e. $z + E_{s,2}$ (crit) $= E_3$

i.e. $z + 2.049 = 2.442$; whence $z = 0.393$ (say 0·4 m)

The upstream depth can be calculated by equating E_1 to E_3 neglecting losses.

i.e. $y_1 + \dfrac{Q^2}{2g \, (by_1)^2} = 2.442$

whence $\quad\quad y_1 = 2.17$ m

At the lower discharge of 3·0 m³/s the depth of uniform flow $= 0.5$ m. Required initial depth for a hydraulic jump of sequent depth 0·5 m

$= 0.29$ m $(= y_3)$

The minimum specific energy required to convey the discharge of 3 m³/s over the sill is that corresponding with critical flow conditions.

$$y_c = \sqrt[3]{\frac{q^2}{g}} = 0.386 \text{ m}; \quad \frac{V_c^2}{2g} = 0.193 \text{ m}$$

$E_{s,c} = 0.579$ m

With the established crest height of 0·4 m the minimum total energy is (E_s)
$= 0.4 + 0.579 = 0.979$ m.

Since this is much greater than 0·5 m, the downstream uniform flow depth, the flow at the crest is certainly critical provided there are no downstream constraints.

Check the existence of a hydraulic jump

$$E_3 = E_2 = 0.979 \text{ m (neglecting losses)}$$

$$y_3 + \frac{Q^2}{2g\,(by_3)^2} = 0 \cdot 979 \text{ m; whence } y_3 = 0 \cdot 191 \text{ m}$$

Since this is less than that required for a jump to form (0·29 m) a hydraulic jump will form in the channel downstream of section 3. The design is therefore satisfactory.

Example 8.19 (The 'critical depth flume')
Using the data of Example 8.15 determine the minimum width of the throat of the venturi flume such that a hydraulic jump will be formed in the downstream channel with a sequent depth equal to the depth of uniform flow. Determine the upstream depth under these conditions. (See fig. 8.33.)

Solution:
Downstream depth (= sequent depth) = depth of uniform flow = 0·8 m.

$$Q = 3 \text{ m}^3/\text{s; channel width} = 2\cdot0 \text{ m}$$

$$V_s = \frac{3}{2 \times 0 \cdot 8} = 1 \cdot 875 \text{ m/s;} \quad F_s = \frac{V_s}{\sqrt{g\,y_s}} = 0 \cdot 67$$

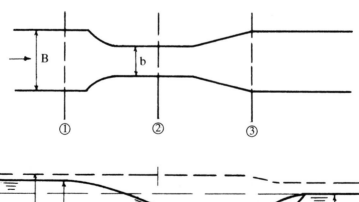

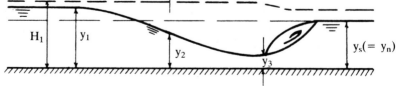

Figure 8.33 Design of venturi flume

Required initial depth for a hydraulic jump to form in the channel, with a sequent depth of 0·8 m,

$$y_i = \frac{y_s}{2} (\sqrt{1 + 8F_s^2} - 1) = 0\cdot456 \text{ m}$$

Thus the maximum value of

$$y_3 = 0\cdot456 \text{ m} \quad \text{and} \quad E_3 = (E_{s,3}) = y_3 + \frac{Q^2}{2g\,(by_3)^2} = 1\cdot007 \text{ m}$$

Conditions in the throat will be critical; if b = throat width,

$$y_{2,c} = \sqrt[3]{\frac{Q^2}{g\,b^2}}$$

Equating energies at 2 and 3.

$$y_{2,c} + \frac{V_{2,c}^2}{2g} = E_3 = 1\cdot007 \text{ m}$$

i.e. $\sqrt[3]{\dfrac{Q^2}{g\,b^2}} + \dfrac{Q^2}{2g\,b^2 \left(\dfrac{Q^2}{g\,b^2}\right)^{2/3}} = 1\cdot007$

i.e. $1\cdot5 \left(\dfrac{Q^2}{g\,b^2}\right)^{1/3} = 1\cdot007 \text{ m}$

whence b = 1·74 m.

(Note that this is narrower than that in Example 8.15 which specifically stated that, in that case, downstream controls did not affect the flow profile.)
The upstream depth can be calculated, neglecting losses, from $E_1 = E_2 = E_3$

i.e. $y_1 + \dfrac{Q^2}{2g\,B^2\,y_1^2} = 1\cdot007 \text{ m}$

whence $y_1 = 0\cdot98$ m.

(Note that this is greater than the uniform flow depth.)

Example 8.20
A trapezoidal concrete-lined channel has a constant bed slope of 0·0015, a bed width of 3 m and side slopes 1:1. A control gate increases the depth immediately upstream to 4·0 m when the discharge is 19·0 m³/s. Compute the water surface profile to a depth 5% greater than the uniform flow depth.

Take n = 0·017 and $\alpha = 1\cdot1$.

Notes: The energy gradient at each depth is calculated as though uniform flow existed at that depth. For hand calculation the Manning equation is much simpler than the Darcy-Colebrook-White equations which could, however, be incorporated in a computer program.

Calculations: Using the Manning equation the depth of uniform flow at $19.0 \, \text{m}^3/\text{s} = 1.75 \, \text{m}$. The systematic calculations are shown in tabular form below. (See equation 8.19 and section 8.11.)

$$\Delta x = \int_{y_1}^{y_2} \frac{dx}{dy} \, dy = x_2 - x_1 \quad \text{where} \quad \frac{dx}{dy} = \frac{1 - \dfrac{\alpha \, Q^2 T}{A^3 \, g}}{S_o - S_f}$$

y (m)	B (m)	A (m²)	R (m)	dx/dy	Δx (m)	x (m)
4·0	11·0	28·0	1·956	677·71		0
3·9	10·8	26·91	1·918	679·10	67·8	67·8
3·8	10·6	25·84	1·880	680·65	68·0	135·8
3·7	10·4	24·79	1·840	682·47	68·2	204·0
3·6	10·2	23·76	1·800	684·59	68·3	272·3
3·5	10·0	22·75	1·760	687·92	68·6	340·9
3·4	9·8	21·76	1·725	690·00	68·9	409·8
3·3	9·6	20·79	1·685	693·45	69·2	472·0
3·2	9·4	19·84	1·646	697·58	69·5	548·5
3·1	9·2	18·91	1·607	702·54	70·0	618·5
3·0	9·0	18·00	1·567	708·56	70·6	689·1
2·9	8·8	17·11	1·527	715·94	71·2	760·3
2·8	8·6	16·24	1·487	725·10	72·0	832·3
2·7	8·4	15·39	1·447	736·64	73·1	905·4
2·6	8·2	14·56	1·406	751·43	74·4	979·8
2·5	8·0	13·75	1·365	770·80	76·1	1056
2·4	7·8	12·96	1·324	796·90	78·4	1134
2·3	7·6	12·19	1·282	833·35	81·5	1216
2·2	7·4	11·44	1·240	886·86	86·0	1302
2·1	7·2	10·71	1·198	971·33	92·9	1395
2·0	7·0	10·00	1·155	1120·92	104·6	1499
1·9	6·8	9·31	1·111	1448·07	128·5	1628
1·8	6·6	8·64	1·068	2667·73	205·8	1834

The surface profile is illustrated by plotting y v.x on the channel bed.

Example 8.21
Using the data of Example 8.20 compute the surface profile using the direct step method.

$$y_o = 4.0 \, \text{m}; \quad S_o = 0.0015; \quad Q = 19 \, \text{m}^3/\text{s}:$$

$$n = 0.017; \quad \alpha = 1.1$$

Solution:
At the control section, depth $= 4.0 \, \text{m}$, $A = 28.0 \, \text{m}^2$, $R = 1.956 \, \text{m}$.

$$\text{Specific energy} \left(y + \frac{Q^2}{2g \, A^2} \right) = 4.0 + \frac{1.1 \times 19^2}{19.62 \times 28^2} = 4.026 \, \text{m}$$

$$S_f = \frac{Q^2 \, n^2}{A^2 \, R^{4/3}} = 5.44 \times 10^{-5}$$

y (m)	A (m²)	R (m)	E (m)	ΔE (m)	S_f	$\bar{S}_f$	Δx (m)	x (m)
4·0	28·0	1·956	4·026	—	$5·44 \times 10^{-5}$	—	0	0
3·9	26·91	1·918	3·928	0·098	$6·05 \times 10^{-5}$	$5·74 \times 10^{-5}$	67·84	67·84
3·8	25·84	1·880	3·830	0·098	$6·74 \times 10^{-5}$	$6·39 \times 10^{-5}$	67·98	135·82
3·7	24·79	1·840	3·733	0·097	$7·52 \times 10^{-5}$	$7·13 \times 10^{-5}$	68·16	203·98
↓	↓	↓	↓	↓	↓	↓	↓	↓
1·8	8·64	1·068	2·0712	0·0623	$1·28 \times 10^{-3}$	$1·16 \times 10^{-3}$	184·94	1809·3

The surface profile is very similar to that calculated by the integration method (Example 8.20).

Example 8.22
Using the standard step method compute the surface profile using the data of Example 8.20.
   The solution is shown in the table on page 244; the intermediate iterations where $H_{(1)} \neq H_{(2)}$ have not been included. It is noted that the result is almost identical with the numerical integration and direct step methods. However, unless a computer is used the calculations in the standard step method are laborious and for prismatic channels with constant bed slopes the other methods would be quicker. The standard step method is particularly suited to natural channels in which the channel geometry and bed elevation at spatial intervals, which are not necessarily equal, have been measured. Variations in roughness coefficient, n, along the channel can also be incorporated.

Example 8.23
A vertical sluice gate, situated in a rectangular channel of bed slope 0·005, width 4·0 m and Manning's n = 0·015, has a vertical opening of 1·0 m and C_c = 0·60. Taking α = 1·1 and β = 1·0 determine the location of the hydraulic jump when the discharge is 20 m³/s and the downstream depth is regulated to 2·0 m.

Solution:
(See fig. 8.34.) Note: $S_f = \dfrac{Q^2 \, n^2}{A^2 \, R^{4/3}}$

Depth at vena-contracta = $1·0 \times 0·6$ = 0·6 m

Depth of uniform flow (from Manning equation) = 1·26 m

Example 8.22

$Q = 19.0$ m^3/s. $n = 0.017$. $S_o = 0.0015$. $\alpha = 1.1$. $b = 3.0$ m. Side slopes $1:1$. $y_n = 1.75$ m.

x (m)	Z	y	A	V	$\dfrac{\alpha V^2}{2g}$	$H_{(1)}$	R	S_f	$\bar{S}_f$	Δx	h_f	$H_{(2)}$
0·00	4·00	4·00	28·00	0·679	0·026	4·026	1·900					
100·00	4·002	3·582	26·39	0·720	0·029	4·031	1·843	0·000064	0·000059	100	0·0059	4·032
200·00	4·005	3·705	24·84	0·765	0·033	4·038	1·786	0·000075	0·000069	100	0·0069	4·038
300·00	4·008	3·558	23·33	0·814	0·037	4·045	1·729	0·000088	0·000082	100	0·0082	4·046
400·00	4·012	3·412	21·88	0·868	0·042	4·054	1·673	0·000105	0·000097	100	0·0097	4·055
500·00	4·017	3·267	20·47	0·928	0·048	4·065		0·000125	0·000115	100	0·0115	4·066
→	→	→	→	→	→	→	→	→	→	→	→	→
1300·00	4·164	2·196	11·41	1·665	0·155	4·302	1·239	0·000602	0·000548	100	0·0548	4·302
1400·00	4·188	2·188	10·62	1·788	0·179	4·367	1·193	0·000731	0·000666	100	0·0666	4·368
1500·00	4·242	1·992	9·94	1·910	0·205	4·447	1·152	0·000874	0·000802	100	0·0802	4·448
1600·00	4·311	1·911	9·38	2·024	0·230	4·541	1·117	0·001022	0·000948	100	0·0948	4·541
1700·00	4·397	1·847	8·95	2·122	0·253	4·650	1·088	0·001162	0·001093	100	0·1093	4·650
1800·00	4·500	1·800	8·64	2·120	0·271	4·771	1·068	0·001280	0·001221	100	0·1221	4·772
1900·00	4·618	1·768	8·43	2·254	0·285	4·903	1·054	0·001369	0·001325	100	0·1325	4·903

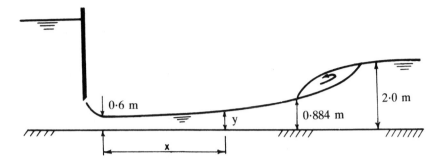

Figure 8.34   Non-uniform flow downstream of sluice gate

Critical depth, y_c = 1·366 m (equation (8·14))
Initial depth at jump (equation (8.16)) = 0·884 m (Note: y_s = 2·0 m.)
Proceeding downstream (in supercritical flow) from the control section, y = 0·6 m and using the direct step method with Δy = 0·04 m, the calculations are shown in the table:

y	A	R	E	ΔE	S_f	$\overline{S}_f$	Δx	x
0·60	2·4	0·461	4·495	—	0·0438	—	—	0
0·64	2·56	0·485	4·063	0·4317	0·0361	0·040	12·36	12·36
0·68	2·72	0·507	3·712	0·3510	0·0300	0·0330	12·50	24·87
0·72	2·88	0·529	3·425	0·2876	0·0253	0·0278	12·67	37·54
0·76	3·04	0·551	3·188	0·2372	0·0216	0·0235	12·85	50·39
0·80	3·20	0·571	2·991	0·1967	0·0185	0·0200	13·07	63·46
0·84	3·36	0·591	2·827	0·1637	0·0160	0·0173	13·31	76·11
0·88	3·52	0·611	2·691	0·1365	0·0140	0·0150	13·61	90·38

Example 8.24
Identify the types of water surface profiles behind (upstream) and between the gate and hydraulic jump using the data of Example 8.23.

Solution:
The solution here is generalised by re-writing the basic water surface slope equation (equation 8.19) as:

$$\frac{dy}{dx} = \frac{S_0 \left(1 - \dfrac{S_f}{S_0}\right)}{\left(1 - \dfrac{\alpha\, Q^2\, B}{g\, A^3}\right)} \tag{i}$$

If the conveyance of a channel is K and its section factor is Z, we can write $Q^2 = K^2\, S_f = K_0^2\, S_0$, $Z^2 = A^3/B$ and $Z_c^2 = \alpha\, Q^2/g$; equation (i) now becomes:

$$\frac{dy}{dx} = \frac{S_0 \left[1 - \left(\frac{K_0}{K}\right)^2 \right]}{\left[1 - \left(\frac{Z_c}{Z}\right)^2 \right]} \tag{ii}$$

Equation (ii) for the case of a wide rectangular channel with the Manning resistance equation reduces to:

$$\frac{dy}{dx} = \frac{S_0 \left[1 - \left(\frac{y_0}{y}\right)^{10/3} \right]}{\left[1 - \left(\frac{y_c}{y}\right)^3 \right]} \tag{iii}$$

Equation (iii) is very convenient to identify the types of surface profile once the normal depth, y_0, and critical depth, y_c, are established in a channel of given slope, S_0.

Case (i): mild channel
In this channel, $S_0 < S_c$ (critical slope i.e. $y_0 = y_c$) and hence $y_0 > y_c$.

> For $y > y_0 > y_c$ (zone 1): dy/dx is positive i.e. increasing depths along x − M_1 profile
> For $y < y_0 > y_c$ (zone 2): dy/dx is negative i.e. decreasing depths along x − M_2 profile
> For $y < y_0 < y_c$ (zone 3): dy/dx is positive i.e. increasing depths along x − M_3 profile

Case (ii): steep channel
Here $S_0 > S_c$ and hence $y_0 < y_c$ and, depending on the level of y, again three profiles (S_1, S_2 and S_3) exist.

Case (iii): critical slope channel
Here $S_0 = S_c$ and $y_0 = y_c$; zone 2 is absent and two profiles, C_1 and C_3, exist.

Case (iv): horizontal channel
Here $S_0 = 0$ and y_0 will not exist and hence zone 1 is absent; again two profiles, H_2 and H_3, exist.

Case (v): adverse slope channel
Now S_0 is negative with no y_0 and two profiles, A_2 and A_3, exist in this case. The discharge $Q = 20$ m^3/s and by the Manning resistance equation $y_0 = 1·26$ m. The critical depth in rectangular channel $y_c = \{ \alpha Q^2/b^2 \ g \}^{1/3}$ from $\alpha Q^2 B/g \ A^3 = 1$. Hence $y_c = 1·366$ m $> y_0$; hence steep channel with gate opening below critical depth. The depth upstream of the gate (by energy balance) $= 4·05$ m $> y_c > y_0$; zone 1 on steep slope; S_1 profile exists immediately upstream of the gate.

The flow downstream of the gate is supercritical (zone 3 of the steep channel) merging with the controlled subcritical flow with the formation of hydraulic jump. Here the S_3 profile forms between the depths, 0·6 m (just d/s of gate) and 0·884 m, sequential to the controlled depth of 2 m.

Example 8.25
Discharge from a natural lake occurs through a very long rectangular channel of bed width 3 m, Manning's n = 0·014 and the bed slope = 0·001. The maximum level of the water surface in the lake above the channel bed at the lake outlet is 3 m. Calculate the discharge in the channel. If the channel slope were to be 0·008, compute the discharge. Also, determine the uniform flow depth and the minimum length of the channel for the uniform flow to establish. Ignore entrance losses.

Solution:
(i) Slope = 0·001; first establish whether this is a mild or steep slope. If the slope were to be assumed critical, the channel will have inlet control with critical depth; critical depth in rectangular channel = ⅔H, H being the energy head (lake level above channel inlet) available.

∴ the critical depth at inlet, y_c = ⅔ × 3 = 2 m

Since the channel is long (with no downstream control) uniform flow with depth y_0 = y_c establishes at the inlet itself; using the Manning resistance equation the corresponding critical slope, S_c, is computed.

At critical depth the discharge is maximum and is computed from y_c = $(q^2/g)^{1/3}$ (from $\alpha Q^2 B/g A^3 = 1$); Hence, Q = qb = 26·58 m³/s and from Q = $An^{-1} R^{2/3} S^{1/2}$ the slope, S = S_c = 0.0047.

If the channel is other than rectangular in cross section, two simultaneous equations (i) the energy equation and (ii) the critical depth criterion, $\alpha Q^2 B/g A^3 = 1$, must be solved for computing y_c.

Since the bed slope $S_0 < S_c$, the channel is of mild slope.

Two equations at the inlet to solve two unknowns, depth and velocity, are required:

(i) energy equation, H = y_0 + $V^2/2g$ and
(ii) the Manning resistance equation, V = $n^{-1} R^{2/3} S^{1/2}$

both are applicable at the inlet (long channel; uniform flow establishes from the inlet). Simultaneous solution of (i) and (ii) gives

y_0 = 2·75 m and V = 2·213 m/s
thus the discharge Q_0 = 18·26 m³/s

The slope S_0 = 0·008 > S_c = 0·0047; the channel is a steep sloped one and the inlet controls the flow. The discharge is maximum (= 26·58 m³/s) with critical depth at the inlet (y_c = 2 m). The corresponding uniform flow depth (y_0) from the Manning resistance equation

$$26·58 = 3 \times y_0 \times (1/0·014) \times [(3 \times y_0)/(3 + 2y_0)]^{2/3} \times (0·008)^{1/2}$$

is computed. Thus $y_0 = 1.64$ m; this unform flow establishes in the channel if its length is at least equal to the length of the surface profile (nonuniform flow) that exists between the inlet depth, y_c, and the uniform flow depth, y_0.

The channel is of steep slope and the flow between these two depths corresponds to region 2 and hence an S_2 profile develops whose length can be computed by any appropriate method. By the step method we obtain $L \approx 80$ m between the two depths; the calculations should commence at a depth slightly less than the critical depth and terminate at a depth slightly higher than the normal (uniform) depth.

Example 8.26
A rectangular channel (b = 15 m, length = 10 km, slope = 1/10 000, Manning's n = 0.015) fed by an upstream lake is discharging into a downstream lake. If the upstream and downstream lake levels are 1.5 m and 2 m (above the channel bed) respectively, determine the discharge rate in the channel.

Solution:
The channel delivery depends upon the following considerations:

(i) Is the channel long (i.e. no downstream control)?
(ii) Is the slope mild or critical or steep?

If we assume a long and mild channel, two equations at its inlet are applicable, i.e. the energy and resistance equations. If we assume a long and critical sloped channel, the two equations are the energy equation and the critical depth criterion. Here it is convenient to assume initially a long and critical sloped channel. Since the channel is of rectangular cross section, $y_c = \frac{2}{3}H = 1$ m. Therefore, from the Manning resistance equation (uniform flow depth $= y_c$) the critical slope, $S_c = 0.00256 > S_0 = 0.0001$ (see Example 8.25). Hence the channel is of mild slope; if we still assume it is long, we now can obtain the corresponding uniform flow rate, Q_0, from the energy and resistance equations.

(iii) Is the channel long enough to satisfy the above assumptions?

Identify the type of water surface profile based on the discharge, uniform flow depth and the downstream lake level and compute its length. If the surface profile length < channel length, the channel is long enough and uniform flow does exist at the inlet i.e. free inlet. If the profile length > channel length, the inlet will be drowned (short channel) and the discharge rate is reduced. To compute the actual discharge rate in a short channel, the following iterative procedure is to be followed:

(a) Assume $Q < Q_0$ (since the inlet is drowned) and compute the corresponding flow depth at the inlet by the energy equation.
(b) Compute the new surface profile length corresponding to this discharge and verify whether it fits between the inlet depth and the downstream lake level.

(c) Repeat (a) and (b) until the profile length matches the channel length.

The problem is best approached on a PC for executing these iterative procedures.

Example 8.27
A concrete twin-box type culvert is proposed to discharge a design flood of $13.0 \, m^3/s$. The following data refer to each opening:

Manning's n	= 0.013
height	= 0.75 m
width	= 1.5 m
length	= 30 m
slope	= 1/100
entrance conditions	= square edge, loss coefficient, K = 0.5
downstream conditions	= free jet

Establish the rating curve (discharge versus headwater elevation above the invert at the entrance) for rising head conditions over a discharge range from 0 to 13.5 m^3/s. Neglect the velocity of approach. Determine the minimum elevation of the road surface assuming a free board of 300 mm to avoid any flooding of the highway.

Solution:
The culvert behaviour is dependent on the headwater level, H, the height of culvert, D, slope, S_0, and length, L; here the outlet is to discharge freely and has no effect on the type of flow through the culvert.

(i) For $H/D \leq 1.2$, open channel flow. If entrance control exists, the depth at the inlet is critical, i.e. the slope is either critical or steep. Assuming entrance control, $y_c = \frac{2}{3} \times H$ (rectangular channel; inlet with no entrance losses) and $V_c = \sqrt{(g \, y_c)}$; hence Q can be computed. Also from the Manning resistance equation, the critical slope, S_c, can be computed and checked against the proposed slope of the culvert; if the slope is then found to be mild, the depth and discharge calculations must be computed by the energy and resistance equations (see Example 8.25).
   The energy equation at the inlet gives $H = 1.75 \, y_c$ assuming entrance control with a loss coefficient of 0.5 i.e. $S_0 \geqslant S_c$.
   For $H = 0.1$ m, $y_c = 0.057$ m and $V_c = 0.748$ m/s. From the Manning's equation, $S_c = 0.00476$ and as $S_0 (= 0.01) > S_c (= 0.00476)$, entrance control exists. The discharge through the culvert of width b can be written as $Q = b y_c \sqrt{g y_c}$ for one box. Various values for y_c and hence the head water levels ($H = 1.75 y_c$) are assumed until $H = 1.2D (= 0.9$ m, the upper limit for open channel flow); a check for $S_0 > S_c$ is necessary for all values.

Headwater level, H (m)	Discharge, Q m^3/s (two boxes)
0·175	0·297
0·525	1·544
0·700	2·377
0·900	3·465

(ii) For H/D $\geq$ 1·2, the culvert entry behaves like an orifice (constriction); if the normal depth in the barrel corresponding to the orifice discharge is less than D, the flow downstream of the inlet is free. The orifice flow equation

$$Q = C_d \times b \times D \times [2g (H - D/2)]^{1/2} \text{ for one box}$$

with the discharge coefficient, C_d = 0·62 (assumed). Computations between H = 1·2D and the value at which y_0 = D are as below:

Headwater level, H (m)	Discharge, Q m^3/s (two boxes)
0·900	4·477
1·300	5·943
1·700	7·113
2·100	8·116
2·500	9·007

(iii) For H/D > 1·2 and $y_0 \geq$ D, pipe flow exists in the culvert. The energy equation between the inlet and outlet of the culvert gives

$$H + S_0L = D + (1 + K)V^2/2g + S_fL$$

where S_f is the friction slope given by the Manning equation, as $S_f = (Vn)^2/R^{4/3}$. The above reduce to

$$Q = 3·41(H - 0·45)^{1/2} \text{ for one box; pipe flow condition with the following:}$$

Headwater level, H (m)	Discharge, m^3/s (two boxes
2·500	9·774
2·900	10·685
3·100	11·112
3·500	11·921
3·900	12·679
4·200	13·219
4·368	13·500 (design discharge)

Elevation of the road surface with a free board of 300 mm = 4·368 + 0·300 = 4·668 m above the culvert invert at its inlet.

Example 8.28

A lateral spillway channel, 120 m long and trapezoidal in section, is designed to carry a discharge which increases at a rate of 3·7 m^3/s per metre length. The cross section has a bed width of 3 m with side slopes of 0·5H : 1V. The bed slope is 0·15 and Manning's n = 0·015. Compute the water surface profile of the design discharge assuming uniform velocity distribution.

Solution:
The type of surface profile depends upon the length of the channel and whether the channel characteristics would permit the existence of a control section, i.e. a section where the depth is critical. The existence of a critical depth section and its longitudinal location are to be examined first; this is achieved by a trial and error process using equation 8.23 and the critical depth criterion, $Q^2 B/g A^3 = 1$. The following table of results is self-explanatory:

x (m)	$Q = qx$ (m³/s)	y_c (m)	A (m²)	P (m)	B (m)	$R = A/P$ (m)	$C = R^{1/6}/n$ (m^{1/2}/s)	x (m)
60	222·0	5·97	35·73	16·37	8·97	2·18	76·00	43·0
43	159·1	4·90	26·70	13·98	7·90	1·91	74·26	56·0
56	207·2	5·57	32·24	15·48	8·57	2·08	75·34	47·6
47	173·9	5·22	29·28	14·69	8·22	1·99	74·79	51·6
52	192·4	5·52	31·79	15·36	8·52	2·07	75·26	48
48	177·6	5·29	29·86	14·85	8·29	2·01	74·90	51·0
51	188·7	5·46	31·28	15·23	8·46	2·05	75·16	48·7
49	181·3	5·35	30·36	14·98	8·35	2·03	75·00	50·0
50	185·0	5·40	30·78	15·09	8·40	2·04	75·07	49
49·5	183·1	5·38	30·61	15·05	8·38	2·03	75·04	49·6

From the table, $x_c = 49·5$ m $<$ L ($= 120$ m). As the length available is greater than x_c, the water surface profile upstream of the control section is in subcritical flow while that of the downstream part is supercritical flow. The surface profile computations are generally carried out by numerical integration combined with trial and error; the procedures are laborious and well described in textbooks of open channel hydraulics (see French[7]; also see worked exampled 15.3).

The nonuniform flow (GVF and SVF) computations may be carried out using the advanced numerical methods. The surface slope dy/dx, given by equations 8.19, 8.22 and 8.25, is a function x and y and can be written as dy/dx $= F(x,y)$ and the Standard Fourth Order Runge-Kutta (SRK) Method uses the following operation:

$$y_{i+1} = y_i + \frac{1}{6} (K_1 + 2K_2 + 2K_3 + K_4) \tag{8.48}$$

in which

$$K_1 = \Delta x\, F(x_i,\ y_i)$$
$$K_2 = \Delta x\, F(x_i + \Delta x/2,\ y_i + K_1/2)$$
$$K_3 = \Delta x\, F(x_i + \Delta x/2,\ y_i + K_2/2)$$
$$K_4 = \Delta x\, F(x_i + \Delta x,\ y_i + K_3)$$

The solution is easily achieved with the help of a PC.

Example 8.29

A rectangular channel of bed width = 2 m, Manning's n = 0·014, is laid on a slope of 1/1000. A side weir is to be designed at a section such that it comes into operation when the discharge in the channel exceeds 0·6 m³/s. A lateral outflow of 0·15 m³/s is expected to be delivered by the side weir when the channel discharge is 0·9 m³/s. Compute the elements of the weir.

Solution:

The sill (crest) height of the side weir is decided by the flow depth corresponding to 0·6 m³/s. The normal depths in the channel (by the Manning resistance equation):

for Q = 0·6 m³/s, y_0 = 0·33 m, and
for Q = 0·9 m³/s, y_0 = 0·44 m.

The crest height is therefore (for the weir to come into operation), s = 0·33 m. For Q = 0·9 m³/s, the critical depth in the channel y_{c1} = $(q^2/g)^{1/3}$ = 0·274 m. Now the sill height s > y_{c1} and y_0 > y_{c1}, the dy/dx of the flow profile over the weir is positive and the flow is subcritical. De Marchi equation assumes $y_1 \approx y_0$ = 0·44 m. With the assumption $E_1 = E_2$ we can write

$$E_1 = 0·493 = y_2 + Q_2^2/[(b\ y_2)^2\ 2g];\ Q_2 = 0·90 - 0·15 = 0·75\ \text{m}^3/\text{s}$$

and hence y_2 = 0·46 m (subcritical flow). The De Marchi functions ϕ_1 = −1·84 and ϕ_2 = −1·42. The De Marchi coefficient C_M = 0·81 − 0·60F_{r1} (assuming a sharp crested weir) where F_{r1} = $V_1/\sqrt{(2g)}$ = 0·49. Hence, C_M = 0·516 and by equation 8·31, the weir length L = 2·442 m.

Recommended reading

1. Ackers, P. (1991) *Hydraulic design of straight compound channels.* Wallingford: Hydraulics Research Ltd, Report SR 281.
2. Ackers, P. (1992) *Hydraulic design of two-stage channels.* Proc Instn Civ. Engrs, Water, Maritime and Energy Board, Vol. 96, pp. 247−57.
3. Bakhmeteff, B.A. (1932) *Hydraulics of Open Channels.* New York: McGraw Hill Book Co.
4. BS 3680: *Methods of measurement of liquid flow in open channels.*
   BS 3680: Part 4A: (1981) *Thin plate weirs.*
   BS 3680: Part 4B: (1969) *Long base weirs.*
   BS 3680: Part 4C: (1981) *Flumes.*
   BS 3680: Part 3A: (1980) *Velocity area methods.* London: British Standards Institute.
5. Chow, Ven Te (1959) *Open-channel Hydraulics.* New York: McGraw Hill Book Co.
6. Fortier, S. and Scobey, F.C. (1926) *Permissible Canal Velocities.* Trans Am Soc Civ Eng, Vol. 89.

7. French, R.H. (1986) *Open Channel Hydraulics*. New York: McGraw-Hill Book Co.
8. Henderson, F.M. (1966) *Open Channel Flow*. New York: The Macmillan Company.
9. Hydraulic Research Station (1977) *Tables for the hydraulic design of pipes*. Dept. of the Environment. London: H.M.S.O.
10. Novak, P., Moffat, A.I.B., Nalluri, C. and Narayanan, R. (1997) *Hydraulic Structures*, 2nd edn. London: E&FN Spon.
11. Ranga Raju, K.G. (1993) *Flow Through Open Channels*, 2nd edn. New Delhi, Tata: McGraw-Hill Book Co.

Problems

1.   Water flows uniformly at a depth of 2 m in a rectangular channel of width 4 m and bed slope 1 : 2000. What is the mean shear stress on the wetted perimeter?

2.   (a) At a measured discharge of 40 m^3/s the depth of uniform flow in a rectangular channel 5 m wide and with a bed slope of 1 : 1000 was 3·05 m. Determine the mean effective roughness size and Manning's roughness coefficient.
(b) Using (i) the Darcy-Weisbach equation (together with the Colebrook-White equation), and (ii) the Manning equation predict the discharge at a depth of 4 m.

3.   Determine the depth of uniform flow in a trapezoidal concrete-lined channel of bed width 3·5 m, bed slope 0·0005 with side slopes at 45° to the horizontal when conveying 36 m^3/s of water.
Manning's roughness coefficient $= 0·014$.

4.   Determine the rate of uniform flow in a circular section channel 3 m in diameter of effective roughness 0·3 mm, laid to a gradient of 1 : 1000 when the depth of flow is 1·0 m. What is the mean velocity and the mean boundary shear stress?

5.   A circular storm water sewer 1·5 m in diameter and effective roughness size 0·6 mm is laid to a slope of 1 : 500. Determine the maximum discharge which the sewer will convey under uniform open channel conditions. If a steady inflow from surface run off exceeds the maximum open channel capacity by 20 per cent show that the sewer will become pressurised (surcharged) and calculate the hydraulic gradient necessary to convey the new flow.

6.   Assuming that a rough turbulent velocity distribution having the form

$$\frac{v}{\sqrt{\dfrac{\tau_o}{\rho}}} = 5\cdot75 \log \frac{30y}{k},$$

exists in a wide river show that the average of current meter measurements taken at 0·2 and 0·8 of the depth from the surface approximates to the mean velocity in a vertical section.

7.   A long, concrete-lined trapezoidal channel with a bed slope 1 : 1000, bed width 3·0 m, side slopes at 45° to the horizontal and Manning roughness 0·014 receives water from a reservoir. Assuming an energy loss of $0\cdot25 \dfrac{V^2}{2g}$ calculate the steady discharge and depth of uniform flow in the channel when the level in the reservoir is 2·0 m above the channel bed at inlet.

8.   A trapezoidal channel with a bed slope of 0·005, bed width 3 m and side slopes 1 : 1·5 (vertical : horizontal) has a gravel bed (n = 0·025) and concrete sides (n = 0·013). Calculate the uniform flow discharge when the depth of flow is 1·5 m using (a) the Einstein, (b) the Pavlovskij, and (c) the Lotter methods.

9.   The figure shows the cross-section of a river channel passing through a flood plain. The main channel has a bank full area of 300 m², a top width of 50 m, a wetted perimeter of 65 m and a Manning roughness coefficient of 0·025. The flood plains have a Manning roughness of 0·035 and the gradient of the main channel and plain is 0·00125. Determine the depth of flow over the flood plain at a flood discharge of 2470 m³/s.

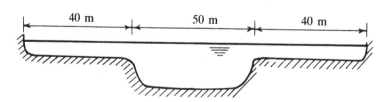

Figure 8.35   Compound channel

10.   A concrete-lined rectangular channel is to be constructed to convey a steady maximum discharge of 160 m³/s to a hydro-power installation. The bed slope is 1 : 5000 and Manning's n appropriate to the type of surface finish is 0·015. Determine the width of the channel and the depth of flow for the 'most economic' section. Give reasons why the actual constructed depth would be made greater than the flow depth.

11.   A concrete-lined trapezoidal channel with a bed slope of 1 : 2000 is to be designed to convey a maximum discharge of 75 m³/s under uniform

flow conditions. The side slopes are at 45° and Manning's n = 0·014. Determine the bed width and depth of flow for the 'most economic' section.

12.   A channel with a bed slope of 1 : 2000 is to be constructed through a stiff clay formation. Compare the relative costs of the alternative design of best rectangular concrete-lined and trapezoidal unlined channels to convey 60 m³/s if the cost of the 100 mm thick lining/m² is twice the cost of excavation/m³. Manning's n for concrete lining = 0·014 and for the unlined channel 0·025. Side slopes (stable) = 1 : 1·5.

13.   An unlined irrigation channel of trapezoidal section is to ber constructed through a sandy formation at a bed slope of 1 : 10 000 to convey a discharge of 40 m³/s. The side slopes are at 25° to the horizontal. The angle of internal friction of the material is 35° and the critical tractive force is 2·5 N/m²; Manning's n = 0·022. Assuming that the maximum boundary shear stress exerted on the bed, due to the water flow, is 0·98 ρg y S_o, and that on the sides is 0·75 ρg y S_o, determine the bed width and flow depth for a non-eroding channel design.

14.   A vertical sluice gate in a long rectangular channel 5 m wide is lowered to produce an opening of 1·0 m. Assuming that free flow conditions exist at the vena contracta downstream of the gate verify that the flow in the vena contracta is supercritical when the discharge is 15 m³/s and determine the depth just upstream of the gate. C_v = 0·98; C_c = 0·6. Take the upstream velocity energy coefficient (Coriolis) to be 1·0 and that at the vena contracta to be 1·2.

15.   A sill is to be constructed on the bed of a rectangular channel conveying a specific discharge of 5 m³/s per metre width. The depth of uniform flow is 2·5 m. Neglecting energy losses:
(a) determine the variation of the depths upstream of the sill (y_1) and over the sill (y_2) for a range of sill heights (z) from 0·1 m to 0·8 m. Take the Coriolis coefficient to be 1·2.
(b) determine the critical depth (y_c), and
(c) the minimum sill height (z_c) to create critical flow conditions at the sill.

16.   A venturi flume with a throat width of 0·5 m is constructed in a rectangular channel 1·5 m wide. The depth of uniform flow in the channel at a discharge of 1·6 m³/s is 0·92 m; α = 1·1. Assuming that downstream conditions do not influence the natural flow profile through the contraction and neglecting losses verify that the flume acts as a 'critical depth flume' and determine the upstream and throat depths.
   Show that for critical flow conditions in the throat the discharge can be obtained from

$$Q = \frac{2}{3}\sqrt{\frac{2g}{3}}\, b\, H_1^{3/2}$$

where b is the throat width and $H_1 = y_1 + \dfrac{\alpha V_1^2}{2g}$.

Calculate the discharge when the upstream depth is 1·0 m. (Verify that critical flow conditions are maintained in the throat.)

17.   A vertical sluice gate in a long rectangular channel 4 m wide, has an opening of 1·0 m and a coefficient of contraction of 0·6. At a discharge of 25 m³/s the depth of uniform flow (y_n) is 3·56 m. Assuming that a hydraulic jump were to occur in the channel downstream with a sequent depth equal to y_n and taking the Boussinesq coefficient to be 1·2 what would be the initial depth of the jump? Hence verify that a hydraulic jump will occur.
   Determine the depth upstream of the gate and the hydrodynamic force on the gate, assuming $C_v = 0·98$ and $\alpha = 1·2$.

18.   If in Problem 17 the gate is raised to give an opening of 1·5 m determine whether, or not, a hydraulic jump will form. Calculate the depths immediately upstream of the gate and at the position of the vena contracta and the force on the gate.

19.   A long rectangular channel 8 m wide, bed slope 1 : 5000 and Manning's roughness 0·015 conveys a steady discharge of 40 m³/s. A sluice gate raises the depth immediately upstream to 5·0 m. Taking the Coriolis coefficient α to be 1·1 determine the uniform flow depth and the distance from the gate at which this depth is exceeded by 10 per cent. What is the depth 5000 m from the gate?

20.   A rectangular channel having a bed width of 4 m, a bed slope of 0·001 and Manning's n = 0·015 conveys a steady discharge of 25 m³/s. A barrage creates a depth upstream of 4·0 m. Compute the water surface profile, taking $\alpha = 1·1$.

21.   A long rectangular channel, 2·5 m wide, bed slope 1 : 1000 and Manning roughness coefficient 0·02 discharges 4·5 m³/s freely to atmosphere at the downstream end. Taking $\alpha = 1·1$ and noting that at a free overfall the depth approximates closely to the critical depth, compute the surface profile to within approximately 10 per cent of the uniform flow depth.

22.   A long trapezoidal channel of bed width 3·5 m, side slopes at 45°, bed slope 0·0003 and Manning's n = 0·018 conveys a steady flow of 50 m³/s. A control structure creates an upstream depth of 5·0 m. Taking $\alpha = 1·1$ determine the distance upstream at which the depth is 4·2 m.

23.   Two reservoirs are connected by a wide rectangular channel length of 1500 m, where Manning's n is 0·02, the bed slope is 4×10^{-4}, the channel entry loss coefficient K = 0·02, the channel invert elevation (u/s) = 101·00 m AOD and the water level in the u/s reservoir = 104·00 mAOD (constant).

(a) Determine the limiting downstream reservoir level to cause uniform flow in the channel.
(b) If the downstream reservoir level is 103·50 m AOD, examine whether or not it will affect the uniform flow rate (submerged inlet and reduced flow).

24.   A culvert is proposed under a highway embankment where the design flood is 15 m^3/s, the width of the highway is 30 m and the natural drainage slope is 0·015. The available pipe barrels are corrugated pipes of diameter in multiples of 250 mm with Manning's n = 0·024; the entry loss coefficient = 0·9.
(a) Compute the proposed culvert barrel size if the maximum permissible headwater level is 4 m above the invert, with the barrel discharging free at its outlet.
(b) If a fare-edged entry (loss coefficient = 0·25) is chosen, calculate the required barrel diameter for the conditions in (a).

25.   A broad crested side weir is to be designed to deliver 10 m^3/s of water into a branch canal. The main channel conveying 100 m^3/s of water is of rectangular cross section with b = 50 m, S_0 = 0·0001 and Manning's n = 0.02. Compute the length and crest height of the weir.

Chapter 9
Dimensional Analysis, Similitude and Hydraulic Models

R.E. Featherstone

9.1 Introduction

Hydraulic engineering structures or machines can be designed using (i) pure theory, (ii) empirical methods, (iii) semi-empirical methods which are mathematical formulations based on theoretical concepts supported by suitably designed experiments or (iv) physical models, (v) mathematical models.

The purely theoretical approach in hydraulic engineering is limited to a few cases of laminar flow, for example the Hagen Poisseuille equation for the hydraulic gradient in the laminar flow of an incompressible fluid in a circular pipeline. Empirical methods are based on correlations between observed variables affecting a particular physical system. Such relationships should only be used under similar circumstances to those under which the data were collected. Due to the inability to express the physical interaction of the parameters involved in mathematical terms some such methods are still in use. One well-known example is in the relationship between wave height, fetch, wind speed and duration for the forecasting of ocean wave characteristics.

A good example of a semi-empirical relationship is the Colebrook-White equation for the friction factors in turbulent flow in pipes (see Chapters 4 and 7). This was obtained from theoretical concepts and experiments designed on the basis of dimensional analysis; it is universally applicable to all Newtonian fluids.

Dimensional analysis also forms the basis for the design and operation of physical scale models which are used to predict the behaviour of their full-sized counterparts called 'prototypes'. Such models, which are generally geometrically similar to the prototype, are used in the design of aircraft, ships, submarines, pumps, turbines, harbours, breakwaters, river and estuary engineering works, spillways, etc.

While mathematical modelling techniques have progressed rapidly due to the advent of high-speed digital computers, enabling the equations of motion coupled with semi-empirical relationships to be solved for complex flow situations such as pipe network analysis, pressure transients in pipelines, unsteady flows in rivers and estuaries, etc., there are many cases, particularly where localised flow patterns cannot be mathematically modelled, when physical models are still needed.

Without the technique of dimensional analysis experimental and computational progress in fluid mechanics would have been considerably retarded.

9.2 Dimensional analysis

The basis of dimensional analysis is to condense the number of separate variables involved in a particular type of physical system into a smaller number of non-dimensional groups of the variables.

The arrangement of the variables in the groups is generally chosen so that each group has a physical significance.

All physical parameters can be expressed in terms of a number of basic dimensions; in engineering the basic dimensions, mass (M), length (L) and time (T) are sufficient for this purpose. For example, velocity = distance/time ($= LT^{-1}$); discharge = volume/time ($= L^3T^{-1}$). Force is expressed using Newton's law of motion (force = mass × acceleration); hence Force = MLT^{-2}.

A list of some physical quantities with their dimensional forms can be seen below.

Physical Quantity	Symbol	Dimensional Form
Length	ℓ	L
Time	t	T
Mass	m	M
Velocity	V	LT^{-1}
Acceleration	a	LT^{-2}
Discharge	Q	L^3T^{-1}
Force	F	MLT^{-2}
Pressure	p	$ML^{-1}T^{-2}$
Power	P	ML^2T^{-3}
Density	ρ	ML^{-3}
Dynamic viscosity	μ	$ML^{-1}T^{-1}$
Kinematic viscosity	ν	L^2T^{-1}
Surface tension	σ	MT^{-2}
Bulk modulus of elasticity	K	$ML^{-1}T^{-2}$

9.3 Physical significance of non-dimensional groups

The main components of force which may act on a fluid element are those due to viscosity, gravity, pressure, surface tension and elasticity. The resultant of these components is called the inertial force and the ratio of this force to each of the force components indicates the relative importance of the force types in a particular flow system.

For example the ratio of inertial force to viscous force,

$$\frac{F_i}{F_\mu} = \frac{\rho L^3 L T^{-2}}{\tau L^2}$$

Now $\tau = \mu \dfrac{dv}{dy} = \mu L T^{-1} L^{-1}$

whence $\dfrac{F_i}{F_\mu} = \dfrac{\rho L^2 T^{-1}}{\mu} = \dfrac{\rho L V}{\mu} = \dfrac{\rho \ell V}{\mu}$

where ℓ is a typical length dimension of the particular system.

The dimensionless term $\dfrac{\rho \ell V}{\mu}$ is in the form of the Reynolds number. Low Reynolds numbers indicate a significant dominance of viscous forces in the system which explains why this non-dimensional parameter may be used to identify the regime of flow, i.e. whether laminar or turbulent.

Similarly it can be shown that the Froude number is the ratio of inertial force to gravity force in the form

$$F_r = \frac{V^2}{g\ell} \quad \left(\text{but usually expressed as } Fr = \frac{V}{\sqrt{g\ell}}\right)$$

The Weber number, We is the ratio of inertial to surface tension force and is expressed by $\dfrac{V}{\sqrt{\sigma/\rho\ell}}$.

9.4 The Buckingham π theorem

This states that the n quantities $Q_1, Q_2, \ldots Q_n$, involved in a physical system can be arranged in $(n-m)$ non-dimensional groups of the quantities where m is the number of basic dimensions required to express the quantities in dimensional form.

Thus $f_1(Q_1, Q_2, \ldots Q_n) = 0$ can be expressed as $f_2(\pi_1, \pi_2, \ldots \pi_{n-m})$ where 'f' means 'a function of ...'. Each π term basically contains m repeated quantities which together contain the m basic dimensions together with one other quantity. In fluid mechanics m = 3 and therefore each π term basically contains four of the quantity terms.

9.5 Similitude and model studies

Similitude, or dynamic similarity, between two geometrically similar systems exists when the ratios of inertial force to the individual force components in the first system are the same as the corresponding ratios in the second system at the corresponding points in space. Hence for absolute dynamic similarity the Reynolds, Froude and Weber numbers must be the same in the two systems. If this can be achieved the flow patterns will be geometrically similar, i.e. kinematic similarity exists.

In using physical scale models to predict the behaviour of prototype systems or designs it is rarely possible (except when only one force type is relevant) to achieve simultaneous equality of the various force ratios. The 'scaling laws' are then based on equality of the predominant force; strict dynamic similarity is thus not achieved resulting in 'scale effect'.

Reynolds modelling is adopted for studies of flows without a free surface such as pipe flow and flow around submerged bodies, e.g. aircraft, submarines, vehicles and buildings.

The Froude number becomes the governing parameter in flows with a free surface since gravitational forces are predominant. Hydraulic structures, including spillways, weirs and stilling basins, rivers and estuaries, hydraulic turbines and pumps and wave-making resistance of ships are modelled according to the Froude law.

Worked examples

Example 9.1
Obtain an expression for the pressure gradient in a circular pipeline, of effective roughness, k, conveying an incompressible fluid of density, ρ, dynamic viscosity, μ, at a mean velocity, V, as a function of non-dimensional groups.

By comparison with the Darcy-Weisbach equation show that the friction factor is a function of relative roughness and the Reynolds number.

Solution:
In full pipe flow gravity and surface tension forces do not influence the flow. Let Δp = pressure drop in a length L.

> Then $f_1(\Delta p, L, \rho, V, D, \mu, k) = 0$ (i)

The repeating variables will be ρ, V and D. Δp clearly is not to be repeated since this variable is required to be expressed in terms of the other variables. If μ or k were to be repeated the relative effect of the parameter would be hidden.

> $f_2(\pi_1, \pi_2, \pi_3, \pi_4) = 0$ (ii)

Then

$$\pi_1 = \rho^\alpha D^\beta V^\gamma \Delta p \qquad \text{(iii)}$$

where α, β and γ are indices to be evaluated.

In dimensional form:

$$\pi_1 = (ML^{-3})^\alpha L^\beta (LT^{-1})^\gamma ML^{-1}T^{-2}$$

The sum of the indices of each dimension must be zero.

Thus for M, $0 = \alpha + 1$, whence $\alpha = -1$;

for T, $0 = -\gamma - 1$, whence $\gamma = -2$;

and for L, $0 = -3\alpha + \beta + \gamma - 1$, whence $\beta = 0$

$$\therefore \pi_1 = \frac{\Delta p}{\rho V^2} \tag{iv}$$

$$\pi_2 = \rho^\alpha D^\beta V^\gamma L \tag{v}$$

The π terms are dimensionless and since D and L have the same dimensions the solution is

$$\pi_2 = \frac{L}{D} \tag{vi}$$

Similarly

$$\pi_3 = \frac{k}{D} \tag{vii}$$

$$\pi_4 = \rho^\alpha D^\beta V^\gamma \mu \tag{viii}$$

$$\pi_4 = (ML^{-3})^\alpha L^\beta (LT^{-1})^\gamma ML^{-1}T^{-1}$$

Indices of M: $0 = \alpha + 1$; $\qquad\qquad \alpha = -1$

Indices of T: $0 = -\gamma - 1$; $\qquad\qquad \gamma = -1$

Indices of L: $0 = -3\alpha + \beta + \gamma - 1$; $\quad \beta = -1$

$$\therefore \pi_4 = \frac{\mu}{\rho D V} \tag{ix}$$

$$\therefore f_2\left(\frac{\Delta p}{\rho V^2}, \frac{L}{D}, \frac{k}{D}, \frac{\mu}{\rho D V}\right) = 0 \tag{x}$$

The π terms can be multiplied or divided and since the pressure gradient is required equation (x) may be reformed thus:

$$f_2\left(\frac{\Delta p}{L}\frac{D}{\rho V^2}, \frac{k}{D}, \frac{\mu}{\rho D V}\right) = 0$$

$$\text{whence } \frac{\Delta p}{L} = \frac{\rho V^2}{D}\, \phi\left[\frac{k}{D}, Re\right]$$

where ϕ means 'a function of' the form of which is to be obtained experimentally.

The hydraulic gradient,

$$\frac{\Delta h}{L} = \frac{\Delta p}{\rho g\, L}$$

$$\text{whence } \frac{\Delta h}{L} = \frac{V^2}{gD}\, \phi\left[\frac{k}{D}, Re\right] \tag{xi}$$

Comparing (xi) with the Darcy-Weisbach equation,

$$\frac{h_f}{L} = \frac{\lambda}{2g} \frac{V^2}{D},$$ it is seen that λ is dimensionless and that

$$\lambda = \phi \left[\frac{k}{D}, \text{Re} \right].$$

This relationship enabled experiments to be designed (as described in Chapter 7) which led eventually to the Colebrook-White equation.

Example 9.2
Show that the discharge of a liquid through a rotodynamic pump having an impeller of diameter, D, and width, B, running at speed, N, when producing a total head, H, can be expressed in the form

$$Q = ND^3 \; \phi \left[\frac{D}{B}, \frac{N^2 D^2}{gH}, \frac{\rho ND^2}{\mu} \right]$$

Solution:

$$f_1(N, D, B, Q, gH, \rho, \mu) = 0$$

Note that the presence of g represents the transformation of pressure head to velocity energy; it is convenient, but not essential, to combine g and H instead of treating them separately.

$$f_2(\pi_1, \pi_2, \pi_3, \pi_4) = 0$$

Using ρ, N and D as the recurring variables

$$\pi_1 = \rho^\alpha N^\beta D^\gamma B$$

$$\pi_1 = \frac{B}{D}$$

$$\pi_2 = \rho^\alpha N^\beta D^\gamma Q$$

$$\pi_2 = (ML^{-3})^\alpha (T^{-1})^\beta L^\gamma L^3 T^{-1}$$

For M, $0 = \alpha$; $\qquad\qquad$ $\alpha = 0$

For T, $0 = -\beta - 1$; $\qquad$ $\beta = -1$

For L, $0 = -3\alpha + \gamma + 3$; $\gamma = -3$

$$\pi_2 = \frac{Q}{ND^3}$$

$$\pi_3 = \rho^\alpha N^\beta D^\gamma (gH)$$

$$\pi_3 = (ML^{-3})^\alpha (T^{-1})^\beta L^\gamma L^2 T^{-2}$$

whence $\pi_3 = \dfrac{gH}{N^2 D^2}$

$$\pi_4 = \rho^\alpha N^\beta D^\gamma \mu$$

$$\pi_4 = (ML^{-3})^\alpha (T^{-1})^\beta L^\gamma ML^{-1}T^{-1}$$

whence $\pi_4 = \dfrac{\mu}{\rho ND^2}$

$$\therefore \ f_1\left(\frac{B}{D}, \frac{Q}{ND^3}, \frac{gH}{N^2D^2}, \frac{\mu}{\rho ND^2}\right)$$

whence $Q = ND^3 \ \phi \left[\dfrac{D}{B}, \dfrac{N^2D^2}{gH}, \dfrac{\rho ND^2}{\mu}\right]$

Note that the π terms may be inverted for convenience and that $\dfrac{\rho ND^2}{\mu}$ is a form of the Reynolds number and $\dfrac{N^2D^2}{gH}$ a form of the square of the Froude number.

Example 9.3
Show that the discharge, Q, of a liquid of density, ρ, dynamic viscosity, μ, and surface tension, σ, over a V-notch under a head, H, may be expressed in the form

$$Q = g^{1/2} H^{5/2} \ \phi \left[\frac{\rho g^{1/2} H^{3/2}}{\mu}, \frac{\rho g H^2}{\sigma}, \theta\right]$$

where θ is the notch angle, and hence define the parameters upon which the discharge coefficient of such weirs is dependent.

Solution:
$\quad$ $f_1(\rho, g, H, Q, \mu, \sigma, \theta) = 0$

$\quad$ i.e. $f_2(\pi_1, \pi_2, \pi_3, \pi_4) = 0$

$\quad$ Using ρ, g and H as the repeating variables,

$\quad$ $\pi_1 = \rho^\alpha g^\beta H^\gamma Q$

$\quad$ $\pi_1 = (ML^{-3})^\alpha (LT^{-2})^\beta L^\gamma L^3 T^{-1}$

For M, $0 = \alpha$; $\qquad\qquad\qquad\ \alpha = 0$

For T, $0 = -2\beta - 1$; $\qquad\qquad \beta = -1/2$

For L, $0 = -3\alpha + \beta + \gamma + 3$; $\quad \gamma = -5/2$

$$\pi_1 = \frac{Q}{g^{1/2} H^{5/2}}$$

$\quad$ $\pi_2 = \rho^\alpha g^\beta H^\gamma \mu$

$\quad$ $\pi_2 = (ML^{-3})^\alpha (LT^{-2})^\beta L^\gamma ML^{-1}T^{-1}$

$$\pi_2 = \frac{\mu}{\rho g^{1/2} H^{3/2}}$$

$$\pi_3 = \rho^\alpha g^\beta H^\gamma \sigma$$

$$\pi_3 = (ML^{-3})^\alpha (LT^{-2})^\beta L^\gamma ML^{-2}$$

$$\pi_3 = \frac{\sigma}{\rho g H^2}$$

$$\pi_4 = \rho^\alpha g^\beta H^\gamma \theta$$

$$\pi_4 = \theta \ (\text{since } \theta \text{ is itself dimensionless})$$

$$\therefore \ f_2 \left[\frac{Q}{g^{1/2} H^{5/2}}, \frac{\mu}{\rho g^{1/2} H^{3/2}}, \frac{\sigma}{\rho g H^2}, \theta \right]$$

Re-arranging: $Q = g^{1/2} H^{5/2} \ \phi \left[\frac{\rho g^{1/2} H^{3/2}}{\mu}, \frac{\rho g H^2}{\sigma}, \theta \right]$

From energy considerations the discharge over a V-notch is expressed as

$$Q = \frac{8}{15} \sqrt{2g} \ C_d \tan \theta/2 \ H^{5/2}$$

Comparing the above two forms it is seen that

$$C_d = f \left(\frac{\rho g^{1/2} H^{3/2}}{\mu}, \frac{\rho g H^2}{\sigma} \right)$$

The group $\dfrac{\rho g^{1/2} H^{3/2}}{\mu}$ has the form of the Reynolds number Re and the

group $\dfrac{\rho g H^2}{\sigma}$ is the square of the Weber number, We.

Hence $C_d = f(\text{Re}, \text{We})$

Surface tension effects, represented by the Weber number may become significant at low discharges.

Example 9.4
Derive an expression for the discharge per unit crest length of a rectangular weir over which a fluid of density ρ, dynamic viscosity μ, is flowing with a head H.

The crest height is P. By comparison with the discharge equation obtained from energy considerations,

$$q = \frac{2}{3} \sqrt{2g} \ C_d H^{3/2}$$

state the parameters on which the discharge coefficient depends for a given crest profile.

Solution:

$$f_1(q, g, H, \rho, \mu, \sigma, P) = 0$$

$$f_2(\pi_1, \pi_2, \pi_3, \pi_4) = 0$$

With ρ, g and H as the repeating variables,

$$\pi_1 = \rho^\alpha g^\beta H^\gamma q$$

$$\pi_1 = \frac{q}{g^{1/2} H^{3/2}}$$

$$\pi_2 = \rho^\alpha g^\beta H^\gamma \mu$$

$$\pi_2 = \frac{\mu}{\rho g^{1/2} H^{3/2}}$$

$$\pi_3 = \rho^\alpha g^\beta H^\gamma \sigma$$

$$\pi_3 = \frac{\sigma}{\rho g H^2}$$

$$\pi_4 = \rho^\alpha g^\beta H^\gamma P$$

$$\pi_4 = P/H$$

$$\therefore f_2 \left(\frac{q}{g^{1/2} H^{3/2}}, \frac{\mu}{\rho g^{1/2} H^{3/2}}, \frac{\sigma}{\rho g H^2}, \frac{P}{H} \right) = 0$$

$$\text{or } q = g^{1/2} H^{3/2} \, \phi \left[\frac{\rho g^{1/2} H^{3/2}}{\mu}, \frac{\rho g H^2}{\sigma}, \frac{P}{H} \right] = 0$$

Hence $C_d = f[Re, We, P/H]$.

In addition, of course, the discharge coefficient will depend on the crest profile, and the influence of this factor together with that of the non-dimensional groups in the above expression can only be found from experiments.

Example 9.5
A spun iron pipeline 300 mm in diameter and 0·3 mm effective roughness is to be used to convey oil of kinematic viscosity $7·0 \times 10^{-5}$ m²/s at a rate of 80 l/s. Laboratory tests on a 30 mm pipeline conveying water at 20°C ($v = 1·0 \times 10^{-6}$ m²/s) are to be carried out to predict the hydraulic gradient in the oil pipeline.

Determine the effective roughness of the 30 mm pipe, the water discharge to be used and the hydraulic gradient in the oil pipeline at the design discharge.

Solution:
It was shown in Example 9.1 that the hydraulic gradient in a pressure pipeline is expressed by

$$S_f = \frac{h_f}{L} = \frac{V^2}{gD} \phi \left[\frac{k}{D}, Re \right] \qquad \text{(i)}$$

For geometrical similarity the relative roughness $\frac{k}{D}$ must be the same in both systems.

$$\left(\frac{k}{D} \right) oil = \frac{0 \cdot 3}{300} = 0 \cdot 001$$

$\therefore$ roughness of water pipe $= 0 \cdot 001 \times 30 = 0 \cdot 03$ mm.

(A uPVC pipeline with chemically cemented joints could be used.)
For dynamic similarity the Reynolds numbers must be the same.

$$\left(\frac{VD}{v} \right)_{water} = \left(\frac{VD}{v} \right)_{oil}$$

$$V_{oil} = 1 \cdot 132 \text{ m/s}; \quad Re_{(oil)} = \frac{1 \cdot 132 \times 0 \cdot 3}{7 \cdot 0 \times 10^{-5}} = 4851 \cdot 0$$

$$\therefore V_{water} = \frac{4851 \cdot 0 \times 1 \cdot 0 \times 10^{-6}}{0 \cdot 03} = 0 \cdot 1617 \text{ m/s}$$

The velocity $0 \cdot 1617$ m/s for water is called the 'corresponding speed' for dynamic similarity.

$\therefore$ Water discharge $= 0 \cdot 114$ l/s.

From (i) $\dfrac{S_{f(oil)}}{S_{f(water)}} = \dfrac{(V^2/gD)_{oil}}{(V^2/gD)_{water}}$ since $\phi \left[\dfrac{k}{D}, Re \right]$

is the same for the two systems at the corresponding speeds.

$$\therefore S_{f(oil)} = \frac{\left(\dfrac{1 \cdot 132^2}{0 \cdot 3} \right)}{\left(\dfrac{0 \cdot 1617^2}{0 \cdot 03} \right)} \times 0 \cdot 0017 = 0 \cdot 00833.$$

Example 9.6
A V-notch is to be used for monitoring the flow of oil of kinematic viscosity $8 \cdot 0 \times 10^{-6}$ m²/s. Laboratory tests using water at 15°C over a geometrically similar notch were used to predict the calibration of the notch when used for oil flow measurement. At a head of $0 \cdot 15$ m, the water discharge was $12 \cdot 15$ l/s. What is the corresponding head when measuring oil flow and what is the corresponding oil discharge?

Solution:
In Example 9.3 it was shown that the discharge over a V-notch is given by

$$Q = g^{1/2} H^{5/2} \phi \left[\frac{\rho g^{1/2} H^{3/2}}{\mu}, \frac{\rho g H^2}{\sigma}, \theta \right] \qquad \text{(i)}$$

θ is the same for both notches and for dynamic similarity the groups,

$$\frac{\rho g^{1/2}\ H^{3/2}}{\mu} \quad \text{and} \quad \frac{\rho g H^2}{\sigma}$$

should be the same respectively for the water and oil systems

i.e. $(Re)_{oil} = (Re)_{water}$ and $(We)_{oil} = (We)_{water}$.

However it will be realised that these two relationships will lead to two different scaling laws but since the surface tension effect will only become significant in relation to the viscous and gravity forces at very low heads this effect can be neglected. The scaling law is therefore obtained by equality of the Reynolds numbers.

Using subscript 'o' for oil and 'w' for water,

$$\left(\frac{H^{3/2}}{\nu}\right)_o = \left(\frac{H^{3/2}}{\nu}\right)_w$$

whence $H_o = 0.15 \left(\dfrac{8 \times 10^{-6}}{1.13 \times 10^{-6}}\right)^{2/3} = 0.553$ m

Using equation (i), and since the ϕ [ ] terms are the same for both systems,

$$\frac{Q_o}{Q_w} = \left(\frac{H_o}{H_w}\right)^{5/2}$$

whence $Q_o = 12.15 \left(\dfrac{0.553}{0.15}\right)^{5/2}$

$$Q_{oil} = 0.3171 \ \text{m}^3/\text{s} \ (317.1 \ \text{l/s}).$$

Example 9.7
(a) Show that the net force acting on the liquid flowing in an open channel may be expressed as

$$F = \rho V^2\ \ell^2\ \phi\ [Re, Fr, We, k/\ell] \tag{i}$$

where ℓ is a typical length dimension.
(b) A $1:50$ scale model of part of a river is to be constructed to investigate channel improvements. A steady discharge of 420 m³/s was measured in the river at a section where the average width was 105 m and water depth 3·5 m.

Determine the corresponding depth, velocity and discharge to be reproduced in the model. Check that the flow in the model is in the turbulent region and discuss how the boundary resistance in the model could be adjusted to produce geometrical similarity of surface profiles.

Solution:
(a) $f(\rho,\ V,\ \ell,\ g,\ \mu,\ \sigma,\ k) = 0$
With ρ, V and ℓ as the repeating variables dimensional analysis yields

$$F = \rho V^2 \, \ell^2 \, \phi \left[\frac{V\ell\rho}{\mu}, \frac{V^2}{g\ell}, \frac{\rho V^2 \ell}{\sigma}, \frac{k}{\ell} \right]$$

i.e. $F = \rho V^2 \, \ell^2 \, \phi \left[\text{Re, Fr, We,} \frac{k}{\ell} \right]$

(b) For dynamic similarity the non-dimensional groups should be equal in the model and prototype. Surface tension effects will be negligible in the prototype and its effect must be minimised in small-scale models. Although the boundary resistance, as reflected in the Re and k/ℓ terms, and the gravity force are both significant, open channel models are operated according to the Froude law,

$$\left(\frac{V}{\sqrt{gy}} \right)_m = \left(\frac{V}{\sqrt{gy}} \right)_p \tag{i}$$

where the length parameter is the depth, y. Subscript 'm' relates to the model and 'p' to the prototype.

$$\frac{y_m}{y_p} = \frac{1}{50} \text{ whence } y_m = \frac{3\cdot5}{50} = 0\cdot07 \text{ m}$$

$$V_p = Q/A = \frac{420}{105 \times 3\cdot5} = 1\cdot14 \text{ m/s.}$$

From (i),

$$V_m = V_p \sqrt{\frac{y_m}{y_p}} = 1\cdot14 \sqrt{\frac{1}{50}} = 0\cdot161 \text{ m/s}$$

$$\frac{Q_m}{Q_p} = \frac{V_m A_m}{V_p A_p} = \frac{V_m}{V_p} \cdot \frac{(by)_m}{(by)_p}$$

$$= \sqrt{\lambda_y} \, \lambda_x \, \lambda_y = \lambda_y^{3/2} \, \lambda_x$$

where λ_y is the vertical scale

and λ_x is the horizontal scale

Note that the term λ is commonly used to indicate a scaling ratio in model studies. It is not to be confused with the Darcy friction factor.

In this case $\lambda_y = \lambda_x$ (undistorted model) whence

$$\frac{Q_m}{Q_p} = \sqrt{\frac{1}{50}} \left(\frac{1}{50} \right)^2$$

$$Q_m = 420 \times \left(\frac{1}{50} \right)^{5/2} = 0\cdot02376 \text{ m}^3\text{/s}$$

$$= 23\cdot76 \text{ l/s.}$$

The Reynolds number in the model for testing the flow regime is best expressed in terms of the hydraulic radius:

$$b_m = \frac{105}{50} = 2\cdot1 \text{ m} \quad (b_m = \text{average width in model} = b_p \times \lambda_x)$$

Wetted perimeter $P_m = 2\cdot1 + 2 \times 0\cdot07 = 2\cdot24$ m

$$R_m = \frac{2\cdot1 \times 0\cdot07}{2\cdot24} = 0\cdot0656 \text{ m}$$

$$Re_m = \frac{0\cdot161 \times 0\cdot0656}{1 \times 10^{-6}} = 10561.6$$

which indicates a turbulent flow.

The discharge in the model has been determined, in relation to the geometrical scale, to correspond with the correct scaling of the gravity forces. In order to scale correctly the viscous resistance forces in the model the discharge ratio should comply with the Reynolds law with the geometrically relative roughness.

In the Froude scaled model therefore, the resistance forces would be underestimated if the boundary roughness were to be modelled to the geometrical scale and, in practice, therefore roughness elements consisting of concrete blocks, wire mesh or vertical rods, are installed and adjusted until the surface profile in the model when operating at the appropriate scale discharge is geometrically similar to the observed prototype surface profile.

Example 9.8
(a) Explain why distorted scale models of rivers are commonly used.
(b) A river model is constructed to a vertical scale of $1:50$ and a horizontal scale of $1:200$. The model is to be used to investigate a flood alleviation scheme. At the design flood of 450 m³/s the average width and depth of flow are 60 m and 4·2 m respectively. Determine the corresponding discharge in the model and check the Reynolds number of the model flow.

Solution:
(a) The size of a river model is determined by the laboratory space available (although in some cases special buildings are constructed to house a particular model, notably the Eastern Schelde model in the Delft Hydraulics Laboratory, the Netherlands). In the case of a long river reach a natural model scale may result in such small flow depths that depth and water elevations cannot be measured with sufficient accuracy, the flow in the model may become laminar, surface tension effects may become significant and sediment studies may be precluded because of the low tractive force.

To avoid these problems geometrical distortion, wherein the vertical scale λ_y is larger than the horizontal scale λ_x, is used, typical vertical distortions being in the range 5 to 100 with a vertical scale not smaller than $1:100$.

(For more detailed discussions on the construction of river and estuary models and scale effects the reader is referred to the literature in this chapter's Recommended reading list.)

(b) Velocity in prototype $= \dfrac{450}{60 \times 4\cdot2} = 1\cdot786$ m/s.

The Froude scaling law is based on the vertical scale ratio

$$V_m = V_p \sqrt{\lambda_y} = 1\cdot786 \sqrt{\dfrac{1}{50}} = 0\cdot25 \text{ m/s}$$

$$\dfrac{Q_m}{Q_p} = \dfrac{V_m}{V_p} \dfrac{(xy)_m}{(by)} = \lambda_y^{3/2}\, \lambda_x$$

$$= \left(\dfrac{1}{50}\right)^{3/2} \times \dfrac{1}{200} = 1\cdot414 \times 10^{-5}$$

$$\therefore Q_m = 450 \times 1\cdot414 \times 10^{-5} \text{ m}^3/\text{s}$$

$$= 6\cdot36 \text{ l/s}$$

Average dimensions of model channel:

Width $= \dfrac{60}{200} = 0\cdot30$ m

Depth $= \dfrac{4\cdot2}{50} = 0\cdot084$ m

Hydraulic radius $\simeq \dfrac{0\cdot3 \times 0\cdot084}{0\cdot468} = 0\cdot0538$

Reynolds number $= \dfrac{VR}{\nu} = \dfrac{0\cdot25 \times 0\cdot0538}{1 \times 10^{-6}} = 13450$

The flow therefore will be turbulent.

Example 9.9
An estuary model is built to a horizontal scale of $1:500$ and vertical scale $1:50$. Tidal oscillations of amplitude $5\cdot5$ m and tidal period $12\cdot4$ h are to be reproduced in the model. What are the corresponding tidal characteristics in the model?

Solution:
The speed of propagation, or celerity, of a gravity wave in which the wavelength is very large in relation to the water depth, y, as in the case of tidal oscillations is given by $c = \sqrt{gy}$. Thus estuary models must be operated according to the Froude law.

The tidal range is modelled according to the vertical scale.

$$\therefore H_m = H_p \times \dfrac{1}{50} = \dfrac{5\cdot5}{50} = 0\cdot11 \text{ m.}$$

Tidal period, $T = \dfrac{L}{c}$ where L is the wavelength. Hence

$$\frac{T_m}{T_p} = \frac{L_m}{L_p} \cdot \frac{c_p}{c_m} = \lambda_x \cdot \sqrt{\frac{y_p}{y_m}} = \frac{\lambda_x}{\sqrt{\lambda_y}}$$

$$\therefore T_m = \frac{12 \cdot 4 \times \dfrac{1}{500}}{\sqrt{\dfrac{1}{50}}} = 0 \cdot 1754 \text{ h}$$

$$= 10 \cdot 52 \text{ minutes.}$$

Example 9.10
The discharge, Q, from a rotodynamic pump developing a total head, H, when running at, N, rev/min is given by

$$Q = ND^3 \, \phi \left[\frac{D}{B}, \frac{N^2 D^2}{gH}, \frac{\rho ND^2}{\mu} \right] \quad \text{(see Example 9.2)} \tag{i}$$

(a) Obtain an expression for the specific speed of a rotodynamic pump and show how to predict the pump characteristics when running at different speeds.
(b) The performance of a new design of rotodynamic pump is to be tested in a $1:5$ scale model. The pump is to run at 1450 rev/min.

The model delivers a discharge of 2·5 l/s of water and a total head of 3 m, with an efficiency of 65 per cent when operating at 2000 rev/min. What is the corresponding discharge head and power consumption in the prototype? Determine the specific speed of the pump and hence state the type of impeller.

Solution:
(a) For geometrically similar machines operating at high Reynolds numbers the term $\dfrac{\rho ND^2}{\mu}$ becomes unimportant and equation (i) may be rewritten:

$$f\left(\frac{Q}{ND^3}, \frac{D}{B}, \frac{N^2 D^2}{gH} \right) = 0$$

The term $\dfrac{D}{B}$ is automatically satisfied by the geometrical similarity and the terms $\dfrac{Q}{ND^3}$ and $\dfrac{N^2 D^2}{gH}$ should have identical values in the model and prototype.

$$\therefore \left(\frac{Q}{ND^3} \right)_m = \left(\frac{Q}{ND^3} \right)_p \tag{ii}$$

$$\text{and} \left(\frac{N^2 D^2}{gH} \right)_m = \left(\frac{N^2 D^2}{gH} \right)_p \tag{iii}$$

The scale

$$\frac{D_m}{D_p} = \frac{N_p}{N_m} \left(\frac{H_m}{H_p}\right)^{1/2} \quad \text{from (iii)}$$

whence, from (ii)

$$\frac{N_m}{N_p} = \sqrt{\frac{Q_p}{Q_m}} \left(\frac{H_m}{H_p}\right)^{3/4} \tag{iv}$$

If H_m and Q_m are made equal to unity then (iv) becomes:

$$N_m = \frac{N_p \sqrt{Q_p}}{H_p^{3/4}} \tag{v}$$

The term $\dfrac{N \sqrt{Q}}{H^{3/4}}$ is called the 'Specific Speed' and is interpreted as the speed at which a geometrically scaled model would run in order to deliver unit discharge when generating unit head. All geometrically similar machines have the same specific speed.

In the case of the same pump running at different speeds, $D_m = D_p$ and (ii) becomes

$$\frac{Q_1}{N_1} = \frac{Q_2}{N_2} \text{ or } Q_2 = Q_1 \left(\frac{N_2}{N_1}\right)$$

and (iii) becomes

$$\frac{N_1}{H_1^2} = \frac{N_2}{H_2^2} \text{ or } H_2 = H_1 \sqrt{\frac{N_2}{N_1}}$$

(b) From (ii)

$$Q_p = Q_m \frac{(ND^3)_p}{(ND^3)_m}$$

$$\therefore Q_p = 2\cdot5 \times 5^3 \times \frac{1450}{2000} = 226 \text{ l/s}$$

From (iii)

$$H_p = H_m \left(\frac{N_p}{N_m} \frac{D_p}{D_m}\right)^2$$

$$= 3\cdot0 \left(\frac{1450}{2000} \times 5\right)^2 = 39\cdot42 \text{ m}$$

$$\text{Power input required} = \frac{9\cdot81 \times 0\cdot226 \times 39\cdot42}{0\cdot65} = 134\cdot5 \text{ kW}$$

$$\text{Specific speed } N_s = \frac{N \sqrt{Q}}{H^{3/4}} = \frac{1450 \sqrt{226}}{(39\cdot42)^{3/4}} = 1385$$

The impeller is of the centrifugal type (see Chapter 6).

Example 9.11
A model of a proposed dam spillway was constructed to a scale of 1 : 25. The design flood discharge over the spillway is 1000 m³/s. What discharge should be provided in the model? What is the velocity in the prototype corresponding with a velocity of 1·5 m/s in the model at the corresponding point?

Solution:
Example 9.4 showed that the discharge per unit crest length of a rectangular weir could be expressed as

$$q = g^{1/2} H^{3/2} \phi \left[\frac{\rho g^{1/2} H^{3/2}}{\mu}, \frac{\rho g H^2}{\sigma}, \frac{P}{H} \right] \tag{i}$$

The governing equation for spillways and weirs is identical. Flood discharges over spillways will result in very high Reynolds numbers and since surface tension effects are also negligible the only factor affecting the discharge coefficient is P/H. In modelling dam spillways, therefore, if the ratios of P/H in the model and prototype are identical and the crest geometry is correctly scaled:

$$\frac{q}{g^{1/2} H^{3/2}} = \text{constant}$$

i.e. $F_{r,m} = F_{r,p}$

Therefore spillway models are operated according to the Froude law and are made sufficiently large that viscous and surface tension effects are negligible.

$$\left(\frac{V}{\sqrt{g\ell}} \right)_m = \left(\frac{V}{\sqrt{g\ell}} \right)_p \tag{ii}$$

$$Q = V\ell^2$$

$$\therefore \left(\frac{Q}{\ell^{5/2}} \right)_m = \left(\frac{Q}{\ell^{5/2}} \right)_p \tag{iii}$$

whence

$$Q_m = 1000 \left(\frac{1}{25} \right)^{5/2} = 0.32 \text{ m}^3/\text{s}$$

From (ii)

$$V_p = V_m \sqrt{\frac{\ell_p}{\ell_m}} = V_m \sqrt{25}$$

$$\therefore V_p = 1.5 \times 5 = 7.5 \text{ m/s}.$$

Recommended reading

1. Allen, J. (1952) *Scale Models in Hydraulic Engineering*. London: Longman.
2. Novak, P and Cǎbélka, J. (1981) *Models in Hydraulic Engineering*. London: Pitman.
3. Webber, N.B. (1971) *Fluid Mechanics for Civil Engineers*. London: Spon.
4. Yalin, M.S. (1971) *Theory of Hydraulic Models*. London: Macmillan.

Problems

1. The head loss of water of kinematic viscosity 1×10^{-6} m²/s in a 50 mm diameter pipeline was 0·25 m over a length of 10·0 m at a discharge of 2·0 l/s. What is the corresponding discharge and hydraulic gradient when oil of kinematic viscosity $8·5 \times 10^{-6}$ flows through a 250 mm diameter pipeline of the same relative roughness?

2. Find the pressure drop at the corresponding speed in a pipe 25 mm in diameter, 30 m long conveying water at 10°C if the pressure head loss in a 200 mm diameter smooth pipe 300 m long in which air is flowing at a velocity of 3 m/s is 10 mm of water. Density of air = 1·3 kg/m³, dynamic viscosity = $1·77 \times 10^{-5}$ Ns/m². Dynamic viscosity of water = $1·3 \times 10^{-3}$ Ns/m².

3. A 50 mm diameter pipe is used to convey air at 4°C (density = 1·12 kg/m³ and dynamic viscosity $1·815 \times 10^{-5}$ Ns/m²) at a mean velocity of 20 m/s.
    Calculate the discharge of water at 20°C for dynamic similarity and obtain the ratio of the pressure drop per unit length in the two cases.

4. If, in modelling a physical system, the Reynolds and Froude numbers are to be the same in the model and prototype determine the ratio of kinematic viscosity of the fluid in the model to that in the prototype.

5. The sequent depth, y_s, of a hydraulic jump in a rectangular channel is related to the initial depth y_i, the discharge per unit width, q, g and p. Express the ratio y_s/y_i in terms of a non-dimensional group and compare with the equation developed from momentum principles:

$$y_s = \frac{y_i}{2} (\sqrt{1 + F_i^2} - 1).$$

6. A 60° V-notch is to be used for measuring the discharge of oil having a kinematic viscosity 10 times that of water. The notch was calibrated using water. When the head over the notch was 0·1 m the discharge was 2·54 l/s.

Determine the corresponding head and discharge when the notch is used for oil flow measurement.

7.   The airflow and wind effects on a bridge structure are to be studied on a 1:25 scale model in a pressurised wind tunnel in which the air density is 8 times that of air at atmospheric pressure and at the same temperature. If the bridge structure is subjected to wind speeds of 30 m/s what is the corresponding wind speed in the wind tunnel? What force on the prototype corresponds with a 1400 N force on the model? (Note the dynamic viscosity of air is unaffected by pressure changes provided the temperature remains constant.)

8.   A rotodynamic pump is designed to operate at 1450 rev/min and to develop a total head of 60 m when discharging 250 l/s.
     The following characteristics of a 1:4 scale model were obtained from tests carried out at 1800 rev/min.
     Obtain the corresponding characteristics of the prototype and state whether, or not, it meets its design requirements.

Q_m (l/s)	0	2	4	6	8
H_m (m)	8	7·6	6·4	4·2	1·0

9. (a) Show that the power output, P, of a hydraulic turbine expressed in terms of non-dimensional groups in the form

$$P = \rho N^3 D^5 \, \phi \left[\frac{Q}{ND^3}, \frac{D}{B}, \frac{N^2 D^2}{gH}, \frac{\rho N D^2}{\mu} \right]$$

Derive an expression for the specific speed of a hydraulic turbine.
   (b) A 1:20 scale model of a hydraulic turbine operates under a constant head of 10 m. the prototype will operate under a head of 150 m at a speed of 300 rev/min. When running at the corresponding speed the model generates 1·2 kW at a discharge of 13·6 l/s. Determine the corresponding speed, power output and discharge of the prototype.

10.   The wave action and forces on a proposed sea wall are to be studied on a 1:10 scale model. The design wave has a period of 9 seconds and a height from crest to trough of 5 m. The depth of water in front of the wall is 7 m.
     Assuming that the wave is a gravity wave in shallow water and that the celerity $c = \sqrt{gy}$ where y is the water depth determine the wave period, wavelength and wave height to be reproduced in the model. If a force of 4 kN due to wave breaking on a 0·5 m length of the model sea wall were recorded, what would be the corresponding force per unit length on the prototype?

Chapter 10
Ideal Fluid Flow and Curvilinear Flow

R.E. Featherstone

10.1 Ideal fluid flow

The analysis of ideal fluid flow is also referred to as 'potential flow'. The concept of an ideal fluid is that of one which is inviscid and incompressible; the flow is also assumed to be irrotational. Since flow in boundary layers is rotational and strongly influenced by viscosity, the analytical techniques of ideal fluid flow cannot be applied in such circumstances. However in many situations the flow of real fluids outside the boundary layer, where viscous effects are small, approximates closely to that of an ideal fluid.

The object of the study of ideal fluid flow is to obtain the flow pattern and pressure distribution in the fluid flow around prescribed boundaries. Examples are the flow over airfoils, through the passages of pump and turbine blades, over dam spillways and under control gates. The governing differential equations of ideal fluid flow have also been successfully applied to oscillatory wave motions, groundwater and seepage flows.

10.2 Streamlines, the stream function

A streamline is a continuous line drawn through the fluid such that it is tangential to the velocity vector at every point. In steady flow the streamlines are identical with the 'pathlines' or tracks of discrete liquid elements. No flow can occur across a streamline and the concept of a 'streamtube' in two-dimensional flow emerges as the flow per unit depth between adjacent streamlines.

From fig. 10.1 (a) $\dfrac{dy}{dx} = \dfrac{v}{u}$; $u\,dy - v\,dx = 0$ (10.1)

The continuity equation for two-dimensional steady ideal fluid flow is

$$\frac{\partial u}{\partial x} + \frac{\partial v}{\partial y} = 0$$ (10.2)

Equation (10.2) is satisfied by the introduction of a stream function denoted by ψ such that

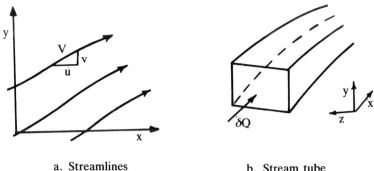

a. Streamlines   b. Stream tube

Figure 10.1   Two-dimensional ideal fluid flow

$$u = \frac{\partial \psi}{\partial y} \text{ and } v = -\frac{\partial \psi}{\partial x} \text{ whence equation (10.2) becomes}$$

$$\frac{\partial^2 \psi}{\partial y \partial x} - \frac{\partial^2 \psi}{\partial x \partial y} \ (= 0)$$

Substitution of u and v in equation (10.1) yields

$$\frac{\partial \psi}{\partial y} \ dy + \frac{\partial \psi}{\partial x} \ dx = 0$$

Now mathematically $d\psi = \dfrac{\partial \psi}{\partial y} \ dy + \dfrac{\partial \psi}{\partial x} \ dx$

whence $\dfrac{d\psi}{ds} = 0$ where s is the direction along a streamline. Thus $\psi =$ constant along a streamline and the pattern of streamlines is obtained by equating the stream function to a series of numerical constants.

Since $\delta Q = Vb$, where b is the spacing of adjacent streamlines and δQ the discharge per unit depth between the streamlines, the velocity vector, V, is universely proportional to the streamline spacing.

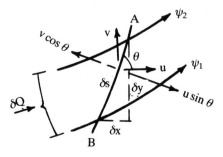

Figure 10.2   Adjacent streamlines

In polar co-ordinates the radial and tangential velocity components, v_r and v_θ respectively are expressed by

$$v_r = \frac{1}{r}\frac{\partial \psi}{\partial \theta}; \quad v_\theta = -\frac{\partial \psi}{\partial r} \tag{10.3}$$

10.3 Relationship between discharge and stream function

Let δQ be the discharge per unit depth between adjacent streamlines (fig. 10.2).

$$\delta Q = u \sin \theta \; \delta s - v \cos \theta \; \delta s$$

$$= u \; \delta y - v \; \delta x$$

$$= \frac{\partial \psi}{\partial y}\; \delta y + \frac{\partial \psi}{\partial x}\; \delta x$$

Now $\delta\psi = \dfrac{\partial \psi}{\partial y}\; \delta y + \dfrac{\partial \psi}{\partial x}\; \delta x;$ whence $\delta Q = \delta\psi = \psi_2 - \psi_1$ $\tag{10.4}$

10.4 Circulation and the velocity potential function

Circulation is the line integral of the tangential velocity around a closed contour, expressed by

$$K = \int v_s \; ds \tag{10.5}$$

Velocity potential is the line integral along the s direction between two points, see fig. 10.3.

$$\phi_A - \phi_B = \int_A^B V \sin \alpha \; ds \tag{10.6}$$

If $\alpha = 0$, $\phi_A - \phi_B = 0$, thus the potential ϕ along AB is constant.

Note that $\dfrac{\partial \phi}{\partial s} = V \sin \alpha$, i.e. $\dfrac{\partial \phi}{\partial s} = v_s$ $\tag{10.7}$

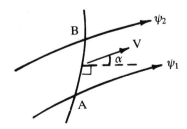

Figure 10.3   Velocity potential

Lines of constant velocity potential are orthogonal to the streamlines and the set of equipotential lines and the set of streamlines form a system of curvilinear squares described as a flow net. (See fig. 10.4.)

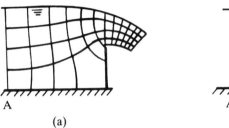

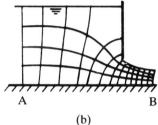

(a)   (b)

Figure 10.4   Flow nets

10.5 Stream functions for basic flow patterns

(a) Uniform rectilinear flow in x-direction (fig. 10.5 (a))

$$u = \frac{\partial \psi}{\partial y}; \quad \psi = uy + f(x) \tag{i}$$

$$v = -\frac{\partial \psi}{\partial x}; \quad \psi = -vx + f(y) \tag{ii}$$

Since $v = 0$ and equations (i) and (ii) are identical

$$f(x) = 0 \quad \text{and} \quad \psi = uy \tag{10.8}$$

(b) Uniform rectilinear flow in y-direction (fig. 10.5 (b))

Similarly since $u = 0$, $\psi = -vx$ \hfill (10.9)

(c) Line source (fig. 10.5 (c))
A line source provides an axi-symmetric radial flow.
Using polar co-ordinates:

$$v_r = \frac{1}{r}\frac{\partial \psi}{\partial \theta}; \quad \psi = rv_r \,\theta + f(r)$$

$$v_\theta = -\frac{\partial \psi}{\partial r}; \quad \psi = -rv_\theta + f(\theta)$$

$v_\theta = 0$, whence $\psi = rv_r \,\theta$

$v_r = \dfrac{q}{2\pi r}$ where q is the 'strength' of the source = discharge per unit depth.

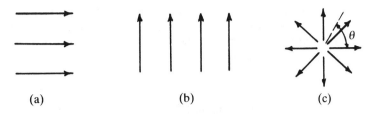

Figure 10.5 Rectilinear flows in x and y directions and line source

$$\therefore \; \psi = \frac{q\theta}{2\pi} \tag{10.10}$$

Similarly a line 'sink' which is a negative source is defined by

$$\psi = -\frac{q\theta}{2\pi} \tag{10.11}$$

10.6 Combinations of basic flow patterns

(a) Source in uniform flow (see fig. 10.6).

The stream function of the resultant flow pattern is obtained by addition of the component stream functions.

Thus $\psi = uy + \dfrac{q\theta}{2\pi}$ (10.12)

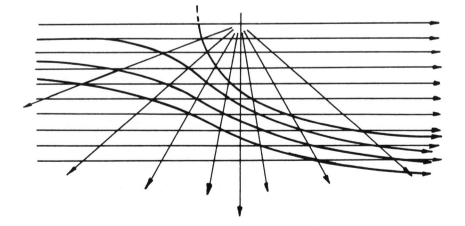

Figure 10.6 Combinations of basic flow patterns

The streamlines are obtained by solving equation (10.12) for a number of values of ψ. Alternatively superimpose the streamlines for the individual flow patterns and algebraically add the values of the stream function where they intersect. Obtain the new streamlines by drawing lines through points having the same value of stream function.

10.7 Pressure at points in the flow field

The pressure, p, at any point (r, θ) in the flow field is obtained from application of the Bernoulli equation:

$$\frac{p_o}{\rho g} + \frac{V_o^2}{2g} = \frac{p}{\rho g} + \frac{V^2}{2g} \text{ where } p_o \text{ and } V_o \text{ are the pressure and velocity}$$

vector in the undisturbed uniform flow and V is the velocity vector at (r, θ). V is obtained from the orthogonal spacing of the streamlines at (r, θ) or analytically from $V = \sqrt{V_r^2 + V_\theta^2}$.

10.8 The use of flow nets and numerical methods

The analytical methods, as described in section 10.6 and illustrated in Example 10.1, can be applied to other combinations of basic flow patterns to simulate, for example, the flow round a cylinder, flow round a corner and vortex flow. The reader is referred to text on fluid mechanics for a more detailed treatment of these applications.

In civil engineering hydraulics however the flows are generally constrained by non-continuous, or complex boundaries. A typical example is the flow under a sluice gate (fig. 10.4 (b)) and such cases are incapable of solution by analytical techniques.

(a) The use of flow nets
One method of solution in such cases is the use of the flow net described in section 10.4. Selecting a suitable number of stream tubes streamlines 'ψ' are drawn starting from equally spaced points where uniform rectilinear flow exists such as section A (fig. 10.4 (a)) and A and B (fig. 10.4 (b)). A system of equipotential lines, 'ϕ', is now added such that they intersect the 'ψ' lines orthogonally. If the streamlines have been drawn correctly to suit the boundary conditions the resulting flow net will correctly form a system of curvilinear 'squares'. As a final test, circles drawn in each 'square' should be tangential to all sides. On the first trial the test will probably fail and successive adjustments are made to the 'ψ' and 'ϕ' lines until the correct pattern is produced.

Local velocities are obtained from the streamline spacings (or the spacing between the equipotential lines since $\Delta \psi$ and $\Delta \phi$ are locally equal) in relation to the rectilinear flow velocities and hence local pressures are calculated from the Bernoulli equation (Example 10.1). The technique has also been widely used in seepage flow problems under water retaining structures.

(b) Numerical methods

Where computer facilities are available the streamline pattern can be obtained for complex boundary problems. With computer graphics a plot of the streamlines can be produced in addition. However the problems can be solved using an electronic calculator. The method involves the solution of the Laplace equation using finite difference methods.

Theory: The assumption of irrotational flow when applied to a liquid element yields

$$\frac{\partial u}{\partial x} - \frac{\partial v}{\partial y} = 0 \tag{10.13}$$

i.e.
$$\frac{\partial^2 \psi}{\partial x^2} + \frac{\partial^2 \psi}{\partial y^2} = 0 \quad \text{or} \quad \nabla^2 \psi = 0 \tag{10.14}$$

Equation (10.14) is known as the Laplace equation.

Since we have seen that the stream function has numerical values at all points over the flow field the streamline pattern can be produced if equation (10.14) can be solved in ψ at discrete points in the field.

Superimpose a square or rectangular mesh of straight lines in the x- and y-directions to generally fit the boundaries of the physical system. (See fig. 10.7.)

Equation (10.14) is to be solved at points such as (x, y) where the grid lines intersect and must first be expressed in finite difference form.

Using the notation

$$f'(x) = \frac{\partial f(x)}{\partial x}; \quad f''(x) = \frac{\partial^2 f(x)}{\partial x^2}, \text{ etc,}$$

Taylor's theorem gives

$$f(x + \Delta x, y) = f(x, y) + \Delta x f'(x, y) + \frac{\Delta x^2}{2!} f''(x, y)$$

$$+ \frac{\Delta x^3}{3!} f'''(x, y) + \frac{\Delta x^4}{4!} f^{iv}(x, y) + \dots$$

and
$$f(x - \Delta x, y) = f(x, y) - \Delta x f'(x, y) + \frac{\Delta x^2}{2!} f''(x, y) - \frac{\Delta x^3}{3!} f'''(x, y)$$

$$+ \frac{\Delta x^4}{4!} f^{iv}(x, y) + \dots$$

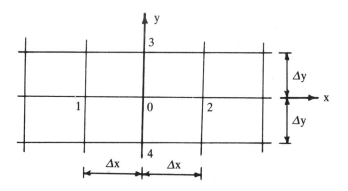

Figure 10.7 Rectangular mesh – numerical method

Adding: $f(x + \Delta x, y) + f(x - \Delta x, y) = 2f(x, y) + \Delta x^2 f''(x, y)$, neglecting Δx^4 and higher order terms.
Thus

$$f''(x, y) = \frac{[f(x + \Delta x, y) - 2f(x, y) + f(x - \Delta x, y)]}{\Delta x^2}$$

Since $\psi = f(x, y)$

$$\frac{\partial^2 \psi}{\partial x^2} = \frac{[\psi(x + \Delta x, y) - 2\psi(x, y) + \psi(x - \Delta x, y)]}{\Delta x^2}$$

Similarly

$$\frac{\partial^2 \psi}{\partial y^2} = \frac{[\psi(x, y + \Delta y) - 2\psi(x, y) + \psi(x, y - \Delta y)]}{\Delta y^2}$$

Using the grid notation of fig. 10.7 for simplicity equation (10.1) becomes

$$\frac{(\psi_1 - 2\psi_0 + \psi_2)}{\Delta x^2} + \frac{(\psi_3 - 2\psi_0 + \psi_4)}{\Delta y^2} = 0 \tag{10.15}$$

(where the point x, y is located at point 0 and point $x + \Delta x$, y is at point 2 etc.).
Or if $\Delta x = \Delta y$,

$$\psi_1 + \psi_2 + \psi_3 + \psi_4 - 4\psi_0 = 0 \tag{10.16}$$

Method of solution
The method of 'relaxation' originally devised by Professor Southwell for the numerical solution of elliptic partial differential equations such as the Laplace, Poisson and Biharmonic equations describing fluid flow and stress distributions in solid bodies is not amenable to automatic computation and is not to be confused with the methods described here.

Boundary conditions
The solid boundaries and free water surfaces are streamlines and therefore have constant values of stream function. The allocation of values to the boundaries can be quite arbitrary: for example in fig. 10.4 the bed can be allocated a value of $\psi = 0$ and the surface, $\psi = 100$.

Where grid points do not coincide with a boundary the following form of equation (10.17) for example, is obtained assuming a linear variation of ψ along the grid lines (fig. 10.8).

$$\psi_o = \frac{\psi_1 + \left(\dfrac{\psi_2}{\lambda_2}\right) + \psi_3 + \psi_4}{3 + \left(\dfrac{1}{\lambda_2}\right)} \qquad (10.17)$$

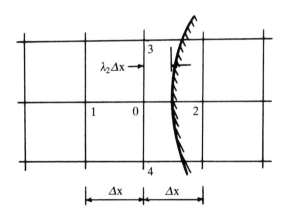

Figure 10.8 Grid points and boundary

(i) Matrix method. Equations (10.15), (10.16) or (10.17) for the interior grid points together with the boundary conditions can be globally expressed in the form

$$[A]\,[\psi] = [B]$$

Thus ψ_1, ψ_2 etc. can be found directly using Gaussian elimination on a computer.
(ii) Method of successive corrections. This method is also amenable to computer solution and is probably quicker than (i) above; it can also be executed using an electronic calculator. Values of ψ are allocated to the boundaries and, in relation to these, estimated values are given to the interior grid points. Considering each interior grid point in turn the allocated value of ψ_o is revised using equations (10.15), (10.16) or (10.17) as appropriate. This procedure is repeated at all grid points as many times as necessary until the differences between the previous and revised values of ψ at EVERY

grid point are less than a prescribed limit. The discrete streamlines are then drawn by interpolation between the grid values of ψ.

10.9 Curvilinear flow of real fluids

The concepts of ideal fluid flow can be used to obtain the velocity and pressure distributions in curvilinear flow of real fluids in ducts and open channels. Curvilinear flow is also referred to as 'Vortex Motion'.

Curvilinear flow is not to be confused with 'rotational' flow; 'rotation' relates to the net rotation of an element about its axis.

Theory: Consider an element of a fluid subjected to curvilinear motion. (See fig. 10.9.)

$$\text{Radial acceleration} = \frac{v_\theta^2}{r}$$

Equating radial forces,

$$dp\,r\,d\theta\,dy = \rho r\,d\theta\,dr\,dy\,\frac{v_\theta^2}{r}$$

whence

$$\frac{dp}{dr} = \frac{\rho\,v_\theta^2}{r}$$

or in terms of pressure head

$$\frac{dh}{dr} = \frac{v_\theta^2}{gr} \tag{10.18}$$

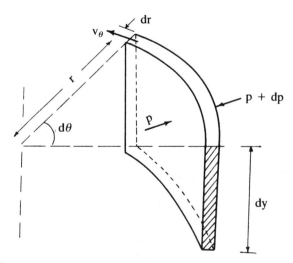

Figure 10.9   Curvilinear flow

10.10 Free and forced vortices

(a) Free vortex motion occurs when there is no external addition of energy; examples occur in bends in ducts and channels, and in the 'sink' vortex. The energy at a point where the velocity is v_θ and pressure head is h is

$$E = h + \frac{v_\theta^2}{2g} \tag{10.19}$$

E is constant across the radius, hence

$$\frac{dh}{dr} + \frac{v_\theta}{g}\frac{d\,(v_\theta)}{dr} = 0$$

From (10.19)

$$\frac{v_\theta^2}{gr} + \frac{v_\theta}{g}\frac{d\,(v_\theta)}{dr} = 0$$

whence

$$\frac{dv_\theta}{v_\theta} = \frac{dr}{r} \quad \text{or } \log_e (v_\theta r) = \text{constant}$$

hence

$$v_\theta r = \text{constant} = K \text{ (circulation)} \tag{10.20}$$

(b) Forced vortex motion is caused by rotating impellers or by rotating a vessel containing a liquid. The equilibrium state is equivalent to the rotation of a solid body where $v_\theta = r\omega$ where ω is the angular velocity (rad/s).

Worked examples

Example 10.1
A line source of strength 180 l/s is placed in a uniform flow of velocity 0·1 m/s.
(a) Plot the streamlines above the x axis.
(b) Obtain the pressure distribution on the streamline denoted by $\psi = 0$.

Solution (a):
It is convenient to consider the uniform flow to be in the-x-direction. This results in a streamline having the value $\psi = 0$ which may be interpreted as the boundary of a solid body.

$$\psi = uy + \frac{q\theta}{2\pi} \quad \text{(equation (10.12))}$$

or

$$\psi = ur \sin \theta + \frac{q\theta}{2\pi}$$

Setting ψ successively to 0, -0.01, -0.02 ... and noting that $u = -0.1$ the co-ordinates (r, θ) along the streamlines can be obtained from equation (10.12).

$\theta°$ \\ ψ	Values of r (m)								
	10	20	40	60	80	100	120	140	160
0	0·29	0·29	0·31	0·35	0·41	0·51	0·69	1·09	2·34
−0·01	0·86	0·58	0·47	0·46	0·51	0·61	0·81	1·24	2·63
−0·02	1·44	0·87	0·62	0·58	0·61	0·71	0·92	1·40	2·92
−0·03	2·02	1·17	0·78	0·69	0·71	0·81	1·04	1·56	3·21

The pairs of co-ordinates, θ and r are plotted to give the streamline pattern. The graphical method gives identical results and is clearly quicker; however the mathematical approach is appropriate in computer applications where computer graph plotting facilities are available. (See fig. 10.10.)

The graphical construction proceeds as follows: the uniform flow is represented by a series of equidistant parallel straight lines (defined by ψ_u in fig. 10.10). Discharge between the streamlines was chosen to be 0·010 m³/s.m in the analytical solution; thus the spacing of the uniform flow streamlines in the diagram represents to scale a distance of 0·10 m. It is convenient to choose the streamlines of the source such that the discharge between them is also 0·010 m³/s.m; the angle $\Delta\theta$ is simply obtained since

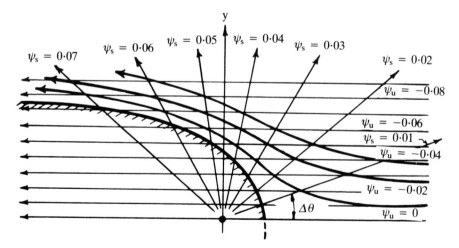

Figure 10.10   Flow net using polar coordinates

$$\Delta \psi_s = \frac{q \Delta \theta}{2\pi}$$

i.e. $0\cdot010 = \dfrac{0\cdot180 \times \Delta\theta}{2\pi}$

whence $\Delta\theta = 20°$

At the points of intersection of the uniform flow and source stream lines the individual stream functions are added algebraically and the resultant flow pattern is obtained by sketching the streamlines through points of equal resultant stream function. The construction and final flow pattern are illustrated in fig. 10.10.

Solution (b):
The pressure distribution in the disturbed flow field at r, θ is obtained from

$$\frac{p_{r,\theta} - p_o}{\rho g} = \frac{V_o{}^2 - V_{r,\theta}^2}{2g} \tag{i}$$

where V_o is the undisturbed velocity and $V_{r,\theta}$ the velocity vector at r, θ.

$$V_{r,\theta} = \sqrt{v_r{}^2 + v_\theta{}^2}$$

$$v_r = \frac{1}{r}\frac{\partial\psi}{\partial\theta} \quad \text{(equation (10.3))}$$

$$= +\frac{ur \cos\theta}{r} + \frac{q}{2\pi r}$$

$$v_\theta = -\frac{\partial\psi}{\partial r} \quad \text{(equation (10.3))}$$

$$= -u \sin\theta$$

$$\therefore\ V_{r,\theta}^2 = u^2 + 2u \cos\theta\frac{q}{2\pi r} + \left(\frac{q}{2\pi r}\right)^2 \tag{ii}$$

Around the 'body', $(\psi = 0)$, $ur \sin\theta + \dfrac{q\theta}{2\pi} = 0$

i.e. $\dfrac{q}{2\pi} = -\dfrac{ur \sin\theta}{\theta}$

Thus $V_{r,\theta}^2 = u^2\left(1 - \dfrac{\sin 2\theta}{\theta} + \dfrac{\sin^2\theta}{\theta^2}\right)$

whence, from (i), since $V_o = u$ in this case,

$$\frac{p_{r,\theta} - p_o}{\rho g} = \frac{u^2}{2g\theta}\left(\sin 2\theta - \frac{\sin^2\theta}{\theta}\right) \tag{iii}$$

Substitution of u = −0·1 m/s into (iii) yields the pressure distribution for a range of θ.

Example 10.2
A discharge of 7 m³/s per metre width flows in a rectangular channel. A vertical sluice gate situated in the channel has an opening of 1·5 m. $C_c = 0.62$, $C_v = 0.95$. Assuming that downstream conditions permit free flow under the gate draw the streamline pattern.

Solution: (See fig. 10.11.)

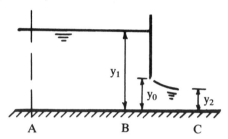

A   B   C

Figure 10.11   Flow under gate

$$y_2 = C_c \times y_o = 0.62 \times 1.5 = 0.93 \text{ m}$$

$$y_1 + \frac{V_1^2}{2g} = y_2 + \frac{V_2^2}{2g} + h_L \qquad\qquad (i)$$

$$\text{and } h_L = \left(\left(\frac{1}{C_v}\right)^2 - 1\right)\frac{V_2^2}{2g}; \quad V_2 = \frac{7}{0.93} = 7.527 \text{ m/s}$$

$$\therefore h_L = 0.31 \text{ m}$$

Whence from (i) y = 3·97 m (say 4·0 m)

Divide the field into a 1·0 m square grid (see fig. 10.11) and allocate boundary values and interior grid values of ψ.

Numbering rows and columns from top left-hand corner, note that at point 4, 7 two of the adjacent grid points do not coincide with the boundary. Using the notation of fig. 10.8, $\lambda_2 = 0.6$ and $\lambda_3 = 0.5$.

The results of successive corrections using equation (10.17) in the form

$$\psi_o = \frac{\psi_1 + \psi_2 + \psi_3 + \psi_4}{4}$$

at all interior grid points with the exception of point 4, 7 where equation (10.17);

$$\psi_o = \frac{\psi_1 + \dfrac{\psi_2}{\lambda_2} + \dfrac{\psi_3}{\lambda_3} + \psi_4}{2 + \dfrac{1}{\lambda_2} + \dfrac{1}{\lambda_3}}$$

is used are shown in the following table:

1st correction

Column		2	3	4	5	6	7
Row	2	80	85	89·5	93·4	97·0	100
	3	60	70	79·4	88·19	93·82	100
	4	35	43·75	52·03	57·55	61.59*	75·57

*Maximum correction = −28·41

2nd correction

Column		2	3	4	5	6	7
Row	2	80	84·88	89·40	93·67	96·87	100
	3	58·75	66·69	74·08	79·78	84·56	100
	4	31·87	37·65	42·32	45·92*	51·51	73·79

*Maximum correction = −11·63

Values (all corrections $< |1\cdot0|$) after 8 iterations

Column		2	3	4	5	6	7
Row	2	76·49	77·86	79·69	83·02	89·39	100
	3	51·90	53·78	56·59	62·20	74·27	100
	4	26·22	27·53	29·69	34·35	45·34	72·71

The above is probably sufficiently accurate for most purposes but computations can be continued if required.

Values after 12 iterations

Column		2	3	4	5	6	7
Row	2	75·78	76·84	78·71	82·33	89·07	100
	3	51·08	52·62*	55·50	61·44	73·91	100
	4	25·76	26·88	29·09	33·92	45·13	72·67

*Maximum correction $|0\cdot125|$

The discrete streamline can now be drawn (see fig. 10.12).

Example 10.3

Water flows under pressure round a bend of inner radius 600 mm in a rectangular duct 600 mm wide and 300 m deep. The discharge is 360 l/s. If the pressure head at the entry to the bend is 3·0 m calculate the velocity and pressure head distributions across the duct at the bend.

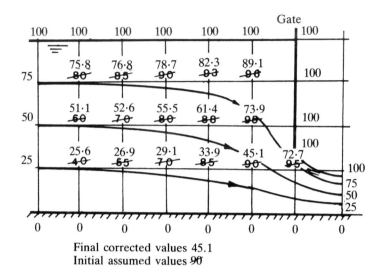

Figure 10.12 Plot of flow net

Solution:
(See fig. 10.13.)

Energy at entry to bend,

$$E = h + \frac{V^2}{2g} \quad \text{where } V = \text{approach velocity } (= 2 \text{ m/s})$$

$$= 3\cdot0 + \frac{2^2}{2g} = 3\cdot204 \text{ m}$$

$$E = \text{const.} = h_A + \frac{v_{\theta.A}^2}{2g} = h_B = \frac{v_{\theta.B}^2}{2g}$$

$$\frac{dh}{dr} = \frac{v_\theta^2}{gr} \quad \text{from (10·18) and } \therefore \text{ the variation of } v_\theta \text{ with radius is required.}$$

Now $v_\theta = \dfrac{K}{r}$ and to evaluate K express the discharge Q as

$$Q = w \int_{r_A}^{r_B} v_\theta \, dr$$

i.e. $Q = w \displaystyle\int_{r_A}^{r_B} \frac{K}{r} \, dr = Kw \left[\log_e \frac{r_B}{r_A} \right]$

$$0\cdot36 = K \times 0\cdot3 \times \log_e \left[\frac{1\cdot2}{0\cdot6} \right] = 0\cdot208 \text{ K}$$

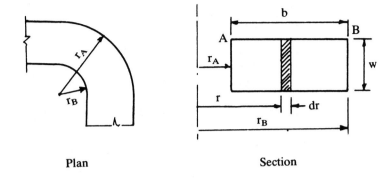

Plan   Section

Figure 10.13   Flow across bend

or

$$K = 1{\cdot}73$$

$$\therefore\ v_{\theta,A} = \frac{1{\cdot}73}{0{\cdot}6} = 2{\cdot}88 \text{ m/s};\quad v_{\theta,C} = \frac{1{\cdot}73}{0{\cdot}9} = 1{\cdot}92 \text{ m/s}$$

$$v_{\theta,B} = \frac{1{\cdot}73}{1{\cdot}2} = 1{\cdot}44 \text{ m/s}$$

$$E\ (= 3{\cdot}204 \text{ m}) = h + \frac{v_\theta^2}{2g};\quad h = 3{\cdot}204 - \frac{v_\theta^2}{2g}$$

whence

$$h_A = 2{\cdot}78 \text{ m};\quad h_C = 3{\cdot}02 \text{ m};\quad h_B = 3.10 \text{ m}.$$

Example 10.4
Water flows round a horizontal 90° bend in a square duct of side length 200 mm, the inner radius being 300 mm. The differential head between the inner and outer sides of the bend is 200 mm of water. Determine the discharge in the duct.

Solution:

$$\frac{dh}{dr} = \frac{v_\theta^2}{gr} \quad (\text{equation } (10.18))$$

$$\Delta h = h_A - h_B = \int_A^B dh = \int_{r_A}^{r_B} \frac{v_{\theta,r}^2 \, dr}{gr}$$

where $v_{\theta,r}$ means the tangential velocity at r.

and $v_{\theta,r} = \dfrac{K}{r}$ (equation (10.20))

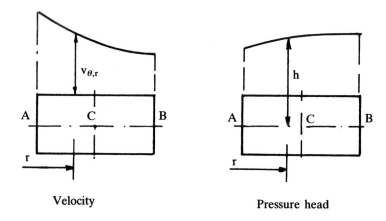

Velocity   Pressure head

Figure 10.14   Velocity and pressure variations across bend

$$\therefore \Delta h \; h_A - h_B = \frac{K^2}{g} \int_{r_A}^{r_B} \frac{dr}{r^3} = \frac{K^2}{2g} \left[-\frac{1}{r^2} \right]_{r_A}^{r_B} \qquad (i)$$

$$Q = w \int_{r_A}^{r_B} v_\theta \, dr = w \int_{r_A}^{r_B} \frac{K}{r} \, dr$$

$$Q = w K \log_e \left(\frac{r_B}{r_A} \right); \quad K = \frac{Q}{w \log_e \left(\frac{r_B}{r_A} \right)}$$

$$\therefore \text{ in (i) } \Delta h = \frac{Q^2}{w^2 \left(\log_e \left(\frac{r_B}{r_A} \right) \right)^2} = \frac{1}{r_A^2} - \frac{1}{r_B^2}$$

$$\text{where } Q = \sqrt{2g} \; \Delta h \; w \log_e \left(\frac{r_B}{r_A} \right) \sqrt{\left(\frac{r_B^2 \times r_A^2}{r_B^2 - r_A^2} \right)}$$

$r_A = 0.3$ m; $r_B = 0.5$ m; $w = 0.2$ m; $\Delta h = 0.2$ m

$\therefore Q = 0.076$ m³/s.

Example 10·5

A cylindrical vessel is rotated at an angular velocity of ω. Show that the surface profile of the contained liquid under equilibrium conditions is parabolic.

Solution:
This is an example of forced vortex motion. (See fig. 10.15.)
For all types of curvilinear flow equation (10.18) is applicable.

$$\frac{dh}{dr} = \frac{v_{\theta,r}^2}{gr} \qquad (i)$$

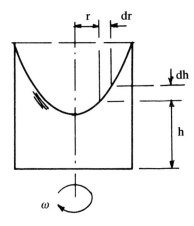

Figure 10.15   Forced vortex motion

and $v_{\theta,r} = r\omega$ (for the forced vortex)   (ii)

$$\therefore h_2 - h_1 = \int_1^2 dh = \frac{1}{g} \int_{r_1}^{r_2} \frac{v_{\theta,r}^2}{r} \cdot dr$$

i.e. $h_2 - h_1 = \dfrac{1}{g} \displaystyle\int_{r_1}^{r_2} r\omega^2 \, dr = \dfrac{\omega^2}{2g} [r_2^2 - r_1^2]$

Taking the origin ($r = 0$, $h = 0$) at 0,

$h = \dfrac{\omega^2}{2g} r^2$ which is a paraboloid of revolution.

Example 10.6
A siphon spillway of constant cross-section 3·0 m wide by 2·0 m deep operates under a head of 6·0 m. The crest radius is 3·0 m and the siphon has a total length of 18·0 m, the length from inlet to crest being 5·0 m along the centre line.
(a) Assuming the inlet head loss to be $0\cdot3\,\dfrac{V^2}{2g}$, the bend loss $0\cdot5\,\dfrac{V^2}{2g}$, the
     Darcy friction factor 0·012 and $\alpha = 1\cdot2$, calculate the discharge through the siphon.
(b) If the level in the reservoir is 0·5 m above the crest of the siphon calculate the pressures at the crest and cowl and comment on the result.

Solution:
Notes: Siphon spillways are used on dams to discharge floodwater. They are particularly useful where the available crest length of a free overfall spillway would be inadequate. Once a siphon spillway has 'primed' it operates like full-bore pipe flow under the head between the reservoir level and the outlet and has a high discharge capacity per unit area. Unlike a free overfall

spillway which provides a gradual increase in discharge (see Example 11.9) the siphon discharge reaches a peak very quickly which may cause a surge to propagate downstream. For this reason a number of siphons may have their crests set at different levels so that their priming times are not simultaneous.

Since siphons incorporate one vertical bend at a high level, the resulting vortex motion can produce very low pressures at the crest which may result in air entrainment, cavitation and vibration. The design of the crest radius is therefore of utmost importance.

In this Example while head losses in the direction of flow are taken into account (i.e. real fluid flow is considered), no energy losses across the flow occur and the curvilinear flow is treated as in ideal fluid flow. (See fig. 10.16.)

Using the principles of resistance to flow in non-circular ducts:

H = entry loss + velocity head + bend loss + friction loss, i.e.

$$H = 0 \cdot 3 \, \frac{V^2}{2g} + \frac{\alpha \, V^2}{2g} + \frac{0 \cdot 5 \, V^2}{2g} + \frac{\lambda \, LV^2}{8gR} \tag{i}$$

R = hydraulic radius = 0·6 m

$$6 = \frac{V^2}{2g} \left(0 \cdot 3 + 1 \cdot 2 + 0 \cdot 5 \, \frac{0 \cdot 012 \times 18}{4 \times 0 \cdot 6} \right)$$

$$6 = \frac{V^2}{2g} \, (2 \cdot 09) \text{ whence } V = 6 \cdot 37 \text{ m/s}$$

Discharge = 6 × 6·37 = 38·22 m³/s

(Note that the discharge may be expressed as $Q = C_d \, A \, \sqrt{2gh}$ where C_d is the overall discharge coefficient.)

Net head at crest, H_A, of bend = 0·5 − losses

$$H_A = 0 \cdot 5 - \frac{0 \cdot 3 \, V^2}{2g} - \frac{\lambda \, L_A \, V^2}{8gR} = 0 \cdot 5 - 2 \cdot 87 \times \left(0 \cdot 3 + \frac{0 \cdot 012 \times 5}{4 \times 0 \cdot 6} \right)$$

(relative to crest)

= −0·433 m

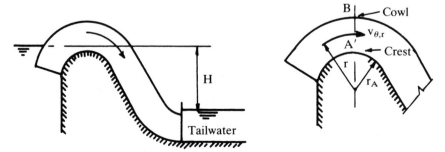

Figure 10.16 Siphon spillway

At the crest, A, $H_A = \dfrac{p_A}{\rho g} + \dfrac{(v_{\theta,A}^2)}{2g}$ (ii)

whence

$\dfrac{p_A}{\rho g} = H_A - \dfrac{(v_{\theta,A}^2)}{2g}$ (iii)

$v_{\theta,A} = \dfrac{K}{r_A}$

The discharge across the section of the duct at the crest,

$$Q = \int_{r_A}^{r_B} v_\theta \, b \, dr \quad \text{where b is the width of the duct.}$$

$$Q = b \int_{r_A}^{r_B} \dfrac{K}{r} \, dr = bK \left[\log_e \dfrac{r_B}{r_A} \right]$$

$$38 \cdot 22 = 3 \times K \left[\log_e \dfrac{5}{3} \right] = 1 \cdot 532 \, K$$

whence $K = 24 \cdot 94$

$$\therefore \; v_{\theta,A} = \dfrac{24 \cdot 940}{3} = 8 \cdot 313 \text{ m/s}; \quad \dfrac{v_{\theta,A}^2}{2g} = 3 \cdot 52 \text{ m}$$

$$\therefore \text{ from equation (iii) } \dfrac{p_A}{\rho g} = -0 \cdot 433 - 3 \cdot 52 = -3 \cdot 953 \text{ m}$$

At B (the cowl) $H_B = \dfrac{p_B}{\rho g} + \dfrac{v_{\theta,B}^2}{2g} + 2 \cdot 0$ (relative to crest)

Since $H_B = H_A$

$\dfrac{p_B}{\rho g} = H_A - \dfrac{v_{\theta,B}^2}{2g} - 2 \cdot 0$

$$v_{\theta,B} = \dfrac{K}{r_B} = \dfrac{24 \cdot 940}{5} = 4 \cdot 988; \quad \dfrac{v_{\theta,B}^2}{2g} = 1 \cdot 268$$

$$\therefore \; \dfrac{p_B}{\rho g} = -0 \cdot 433 - 1 \cdot 268 - 2 \cdot 0 = -3 \cdot 701 \text{ m}$$

Comment: Neither $\dfrac{p_A}{\rho g}$ nor $\dfrac{p_B}{\rho g}$ are particularly low and there should be no danger of cavitation. The spillway could satisfactorily operate under a larger gross head.

The reader should determine the maximum operating head if the crest pressure is not to fall below — 7 m below atmospheric pressure head.

Example 10.7
At a discharge of 10 m³/s the depth of uniform flow in a rectangular channel 3 m wide is 2·2 m. The water flows round a 90° bend of inner radius 5·0 m. Assuming no energy loss at the bend calculate the depth of water at the inner and outer radii of the bend.

Solution:
(See fig. 10.17.)
   Since there is no energy loss between the entry to the bend and points within the bend, $E_o = E_A = E_B = E$
i.e.

$$y_o + \frac{V_o^2}{2g} = y_A + \frac{v_{\theta,A}^2}{2g} = y_B + \frac{v_{\theta,B}^2}{2g} \qquad (i)$$

Hence if $v_{\theta,A}$ and $v_{\theta,B}$ can be determined y_A and y_B can be calculated.

$$y_o = 2\cdot2 \text{ m}; \quad V_o = \frac{10}{3 \times 2\cdot2} = 1\cdot515 \text{ m/s}; \quad \frac{V_o^2}{2g} = 0\cdot117 \text{ m}$$

$$\therefore E_o = 2\cdot2 + 0\cdot117 = 2\cdot317 \text{ m}$$

$$v_{\theta,r} = \frac{K}{r} \text{ and } Q = \int_{r_A}^{r_B} v_{\theta,r} \, y_r \, dr \qquad (ii)$$

$$y_r = E - \frac{v_{\theta,r}^2}{2g} = E - \frac{K^2}{2g \, r^2} \text{ and substituting in (ii):}$$

$$\therefore Q = K \int_{r_A}^{r_B} \frac{1}{r} \left(E - \frac{K^2}{2g \, r^2} \right) dr = K \int_{r_A}^{r_B} \left(\frac{E}{r} - \frac{K^2}{2g \, r^3} \right) dr$$

$$Q = K \left[E \log_e \left(\frac{r_B}{r_A} \right) + \frac{K^2}{4g} \left(\frac{1}{r_B^2} - \frac{1}{r_A^2} \right) \right]$$

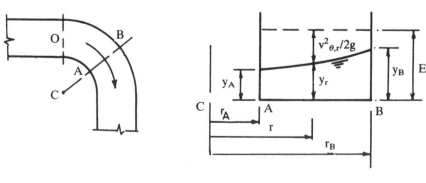

Plan   Section

Figure 10.17   Flow across rectangular channel bend

$$10 = K \left[2 \cdot 317 \log_e \left(\frac{8}{5} \right) + \frac{K^2}{4g} \left(\frac{1}{8^2} - \frac{1}{5^2} \right) \right]$$

By trial $K = 9 \cdot 71$

$$\therefore v_{\theta,A} = \frac{9 \cdot 71}{5} = 1 \cdot 942; \quad \frac{v_{\theta,A}^2}{2g} = 0 \cdot 192 \text{ m}$$

$$v_{\theta,B} = \frac{9 \cdot 71}{5} = 1 \cdot 2137; \quad \frac{v_{\theta,B}^2}{2g} = 0 \cdot 075 \text{ m}$$

$$\therefore y_A = 2 \cdot 317 - 0 \cdot 192 = 2 \cdot 125 \text{ m}$$

$$y_B = 2 \cdot 317 - 0 \cdot 075 = 2 \cdot 242 \text{ m}.$$

Recommended reading

1. Featherstone, R.E. (1981) 'Hydraulics (Mechanics of Fluids)' in *Kempe's Engineers Year Book*, Vol. 1. London: Morgan-Grampian.
2. *Design of Small Dams* (1987), 3rd edn. US Department of the Interior, Bureau of Reclamation, US Government Printing Office, Washington.
3. Vallentine, H.R. (1969) *Applied Hydrodynamics*. London: Butterworth.

Problems

1. Determine the stream function for a uniform rectilinear flow of velocity V inclined at α to the x axis (a) in Cartesian, and (b) in polar forms.

2. A stream function is defined by $\psi = x \, y$. Determine the flow pattern and the velocity potential function.

3. Draw the streamlines defining 'streamtubes' conveying 1 m^3/s per metre depth for a source of strength 12 m^3/s.m in a rectilinear uniform flow of velocity $-1 \cdot 0$ m/s.

4. In the system described in Problem 3 a sink of strength 12 m^3/s.m is situated 5 m downstream from the origin of source in the direction of the x axis. Draw the streamlines and determine the shape of the 'body' defined by the streamline $\psi = 0$.

5. A pollutant is released steadily from the vertical outlet of an outfall into a river (fig. 10.18) such that it rises vertically without radial flow (neglecting the effects of entrainment). The pollutant is carried downstream by a uniform current of 1·0 m/s. A water abstraction is situated 30 m downstream of the outfall; this may be considered as a line sink the total

inflow over the 2 m of the stream being 6 m³/s. Neglecting the effects of dispersion investigate the possibility of the pollutant entering the intake. (See fig. 10.18.)

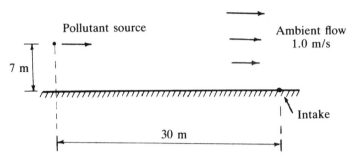

Figure 10.18   Source release of pollutant

6.   Water flows through a rectangular duct 4 m wide and 1 m deep at a rate of 20 m³/s. The flow passes through the side contraction shown in fig. 10.19. Assuming ideal fluid flow and using either a flow net construction or a numerical method of streamline plotting determine the pressure distribution through the transition between AA and BB (a) along the centre line of the duct, and (b) along the boundary if the pressure head at AA is 10 m of water. (See fig. 10.19.)

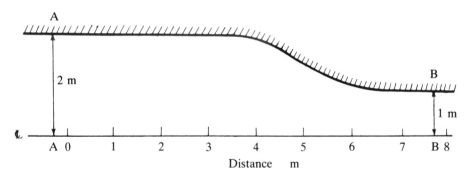

Figure 10.19   Flow through contracted rectangular duct

7.   A discharge of 6 m³/s per unit width approaches a vertical sluice gate in a rectangular channel at a depth of 4 m. The vertical gate opening is 1·5 m and $C_d = 0·6$. Assuming that downstream conditions do not affect the natural flow through the gate opening verify that the flow through the opening is supercritical, draw the streamline pattern and determine the pressure distribution and force per unit width on the gate. Compare the value of force obtained with that obtained using the momentum equation.

8. Water flows under pressure in a horizontal rectangular duct of width b and depth w. The pressure head difference across the duct, at a horizontal bend of radius R to the centre line, is h. Show that if R = 1·5 w the discharge is obtained from the expression, $Q = 2 \cdot 5 \, dw \, \sqrt{h}$.

9. Water flows round a horizontal 90° bend in a square duct of side length 200 mm, the inner radius being 400 mm. The differential head between the inner and outer sides of the bend is 150 mm of water. Determine the discharge in the duct.

10. The depth of uniform flow of water in a rectangular channel 5 m wide conveying 40 m³/s is 2·5 m. The water flows round a 90° bend of inner radius 10 m. Assuming no energy loss through the bend determine the depth of water at the inner and outer radii of the bend at the discharge of 150 m³/s.

11. A proposed siphon spillway is of uniform rectangular section 5 m wide and 3 m deep. The crest radius is 3 m and the length of the approach to the crest along the centre line is 7 m. The overall discharge coefficient is 0·65, the entry head loss = $\dfrac{0 \cdot 2 \, V^2}{2g}$, where V is the mean velocity, and the Darcy friction factor is 0·015. Determine the discharge when the siphon operates under a head of 6 m with the upstream water level 0·5 m above the crest and determine the pressure heads at the crest and cowl.

12. Water discharges from a tank through a circular orifice 25 mm in diameter in the base. The discharge coefficient under conditions of radial flow towards the orifice in the tank is 0·6. A free vortex forms when water is discharging under a head of 150 mm; at a horizontal distance of 10 mm from the centre line of the orifice the water surface is 50 mm below the top water level. Determine the discharge through the orifice.

13. A cylindrical vessel 0·61 m in diameter and 0·97 m deep, open at the top, is rotated about a vertical axis at 105 rev/min. If the vessel was originally full of water how much water will remain under equilibrium conditions?

Chapter 11
Gradually Varied Unsteady Flow from Reservoirs

R.E. Featherstone

11.1 Discharge between reservoirs under varying head

Figure 11.1 shows two reservoirs, of constant area, interconnected by a pipeline, water transfer occurring under gravity. Reservoir A receives an inflow I_1 while a discharge Q_2 is withdrawn from B; I_1 and Q_2 may be time variant.

In the general case the head, h, and hence the transfer discharge, will vary with time and the object is to obtain a differential equation describing the rate of variation of head with time, $\dfrac{dh}{dt}$. The corresponding rate of change of discharge is considered to be sufficiently small that the steady state discharge v. head relationship for the pipeline flow may be applied at any instantaneous head; compressibility effects are also neglected. Such unsteady flow situations are therefore sometimes referred to as 'quasi-steady' flow.

Let h = gross head at any instant

Δh_1 = change in level in A during a small time interval Δt

Δh_2 = change in level in B in time Δt

Then change in total head = $\Delta h = \Delta h_1 - \Delta h_2$ (11.1)

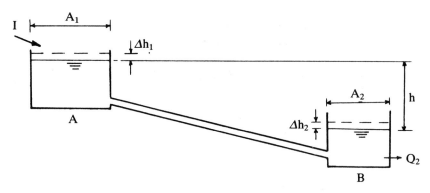

Figure 11.1   Flow through pipe with varying head

Continuity equation for A: $I - Q_1 = A_1 \dfrac{\Delta h_1}{t}$ (11.2)

Continuity equation for B: $Q_1 - Q_2 = A_2 \dfrac{\Delta h_2}{t}$ (11.3)

Note that Q_1 is the discharge in the pipeline and hence is the inflow rate into B.

From the steady state head v. discharge relationship for the pipeline

$$h = \left(K_m + \frac{\lambda L}{D}\right) \frac{Q_1^2}{2 g A_p^2}$$

where A_p = area of pipe and K_m the minor loss coefficient.
whence

$$Q_1 = K h^{1/2} \tag{11.4}$$

in which

$$K = A_p \sqrt{\frac{2g}{K_m + \dfrac{\lambda L}{D}}}$$

From equations (11.1), (11.2) and (11.3)

$$\Delta h = \left(\frac{I - Q_1}{A_1} - \frac{Q_1 - Q_2}{A_2}\right) \Delta t$$

Introducing (11.4),

$$\Delta h = \left(\frac{I}{A_1} - K h^{1/2}\left(\frac{1}{A_1} + \frac{1}{A_2}\right) + \frac{Q_2}{A_2}\right) \Delta t$$

and in the limit as $\Delta t \to 0$,

$$dt = \frac{dh}{\left(\dfrac{I}{A_1} + \dfrac{Q_2}{A_2}\right) - K h^{1/2}\left(\dfrac{1}{A_1} + \dfrac{1}{A_2}\right)} \tag{11.5}$$

In a similar dynamic system where the upper reservoir discharges to atmosphere through a pipeline, orifice or valve (fig. 11.2), the term Δh_2 in equations (11.1) and (11.3) disappears and a similar treatment, or alternatively removing the irrelevant terms A_2 and Q_2 from equation (11.5), yields

$$dt = \frac{A_1 \, dh}{I - K h^{1/2}} \tag{11.6}$$

or

$$I - K h^{1/2} = A_1 \frac{dh}{dt} \tag{11.7}$$

where $K\ h^{1/2}$ represents the appropriate steady state discharge v. head relationship for the outlet device.

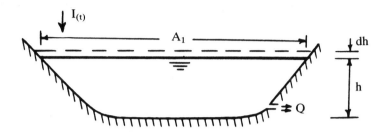

Figure 11.2   Flow through orifice under varying head

11.2 Unsteady flow over a spillway

The computation of the time variation in reservoir elevation and spillway discharge during a flood inflow to the reservoir is essential in design of the spillway to ensure safety of the impounding structure. (See fig. 11.3.)

The continuity equation is

$$I_{(t)} - Q_{(t)} = \frac{dS}{dt} \tag{11.8}$$

where $\dfrac{dS}{dt}$ is the rate of change of storage, or volume, S.

$\dfrac{dS}{dt}$ may be expressed as $A\ \dfrac{dh}{dt}$ where A is the instantaneous plan area of the reservoir and $\dfrac{dh}{dt}$ the instantaneous rate of change of depth.

Assuming that in the case of a fixed crest free overfall spillway, the discharge rate Q may be expressed by the steady state relationship

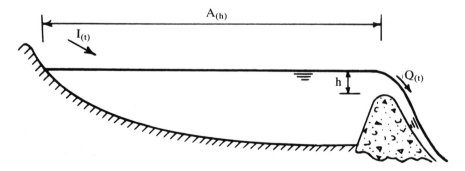

Figure 11.3   Reservoir routeing

$$Q = \frac{2}{3} \sqrt{2g}\; C_d\; L\; h^{3/2} \tag{11.9}$$

where L is the crest length and C_d the discharge coefficient,

i.e. $Q = K\, h^{3/2}$ (11.10)

Equation (11.7) becomes

$$I_{(t)} - K\, h^{3/2} = \frac{dS}{dt} = A\, \frac{dh}{dt} \tag{11.11}$$

$I_{(t)}$ is the known time variant inflow rate. Except in the rather special case where A does not vary with depth and $I_{(t)}$ is constant, equation (11.11) is not directly integrable and in general must be evaluated numerically.

11.3 Flow establishment

(See fig. 11.4.)

Figure 11.4 shows a constant head tank discharging to atmosphere through a pipeline terminated by a control valve. If the valve, which is initially closed, is suddenly opened the discharge will not instantaneously attain its equilibrium value and the object is to determine the time taken for this to be attained, i.e. the time of steady flow establishment.

Initially the total head, H, is available to accelerate the flow but this decreases as the velocity increases due to the associated head losses. At any instant the available head $= H - K\, V^2$ where K is the head loss coefficient,

$$= \left(\frac{\lambda L}{D} + K_m \right) \frac{1}{2g}$$

where K_m is the minor loss coefficient.

The equation of motion (Force = mass × acceleration)

$$\rho g\, A_p\, (H - K\, V^2) = \rho\, A_p\, L\, \frac{dV}{dt} \tag{11.12}$$

where A_p = area of pipe cross-section

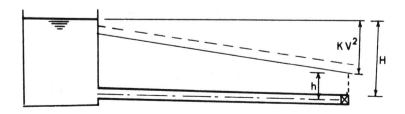

Figure 11.4 Flow establishment through pipeline

whence

$$T = \int_{t_1}^{t_2} dt = \frac{L}{g} \int_{V_1}^{V_2} \frac{dV}{H - K V^2} \tag{11.13}$$

Assuming λ = constant in the interval between velocities V_1 and V_2 and writing $a^2 = H$, and $b^2 = K$, (constant), equation (11.13) becomes

$$\begin{aligned}
T &= \frac{L}{g} \int_{V_1}^{V_2} \frac{dV}{a^2 - b^2 V^2} \\[2mm]
&= \frac{L}{2ag} \int_{V_1}^{V_2} \left(\frac{1}{a + b V} + \frac{1}{a - b V} \right) dV \\[2mm]
&= \frac{L}{2gab} \left[\log_e \frac{(a + b V)}{(a - b V)} \right]_{V_1}^{V_2} \\[2mm]
&= \frac{L}{2g \sqrt{K H}} \left[\log_e \frac{(\sqrt{H} + \sqrt{K} V)}{(\sqrt{H} - \sqrt{K} V)} \right]_{V_1}^{V_2}
\end{aligned} \tag{11.14}$$

Hence the time for flow establishment is the value of the integral (11.13) between $V = 0$ and $V = V_o$ where V_o is the steady state velocity in the pipeline under the head H. Therefore, since $H = K V_o^2$, equation (11.14) becomes:

$$T = \frac{L V_o}{2g H} \log_e \left[\frac{V_o + V}{V_o - V} \right]_{V=0}^{V_o} \tag{11.15}$$

If the variation of λ with discharge is taken into account during the accelerating period the time of flow establishment must be evaluated by numerical or graphical integration of equation (11.13).

Worked examples

Example 11.1
A circular orifice 20 mm in diameter with a discharge coefficient $C_d = 0.6$ is fitted to the base of a tank having a constant cross-sectional area of 1.5 m. Determine the time taken for the water level to fall from 3.0 m to 0.5 m above the orifice.

Solution:
Since there is no inflow (I = 0), equation (11.7) becomes

$$-K h^{1/2} = A_1 \frac{dh}{dt}$$

whence

$$dt = -A_1 \frac{dh}{K\,h^{1/2}}$$

Integrating:

$$\text{Time } T = \int_{t=0}^{t_1} dt = -\frac{A_1}{K} \int_{h_o}^{h_1} \frac{dh}{h^{1/2}}$$

$$T = -\frac{A_1}{K} \left[2\,h^{1/2} \right]_{h_o}^{h_1} = \frac{2\,A_1}{K} \left[h^{1/2} \right]_{h_1}^{h_o}$$

$$K = C_d\,A_o\,\sqrt{2g} = 0{\cdot}6 \times 3{\cdot}142 \times 4{\cdot}43 = 8{\cdot}35 \times 10^{-4}$$

whence

$$T = \frac{2 \times 1{\cdot}5}{8{\cdot}35 \times 10^{-4}} \left[h^{1/2} \right]_{0{\cdot}5}^{3{\cdot}0} = 3682{\cdot}4 \text{ sec.}$$

Example 11.2
If the tank in Example 11.1 has a variable area expressed by $A_1 = 0{\cdot}0625\,(5+h)^2$ calculate the time for the head to fall from 4 m to 1 m.

Solution:

$$dt = \frac{A_1\,dh}{K\,h^{1/2}}$$

whence

$$T = -\frac{0{\cdot}0625}{0{\cdot}000835} \int_{3{\cdot}0}^{0{\cdot}5} \frac{(5+h)^2}{h^{1/2}}\,dh$$

$$T = 74{\cdot}85 \int_{0{\cdot}5}^{3{\cdot}0} (25\,h^{-1/2} + 10\,h^{1/2} + h^{3/2})\,dh$$

$$= 74{\cdot}85 \left[50\,h^{1/2} + \frac{20}{3}\,h^{3/2} + \frac{2}{5}\,h^{5/2} \right]_{0{\cdot}5}^{3{\cdot}0}$$

$$= 6713{\cdot}75 \text{ sec.}$$

Example 11.3
A pipeline 1000 m long, 100 mm in diameter with a roughness size of 0·03 mm discharges water to atmosphere from a tank having a cross-sectional area of 1000 m². Find the time taken for the water level to fall from 20 m to 15 m above the pipe outlet.

Solution:

$$\text{Continuity } I - Q = A_1 \frac{dh}{dt} \qquad\qquad\qquad (i)$$

The pipeline discharge Q may be expressed by the Darcy-Colebrook-White equation

$$Q = -2 A_p \sqrt{2g\, D\, \frac{h_f}{L}}\, \log\left[\frac{k}{3 \cdot 7\, D} + \frac{2 \cdot 51\, v}{D\sqrt{2g\, D\, \frac{h_f}{L}}}\right]$$ (ii)

However if this were substituted into (i) the resulting equation would need to be evaluated using graphical or numerical integration methods. Such a procedure is fairly straightforward but if constant value of λ over the range of discharges is adopted a direct solution is obtained. In this latter case Q is expressed by

$$Q = \frac{A_p \sqrt{2g}\, h^{1/2}}{\sqrt{K_m + \frac{\lambda\, L}{D}}} = K\, h^{1/2} \quad \text{(equation (11.4))}$$

Thus (i) reduces to equation (11.6): $dt = \dfrac{A_1\, dh}{I - K\, h^{1/2}}$.

To evaluate K calculate the pipe velocities at values of $h_f = 20$ m and 15 m using equation (ii), i.e. neglecting minor losses, and evaluate λ from the Darcy-Weisbach equation:

$$\lambda = \frac{2g\, D}{V^2}\, \frac{h_f}{L}$$

(a) when $h_f = h = 20$ m, $V = 1 \cdot 46$ m/s and $\lambda = 0 \cdot 0184$

(b) when $h_f = h = 15$ m, $V = 1 \cdot 25$ m/s and $\lambda = 0 \cdot 0188$

Adopting $\lambda = 0 \cdot 0186$ and $K_m = 1 \cdot 5$, $K = 0 \cdot 00254$.

whence

$$T = \int_{h_1}^{h_2} \frac{A_1\, dh}{K\, h^{1/2}} \quad \text{(since I = 0)}$$

$$T = \frac{2 \times 1000}{0 \cdot 00254}\, [20^{1/2} - 15^{1/2}] = 471\,660 \text{ s}$$

$$= 131 \cdot 02 \text{ h.}$$

Example 11.4
If in Example 11.3 a constant inflow of 5 l/s enters the tank determine the time for the head to fall from 20 m to 18 m.

$$dt = \frac{A_1\, dh}{I - K\, h^{1/2}}$$ (i)

$$T = A_1 \int_{h_o}^{h_1} \frac{dh}{I - K h^{1/2}} \tag{ii}$$

Write $y = K h^{1/2} - I$ (iii)

whence $h = \frac{1}{K^2} (y^2 + 2 I y + I^2)$

$$dh = \frac{1}{K^2} (2 y + 2 I) dy$$

(ii) becomes: $T = -2 \frac{A_1}{K^2} \int_{y_o}^{y_1} \frac{(y + I)}{y} dy$

i.e. $T = -2 \frac{A_1}{K^2} \int_{y_o}^{y_1} \left(1 + \frac{I}{y}\right) dy = -2 \frac{A_1}{K^2} \left[y + I \log_e (y)\right]_{y_o}^{y_1}$

Substituting for y from (iii)

$$T = 2 \frac{A_1}{K^2} \left[K h^{1/2} - I + I \log_e (K h^{1/2} - I)\right]_{18}^{20}$$

Using $K = 0.00254$ (as in Example 11.3),

$$T = \frac{2 \times 1000}{0.00254^2} \{0.00254 (\sqrt{20} - \sqrt{18}) + 0.005 \times$$

$$[\log_e (0.00254 \sqrt{20} - 0.005) - \log_e (0.0254 \sqrt{18} - 0.005)]\}$$

$$= 329579.8 \text{ s}$$

$$= 91.55 \text{ h.}$$

Example 11.5
Reservoir A with a constant surface area of 10 000 m² delivers water to reservoir B with a constant area of 2500 m² through a 10 000 m long, 200 mm diameter pipeline of roughness 0·06 mm. Minor losses including entry and velocity head, total 20 $V^2/2g$. A steady inflow of 10 l/s enters reservoir A and a steady flow of 20 l/s is drawn from B.

  If the initial level difference is 100 m determine the time taken for this to become 90 m.

Solution:
Refer to section 11.2 and fig. 11.1.
  From equation (11.5)

$$dt = \frac{dh}{\left(\dfrac{I}{A_1} + \dfrac{Q_2}{A_2}\right) - K h^{1/2} \left(\dfrac{1}{A_1} + \dfrac{1}{A_2}\right)} \tag{i}$$

  Since, in this case I, Q_2, A_1 and A_2 are constant equation (i) can be directly integrated

Let $W = \left(\dfrac{I}{A_1} + \dfrac{Q_2}{A_2}\right)$ and $Z = K\left(\dfrac{1}{A_1} + \dfrac{1}{A_2}\right)$

Then in (i)

$$dt = \frac{dh}{W - Z\,h^{1/2}}$$

and $$T = \int_{h_1}^{h_2} \frac{dh}{W - Z\,h^{1/2}} \tag{ii}$$

Using the mathematical technique of Example 11.4 this integral (ii) becomes

$$T = \frac{2}{Z^2}\left[Z\,(h_1^{1/2} - h_2^{1/2}) - W\,\log_e\left(\frac{Z\,h_1^{1/2} - W}{Z\,h_2^{1/2} - W}\right)\right] \tag{iii}$$

Now $K = \dfrac{A_p\,\sqrt{2g}}{\sqrt{K_m + \dfrac{\lambda L}{D}}}$ and λ is evaluated as in Example 11.4.

Adopting $\lambda = 0.0173$, $K = 4.678 \times 10^{-3}$

$W = 9.0 \times 10^{-6}$ and $Z = 2.339 \times 10^{-6}$

Thus from equation (iii), $T = 725\,442$ s
$\qquad\qquad\qquad\qquad\quad = 201.51$ h.

Example 11.6
For the system described in Example 11.5 find the time taken for the level in reservoir A to fall by 1.0 m.

Solution:
Continuity equation for reservoir A (from equation (11.1)):

$$I - Q_1 = A_1\,\frac{dh_1}{dt} \tag{i}$$

Since $Q_1 = K\,h^{1/2}$, $I - K\,h^{1/2} = A_1\,\dfrac{dh_1}{dt}$ \qquad (ii)

Thus the rate of change of level in A is related to the instantaneous gross head h. A numerical or graphical integration method must be used to evaluate the time taken for the level in A to change by a specified amount since, in equation (ii) the head h, also varies with time.
Thus

$$\Delta h_1 = \frac{I - K\,h^{1/2}}{A_1}\,\Delta t \tag{iii}$$

By taking a series of values of h the variation of h with time is evaluated using equation (iii) in Example 11.5.

h (m)	time (s)
99	69786·6
98	140148·3
97	211100·0
96	282656·0
95	354827·0

Plotting h v. time (fig. 11.5 (a)) values of h at discrete time intervals Δt, say 100 000 s, are obtained. Equation (iii) is then evaluated for Δh_1 in the time interval Δt.

The following values are obtained from fig. 11.5 (a) and equation (iii). $\bar{h}$ is the average head in the time interval. The negative sign for Δh_1 indicates a falling level in A.

Time (s × 10^5)	0	1	2	3
h (m)	99	98·5	97·2	95·0
$h^{1/2}$		9·94	9·89	9·89
Δh_1 (m)		−0·355	−0·353	−0·349
$\Sigma\Delta h_1$ (m)		−0·355	−0·708	−1·057

From the graph of $\Sigma\Delta h_1$ v. time (graph (b)) t = $2·95 \times 10^5$ s when $\Sigma\Delta h_1 = -1·0$ m.

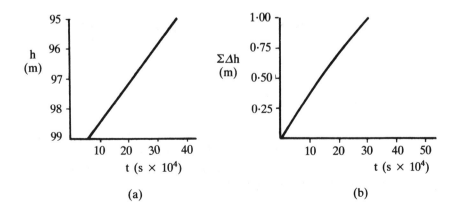

Figure 11.5 Variation of reservoir levels with time

Example 11.7

An impounding reservoir is to be partially emptied from the level of the spillway crest through three 1·0 m diameter valves with $C_d = 0·95$, and set at the same level with their axes 20 m below the spillway crest. The variation of reservoir surface area (A_1) with height (h), above the valve centre line is tabulated (see table below).

   Assuming that a continuous inflow of 0·5 m³/s enters the reservoir determine the time required to lower the level from the spillway crest to 1·0 m above the valves.

h (m)	0	5	10	15	20
A_1 (m² × 10⁶)	2·5	6·0	11·5	15·0	17·0

Solution:

$$I - Q = A_1 \frac{dh}{dt} \tag{i}$$

Q (the outflow) $= 3 \times C_d \, A_o \, \sqrt{2g} \, h^{1/2} = K \, h^{1/2}$

where $\quad K = 3 \times 0·95 \times 0·7854 \sqrt{2g} = 9·915$

From (i) $T = \displaystyle\int_{t_o}^{t_1} dt = \int_{h_o}^{h_1} \frac{A_1 \, dh}{I - K \, h^{1/2}} \tag{ii}$

Since A_1 is not readily expressible as a continuous function of h, equation (ii) is not directly integrable, but may be evaluated graphically or numerically.

Writing $X(h) = \dfrac{A_1}{I - K \, h^{1/2}}, \; T = \displaystyle\int_{h_o}^{h_1} X(h) \, dh$

$X(h)$ is evaluated for a number of discrete values of h and the integral

$$\int_{h_o}^{h_1} X(h) \, dh$$

obtained from the area under the $X(h)$ v. h curve.

h (m)	20	15	10	5	1
A_1 (h) (m² × 10⁶)	17·0	15·0	11·5	6·0	3·0
$X(h)$ (s m⁻¹ × 10⁶)	−0·38776	−0·39571	−0·37272	−0·27687	−0·318640

The -ve sign for $X(h)$ indicates that h is decreasing with time. Plotting $X(h)$ v. the area between h = 20·0 and h = 1·0 yields

$\qquad T = 6·61 \times 10^6 \text{ s} = 1836 \text{ h.}$

Example 11.8

Flood water discharges from an impounding reservoir over a fixed crest spill-way 100 m long; $C_d = 0.7$. The variation of surface area with head above the crest (h) is shown in the two tables on page 312.

Calculate the outflow hydrograph and state the maximum water level and peak outflow, assuming that an outflow of 20 m^3/s exists at t = 0 h.

Solution:

Refer to fig. 11.3.

The continuity equation (equation (11.8)) is

$$I_{(t)} - Q_{(t)} = \frac{dS}{dt} \tag{i}$$

where S = volume of storage

$$Q_{(t)} = \text{outflow rate over spillway crest} = \frac{2}{3} \sqrt{2g} \, C_d \, L \, h^{3/2}$$

$$= K \, h^{3/2} \tag{ii}$$

Thus (i) becomes

$$I_{(t)} - K \, h^{3/2} = \frac{dS}{dt} = \frac{A \, dh}{dt} \tag{iii}$$

Since $I_{(t)}$ is not a function of h and S is not a readily expressible function of h equation (iii) is best evaluated using a numerical method. Taking small time intervals Δt (iii) may be written:

$$\bar{I} - K \, (\bar{h})^{3/2} = \bar{A} \, \frac{\Delta h}{\Delta t} \tag{iv}$$

where $\bar{I} = \dfrac{I_1 + I_2}{2}$; $(\bar{h})^{3/2} = \dfrac{(h_1 + \Delta h)^{3/2} + h_1^{3/2}}{2}$

$$\bar{A} = \frac{A_1 + A_2}{2}; \quad A_1 = A_{(h_1)}; \quad A_2 = A_{(h_1 + \Delta h)}$$

Subscripts 1 and 2 indicate values at the beginning and end of each time interval respectively.

Δh is estimated and adjusted until equation (iv) is satisfied and the computation proceeds to the next time interval. A curve, or numerical relationship, relating A with h is required. The above procedure is best carried out on a digital computer.

A simpler, explicit method, is obtained by writing equation (i) in finite difference form taking discrete time intervals Δt.

$$\frac{I_1 + I_2}{2} - \frac{(Q_1 + Q_2)}{2} = \frac{S_2 - S_1}{\Delta t} \tag{v}$$

$$\text{i.e. } I_1 + I_2 + \frac{2S_1}{\Delta t} - Q_1 = \frac{2S_2}{\Delta t} + Q_2 \qquad \text{(vi)}$$

At each time step the values of S_1 and Q_1 are known; hence the value of the left-hand side (L H S) is known and hence S_2 and Q_2 can be found from curves relating $\left(\dfrac{2S}{\Delta t} + Q\right)$ v. h and Q v. h.

Taking $S = \Sigma\,(A_{(h)} \times \Delta h)$ and $\Delta t = 1\text{ h} = 3600\text{ s}$, (see bottom table on page 315).

Table of calculations of outflow rate and head

Time (h)	I (m³/s)	$\dfrac{2S}{\Delta t} - Q$	L H S	Q (m³/s)	h (m)
0	20	1160		20	0·21
1	40	1180	1220	20	0·21
2	70	1242	1290	24	0·23
3	100	1360	1412	26	0·25
4	128	1528	1588	30	0·27
5	150	1730	1806	38	0·32
6	155	1949	2035	43	0·35
7	140	2144	2244	50	0·39
8	112	2286	2396	55	0·41
9	73	2359	2471	56	0·42
10	46		2478	56	0·42

Notes: At $t = 0$, $Q = 20\text{ m}^3/\text{s}$, hence $h = 0\cdot 21\text{ m}$

and $\dfrac{2S}{\Delta t} + Q = 1200$; $\dfrac{2S}{\Delta t} - Q = \dfrac{2S}{\Delta t} + Q - (2Q)$

hence column 3 is completed.

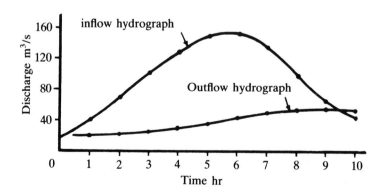

Figure 11.6   Reservoir flood routeing

h (m)	0	0·1	0·2	0·3	0·4	0·5	0·6	0·7	0·8	0·9	1·0
A (m² × 10⁶)	10·00	10·10	10·20	10·34	10·46	10·60	10·75	10·92	11·10	11·30	11·50

Inflow hydrograph

Time (h)	0	1	2	3	4	5	6	7	8	9	10
Inflow (m³/s)	20	40	70	100	128	150	155	140	111	73	46

Table of calculations of $\frac{2S}{\Delta t} + Q$ v. h

	0·1	0·2	0·3	0·4	0·5	0·6	0·7	0·8	0·9	1·0
$Q = K(h)^{3/2}$	6·54	18·49	33·97	52·29	73·08	96·07	121·06	147·91	176·49	206·71
$A_{(h)}$ (m² × 10⁶)	10·1	10·2	10·34	10·46	10·60	10·75	10·92	11·10	11·30	11·50
$\left(\frac{2S}{\Delta t} + Q\right)$ (m³/s)	567	1146	1736	2335	2945	3565	4197	4840	5497	6166

At t = 1 h, $I_1 + I_2 + \dfrac{2S}{\Delta t} - Q = 20 + 40 + 1160 = 1220$

$\dfrac{2S}{\Delta t} + Q = 1220$ whence h = 0·21 m and Q = 20 m³/s

Peak outflow = 58 m³/s at t = 9·5 hrs; h_{max} = 0·42 m

Plotting the inflow and outflow hydrographs (see fig. 11.6).
Note that Q_{max} coincides with the falling limb of the inflow hydrograph.

Example 11.9
A pipeline 5000 m long, 300 mm in diameter and roughness size 0·03 mm
discharges water from a reservoir to atmosphere through a terminal valve.
The difference in level between the reservoir and valve is 20 m which may
be assumed constant. If the valve, which is initially closed, is suddenly
opened, determine the time for steady flow to become established neglecting
compressibility effects. Assume minor losses = $\dfrac{5\,V^2}{2g}$ including velocity head
and entry loss.

Solution:
From equation (11.15), section 11.3, the theoretical time for flow establishment is

$$T = \frac{L\,V_o}{2g\,H}\,\log_e\left[\frac{V_o + V}{V_o - V}\right]^{V_o}_{V=0} \qquad\qquad (i)$$

Using the techniques of Chapter 4 the steady state velocity, V_o, under the
head of 20 m is calculated to be 1·23 m/s. Theoretically the time taken to
attain steady flow is infinite (substituting $V = V_o$ in equation (i)); therefore
adopt V_o = 0·99 × 1·23 = 1·2 m/s (say) whence in (i)

$$T = \frac{5000 \times 1·23}{19·62 \times 20}\,\log_e\left[\frac{2·43}{0·03}\right] = 68·9 \text{ s}$$

The above solution assumes a constant value of λ. The reader should
evaluate T, taking account of the variation of λ with velocity, using a
graphical or numerical integration of equation (11.13).

Problems

1.   A reservoir with a constant plan area of 20000 m² discharges water to
atmosphere through a 2000 m long pipeline of 300 mm diameter and roughness 0·15 mm. The reservoir receives a steady inflow of 20 l/s. If the head
between the reservoir surface and the pipe outlet is initially 40 m determine
(neglecting minor losses) the time taken for the head to fall to 35m.

2.   An impounding reservoir, of constant area 100 000 m² discharges to a service reservoir through 20 km of 400 mm diameter pipeline of roughness 0·06 mm. Minor losses amount to 20 V²/2g. The impounding reservoir, of constant plan area 10 000 m², receives a steady inflow of 30 l/s while a steady outflow of 10 l/s takes place from the service reservoir. If the initial level difference is 50 m determine the time taken for the head to become 48 m and the time for the level in the upper reservoir to fall by 0·5 m.

3.   An impounding reservoir delivers water to a hydro-electric plant through four pipelines each 2·0 m in diameter and 1000 m long with a roughness of 0·3 mm. The reservoir is to be drawn down using the four pipelines with by-passes round the turbines to discharge into the tailrace which has a constant level. Allowing $\dfrac{5\,V^2}{2g}$ for local losses including velocity head, entry and by-pass losses determine the time taken for the level in the reservoir to fall from 50 m above the tail-race to 20 m above the tailrace assuming a constant inflow of 1·0 m³/s.

Level above tailrace (m)	20	30	40	50
Surface area of reservoir (m² × 10⁶)	2·0	4·0	6·8	12·2

4.   Using the data given for the reservoir in Example 11.7 obtain the outflow hydrograph resulting from the following inflow hydrograph if the spillway is 50 m long. (Assume the outflow rate at t = 0 to be 10 m³/s.)

Time (h)	0	1	2	3	4	5	6	7	8	9	10	11	12
Inflow (m³/s)	20	40	80	130	216	250	228	176	120	80	52	44	20

Chapter 12
Mass Oscillations and Pressure Transients in Pipelines

R.E. Featherstone

12.1 Mass oscillation in pipe systems – surge chamber operation

When compressibility effects are not significant the unsteady flow in pipelines is called 'surge'. A typical example of surge occurs in the operation of a medium to high head hydro-electric scheme (fig. 12.1).

If, while running under steady power conditions, the turbine is required to be closed down, values in the inlet to the turbine runner passages will be closed slowly. This will result in pressure transients, which involve compressibility effects, occurring in the penstocks between the turbine inlet valve and the surge chamber. (For details of pressure transients see sections 12.5 et seq.) The pressure transients do not proceed beyond the surge chamber and hence high pressures in the tunnel are prevented resulting in a reduced cost of construction.

Due to the presence of the surge chamber the momentum of the water in the tunnel is not destroyed quickly and water continues to flow, passing into the surge chamber the level in which stops rising when the pressure in the tunnel at the surge chamber inlet is balanced by the pressure created by the head in the chamber.

At this time the level in the chamber will be higher than that in the reservoir and reversed flow will occur, setting up a long period oscillation

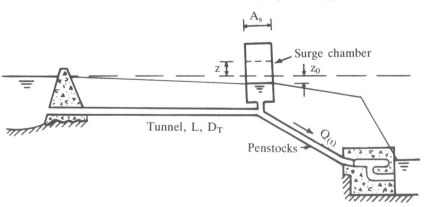

Figure 12.1 Medium head hydro-power scheme

between the two. **Figure 12.2** shows a typical time variation of level in a surge chamber, the oscillations being eventually damped out by friction in the tunnel, losses at the inlet to the chamber and in the chamber itself. Since there are many different types of surge chamber and types of inlet the reader is referred to the Recommended reading for such details. (See fig. 12.2.)

The governing equations describing the mass oscillations in the reservoir-tunnel-surge chamber system are:

(a) The dynamic equation: $\dfrac{L}{g}\dfrac{dV}{dt} + z + F_s\,V_s\,|V_s| + F_T\,V\,|V| = 0$

$$(12.1)$$

(b) The continuity equation: $V\,A_T = A_s\,\dfrac{dz}{dt} + Q$ $\qquad(12.2)$

L = length of tunnel

z = elevation of water in surge chamber above that in the reservoir

F_s = head loss coefficient at surge chamber inlet (throttle)

V_s = velocity in surge chamber $\left(= \dfrac{dz}{dt}\right)$

F_T = head loss coefficient for tunnel friction head loss $\left(= \dfrac{\lambda\,L}{2g\,D_T}\right)$

D_T = diameter of tunnel

V = velocity in tunnel

A_T = area of cross-section of tunnel

A_s = area of cross-section of surge chamber

Q = discharge to turbines.

Equations (12.1) and (12.2) can only be integrated directly for cases of sudden load rejection; tunnel friction and throttle losses may be included.

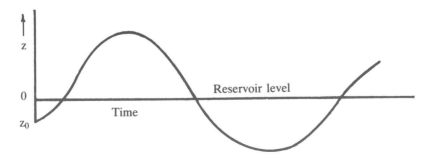

Figure 12.2 Surge chamber oscillations

12.2 Solution, neglecting tunnel friction and throttle losses for sudden discharge stoppage

F_s, F_T and $Q = 0$ and equations (12.1) and (12.2) become:

$$\frac{L}{g}\frac{dV}{dt} + z = 0 \tag{12.3}$$

$$V\,A_T = A_s\,\frac{dz}{dt} \tag{12.4}$$

Differentiating (12.4): $\quad \dfrac{dV}{dt} = \dfrac{A_s}{A_T}\dfrac{d^2z}{dt^2}$ $\tag{12.5}$

Hence $\quad \dfrac{L}{g}\dfrac{A_s}{A_T}\dfrac{d^2z}{dt^2} + z = 0$ $\tag{12.6}$

This is a linear homogeneous second order differential equation with constant coefficients the solution to which is

$$z = C_1 \cos\frac{2\pi t}{T} + C_2 \sin\frac{2\pi t}{T} \tag{12.7}$$

where T is the period of oscillation. Since the tunnel is assumed to be frictionless $z = 0$ at $t = 0$

Hence $\quad z = C_2 \sin\dfrac{2\pi t}{T}$ $\tag{12.8}$

$$T = 2\pi\sqrt{\frac{L}{g}\frac{A_s}{A_T}} \tag{12.9}$$

$$\frac{dz}{dt} = C_2\frac{2\pi}{T}\cos\frac{2\pi t}{T} \quad \text{(from (12.8))} \tag{12.10}$$

and $\quad \dfrac{dz}{dt} = V\dfrac{A_T}{A_s}$ (from (12.4)) $\tag{12.11}$

Hence $\quad V = \dfrac{A_s}{A_T}\,C_2\,\dfrac{2\pi}{T}\cos\dfrac{2\pi t}{T}$

when $\quad t = 0$, $V = V_o$ and $V_o = \dfrac{A_s}{A_T}\,C_2\,\dfrac{2\pi}{T}$ $\tag{12.12}$

Substituting for T from (12.9), $C_2 = V_o\sqrt{\dfrac{L}{g}\dfrac{A_T}{A_s}}$ $\tag{12.13}$

whence (from (12.8)), $\qquad\qquad z = V_o\sqrt{\dfrac{L}{g}\dfrac{A_T}{A_s}}\sin\dfrac{2\pi t}{T}$ $\tag{12.14}$

12.3 Solution, including tunnel and surge chamber losses for sudden discharge stoppage

Since $Q = 0$, $V_s A_s = V_T A_T$ whence equation (12.1) becomes:

$$\frac{L}{g} \frac{dV}{dt} + z + \left(F_s \left(\frac{A_T}{A_s} \right)^2 + F_T \right) V \, |V| = 0$$

i.e. $\dfrac{L}{g} \dfrac{dV}{dt} + z + F_R V \, |V| = 0$ \hfill (12.15)

From equations (12.4), (12.5) and (12.15) the following relationships are obtainable:

$$V^2 = -\frac{z}{F_R} + \frac{L \, A_T}{2g \, A_s \, F_R^2} + C \exp \left(-2g \, A_s \, F_R \, z/L \, A_T \right) \hfill (12.16)$$

when $t = 0$, $V = V_o$ and $z_o = F_T V_o^2$ whence we obtain:

$$\frac{V^2 + \dfrac{z}{F_R} - \dfrac{L \, A_T}{2g \, A_s \, F_R^2}}{V_o^2 + \dfrac{F_T V_o^2}{F_R} - \dfrac{L \, A_T}{2g \, A_s \, F_R^2}} = \exp \left(-2g \, A_s \, F_R \, (z + F_T V_o^2)/L \, A_T \right)$$

$$\hfill (12.17)$$

z max occurs when $V = 0$

Note that since $\dfrac{dz}{dt} = V_s$, $dt = \dfrac{dz}{V_s}$

and the time corresponding with any value of z is

$$T_z = \int_0^t dt = \int_o^z \frac{dz}{V_s} = \int_o^z \frac{dz}{\dfrac{V \, A_T}{A_s}} \hfill (12.18)$$

Equation (12.18) can be evaluated by using graphical integration or numerical integration method, in the latter case taking small intervals of z.

12.4 Finite difference methods in the solution of the surge chamber equations

Numerical methods of analysis using digital computers provide solutions to a wide range of operating conditions, and types and shapes of surge chambers.

Considering the general case of a surge chamber with a variable area and taking a finite interval Δt during which V changes by ΔV and z changes by Δz equations (12.1) and (12.2) become

$$\frac{L}{g} \frac{\Delta V}{\Delta t} + z_m + F_T V_m \, |V_m| + F_s \, V_s \, |V_s| = 0 \hfill (12.19)$$

$$V_m \, A_T = A_{s.m} \frac{\Delta z}{\Delta t} + Q_m \tag{12.20}$$

where subscript m indicates the average value in the interval and $A_{s.m}$ is the average area of the surge chamber between z and $z + \Delta z$.

(a) Solution by successive estimates. In each time interval estimate ΔV. Then, $V_m = V_i + \dfrac{\Delta V}{2}$, and from equation (12.19) calculate $z_m \left(= z_i + \dfrac{\Delta z}{2} \right)$ whence z is calculated, noting that $V_s = \Delta z / \Delta t$. Subscript i indicates values at the beginning of the time interval and which are therefore known. Q_m is known since the time variation of discharge to the turbines will be prescribed and substitution of Δz into (12.20) yields V_m. If the two values of V_m agree, the estimated value of ΔV is correct; otherwise adjust ΔV and repeat until agreement is achieved and proceed to the next time interval.

Alternatively, estimate Δz and proceed in similar fashion; this is preferable if the chamber has a variable area. In both cases the time variation of z is obtained.

Such calculations are ideally carried out on a digital computer or programmable calculator, due to the repetitive nature of the computations. However the calculations are simple and since the time interval can be fairly lengthy in such cases (e.g. 10 sec) basic electronic calculators can be used.

(b) Direct solution of equations (12.19) and (12.20). From equation (12.20)

$$\Delta z = \frac{\Delta t}{A_{s.m}} \left(V_i \, A_T + \frac{A_T}{2} \Delta V - Q_m \right) \tag{12.21}$$

where $V_m = V_i + \dfrac{\Delta V}{2}$

Also $z_m = z_i + \dfrac{\Delta z}{2}$ and $V_s = \dfrac{\Delta z}{\Delta t}$; equation (12.19) becomes

$$\frac{L}{g} \frac{\Delta V}{\Delta t} + z_i + \frac{\Delta t}{2A_s} \left(V_i \, A_T + \frac{A_T}{2} \Delta V - Q_m \right) \pm F_T \left(V_i + \frac{\Delta V}{2} \right)^2$$

$$\pm \frac{F_s}{A_s^2} \left(A_T^2 \left(V_i^2 + V_i \, \Delta V + \frac{\Delta V^2}{4} \right) - 2A_T \left(V_i + \frac{\Delta V}{2} \right) Q_m + Q_m^2 \right) = 0$$

Rearranging

$$\pm \frac{F_R}{4} \Delta V^2 + \left(\frac{L}{g\Delta t} + \frac{A_T}{4A_{s.m}} \Delta t \pm \left(F_R \, V_i - \frac{F_s \, A_T \, Q_m}{A_s^2} \right) \right) \Delta V + z_i$$

$$+ \frac{A_T}{2A_{s.m}} V_i \, \Delta t - \frac{Q_m}{2A_{s.m}} \Delta t \pm \left(F_R \, V_i^2 + \frac{F_s}{A_s^2} Q_m (-2V_i \, A_T + Q_m) \right) = 0 \tag{12.22}$$

which is of the form:

$$a \, \Delta V^2 + b \, \Delta V + c = 0 \tag{12.23}$$

$$\text{whence } \Delta V = \frac{-b + \sqrt{b^2 - 4ac}}{2a} \tag{12.24}$$

ΔV is therefore determined explicitly in each successive time step Δt and the corresponding change in z is obtained from equation (12.21). Note that if V becomes -ve (i.e. on the downswing) the -ve value of F_R is used. As with most finite difference methods, in this case Δt should be small since the use of average values of the variables implies a linear time variation. A ten second time interval gives a sufficiently accurate solution.

12.5 Pressure transients in pipelines (waterhammer)

Changes in the discharge in pipelines, caused by valve or pump operation, either closure or opening, result in pressure surges which are propagated along the pipeline from the source. If the changes in control are gradual the time variation of pressures and discharge may be achieved by assuming the liquid to be incompressible and neglecting the elastic properties of the pipeline such as in the problems on surge analysis dealt with in sections 12.1 to 12.4.

In the case of rapid valve closure or pump stoppage the resulting deceleration of the liquid column causes pressure surges having large pressure differences across the wave front. The speed (celerity) of the pressure wave is dependent on the compressibility of the liquid and the elasticity of the pipeline and these parameters are therefore incorporated in the analysis.

The simplest case of waterhammer, that due to an instantaneous valve closure, can be used to illustrate the phenomenon (fig. 12.3).

At time Δt after closure the pressure wave has reached a point $x = c\Delta t$ where c is the celerity of the wave. In front of the wave the velocity is Vo and behind it the water has come to rest. The pressure within the region $0 - x$ will have increased significantly and the pipe diameter will have increased due to the increased stress. The density of the liquid will increase due to its compressibility. Note that it takes a time $t = \dfrac{L}{c}$ for the whole column to come to rest. At this time the wave has reached the reservoir where the energy is fixed at HR (fig. 12.3 (c)). Thus the increased stored strain energy in the pipeline cannot be sustained and in its release water is forced to flow back into the reservoir in the direction of the pressure gradient. The wave front retreats to the valve at celerity, c (fig. 12.3 (d)) and arrives at $t = \dfrac{2L}{c}$ $(= T)$ (fig. 12.3 (e)). Due to the subsequent arrest of retreating column a reduced pressure wave is propagated to the reservoir

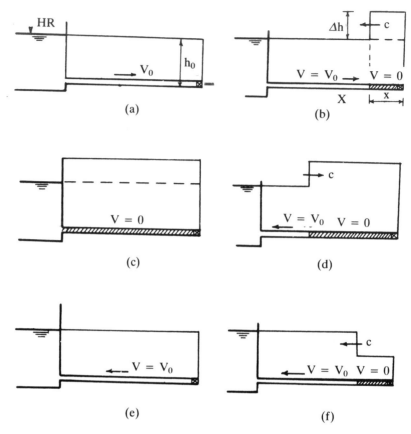

Figure 12.3 Pressure transients in uniform pipeline due to sudden valve closure

and the whole sequence repeated. In practice friction eventually damps out the oscillations.

It will be shown later that in the case of an instantaneous stoppage the pressure head rise $\Delta h = \dfrac{cVo}{g}$ where Vo is the initial steady velocity. c may be of the order of 1300 m/s for steel or iron pipelines. For example if Vo = 2 m/s, $\Delta h = \dfrac{1300 \times 2}{9 \cdot 81} = 265$ m thus giving some idea of the potentially damaging effects of waterhammer.

12.6 The basic differential equations of waterhammer

The continuity and dynamic equations applied to the element of flow δx (fig. 12.4) yield

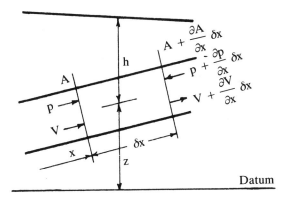

Figure 12.4 Unsteady flow field

continuity equation: $\dfrac{\partial h}{\partial t} + V \dfrac{\partial h}{\partial x} + \dfrac{c^2}{g} \dfrac{\partial V}{\partial x} = 0$ (12.25)

dynamic equation: $\dfrac{\partial h}{\partial x} + \dfrac{V}{g} \dfrac{\partial V}{\partial x} + \dfrac{1}{g} \dfrac{\partial V}{\partial t} + \dfrac{\lambda}{2Dg} V \, |V| = 0$ (12.26)

c is the speed of propagation of the pressure wave given by

$$ c = \sqrt{\dfrac{1}{\rho\left(\dfrac{1}{K} + \dfrac{C_1 D}{TE}\right)}} \tag{12.27} $$

where K = bulk modulus of liquid

ρ = density of liquid

E = elastic modules of pipe material

T = pipe wall thickness

D = pipe diameter

C_1 is a constant depending on the method of pipeline anchoring.

For a thin-walled pipe fixed at the upper end, containing no expansion joints, but free to move in the longitudinal direction $C_1 = \dfrac{5}{4} - \eta$ where η is Poisson's ratio for the pipe wall material. For steel and iron $\eta = 0 \cdot 3$ $C_1 = 1 - \eta^2$ for a pipe without expansion joints and anchored throughout its length.

$C_1 = 1 - \dfrac{\eta}{2}$ for a pipe with expansion joints throughout its length.

Equation (12.27) can be expressed in the form

$$c = \sqrt{\frac{K^1}{\rho}}$$

where K^1 is the effective bulk modulus of the fluid in the flexible pipeline. Since the speed of propagation of a pressure wave (or speed of sound) in an infinite fluid or in a rigid pipeline is $c = \sqrt{\dfrac{K}{\rho}}$ the effect of the term $\dfrac{C_1 D}{TE}$ is to reduce the speed of propagation.

For water $K = 2\cdot1 \times 10^9$ N/m^2

For steel $E = 2\cdot1 \times 10^{11}$ N/m^2

12.7 Solutions of the waterhammer equations

Equations (12.25) and (12.26) can only be solved analytically if certain simplifying assumptions are made, such as the neglect of certain terms, and for simple boundary conditions such as reservoirs and valves. Some methods neglect friction losses which can lead to serious errors in the calculation of pressure transients.

Since the advent of the digital computer equations (12.25) and (12.26) when expressed in discretised form can be readily evaluated for a whole range of boundary conditions and pipe configurations including networks. The computations proceed in small time steps and the pipeline is 'divided' into equal sections. At every 'station' between adjacent sections the transient pressure head and velocity are calculated at each time step. In this way the complete history of the waterhammer in space and time is revealed. No simplifying assumptions to the basic equations need be made. Indeed if interior boundary conditions such as low pressures causing cavitation arise, especially in sloping pipelines, these can be incorporated in the analysis.

The most commonly used numerical method is probably the method of 'characteristics' which reduces the partial differential equations to a pair of simultaneous ordinary differential equations. These, when expressed in numerical (finite difference) form can be programmed for automatic evaluation on a digital computer. Such methods are, however, beyond the scope of this text.

In the present day, therefore, it hardly seems justifiable to use simplified methods. However, in the following sections some of these methods will be illustrated. These were developed before the advent of computers and represent examples of classic analytical techniques. In some cases friction can be included and the results, for example for the pressure transients at closing valves, are very similar to those using the 'characteristics' method. Friction losses are often simplified by assuming them to be localised, for example at an 'orifice' at the outlet from a reservoir. Such losses may also be discretely distributed along the pipeline but this makes the analysis more laborious.

The analytical and graphical methods which are included hereafter illustrate some aspects of the waterhammer phenomenon. The inclusion of additional features has been kept to a minimum in view of the superiority of numerical analysis for practical problems.

12.8 The Allievi equations

The differential equations (12.25) and (12.26) cannot be solved analytically unless certain simplifications are carried out.

The term $V \dfrac{\partial h}{\partial x}$ is of the order of $\dfrac{V}{(V + c)} \dfrac{\partial h}{\partial t}$ and may, therefore, be small.

$\dfrac{\partial V}{\partial x}$ is small compared with $\dfrac{\partial V}{\partial t}$ since $\dfrac{dV}{dt} = -\dfrac{V \partial V}{\partial x} + \dfrac{\partial V}{\partial t}$

$$= \frac{\partial V}{\partial t} \left(1 - \frac{V \partial V}{\partial x} \frac{\partial t}{\partial V} \right)$$

$$= \frac{\partial V}{\partial t} \left(1 - \frac{V}{\partial x / \partial t} \right)$$

in which the last term is small.
Neglecting the friction loss term, in addition, yields

$$\frac{\partial V}{\partial x} = -\frac{g}{c^2} \frac{\partial h}{\partial t} \quad \text{continuity equation} \tag{12.28}$$

$$\frac{\partial V}{\partial t} = -g \frac{\partial h}{\partial x} \quad \text{dynamic equation} \tag{12.29}$$

Riemann obtained a solution to such simultaneous differential equations which may be expressed in the form:

$$h_{tx} = h_o + F \left(t - \frac{x}{c} \right) + f \left(t + \frac{x}{c} \right) \tag{12.30}$$

$$V_{tx} = V_o - \frac{g}{c} \left(F \left(t - \frac{x}{c} \right) - f \left(t + \frac{x}{c} \right) \right) \tag{12.31}$$

where F and f mean 'a function of' and + x is measured as in fig. 12.5 so that the signs in (12.28) and (12.29) become positive.
The functions F and f have the dimension of head.

An observer travelling along the pipeline in the + x direction will be at a position $X_1 = ct + X_o$ at time t, where X_o is his position at $t = 0$.

For the observer the function $F \left(t - \dfrac{x}{c} \right) = F \left(t - \dfrac{ct + X_o}{c} \right) = F \left(\dfrac{X_o}{c} \right)$

$$= \text{const.}$$

Thus the function $F\left(t - \dfrac{x}{c}\right)$ is a pressure (head) wave which is propagated upstream at the wavespeed c.

By similar argument the function $f\left(t + \dfrac{x}{c}\right)$ is a pressure wave propagated in the $-$ x direction at the wavespeed c (fig. 12.5).

An $F\left(t + \dfrac{x}{c}\right)$ wave generated, for example by valve operation at x = 0, will propagate upstream towards the reservoir at which it will be completely and negatively reflected as a f wave at time $\dfrac{T}{2} = \left(\dfrac{L}{c}\right)$, where T is the waterhammer period $\left(\dfrac{2L}{c}\right)$.

Denoting i as the discrete time period with interval T the equations (12.30) and (12.31) can be written for the downstream control, e.g. valve.

$$h_i = h_o + F_i + f_i \tag{12.32}$$

$$V_i = V_o - \frac{g}{c}(F_i - f_i) \tag{12.33}$$

At i = 0, t = 0 then $h_o = h_o + F_o + f_o$

$$V_o = V_o - \frac{g}{c}(F_o - f_o)$$

At i = 1, t = T $h_1 = h_o + F_1 + f_1$

$$V_1 = V_o - \frac{g}{c}(F_1 - f_1)$$

and so on.

Since f is the reflected F wave, $f_o = 0$; $f_1 = -F_o : f_2 = -F_1$ etc. Thus

$$h_1 = h_o + F_1 - F_o$$

$$V_1 = V_o - \frac{g}{c}(F_1 + F_o)$$

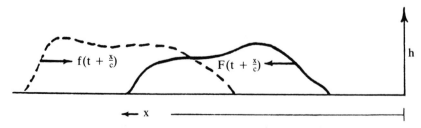

Figure 12.5 Propagation of pressure wave

and $h_2 = h_o + F_2 - F_1$

$$V_2 = V_o - \frac{g}{c}(F_2 + F_1)$$

Adding successive pairs of head equations and subtracting successive pairs of velocity equations

$$h_1 + h_o = 2h_o + F_1$$

$$h_2 + h_1 = 2h_o + F_2 - F_o$$

$$V_o - V_1 = \frac{g}{c}(F_1 + F_o)$$

$$V_1 - V_2 = -\frac{g}{c}(F_o - F_2)$$

whence, in general $F_i - F_{i-2} = \frac{c}{g}(V_{i-1} - V_i)$

$$h_i + h_{i-1} - 2h_o = \frac{c}{g}(V_{i-1} - V_i) \tag{12.34}$$

Boundary conditions.
Valve at downstream end.
  The discharge through a valve of area $A_{v,i}$ at time i when the pressure head behind it is h_i is given by

$$Q = C_{d,i} A_{v,i} \sqrt{2g\, h_i} \tag{12.35}$$

and the velocity in the pipe immediately upstream is

$$V_i = \frac{C_{d,i} A_{v,i}}{Ap} \sqrt{2g\, h_i} \tag{12.36}$$

where Ap is the area of the pipe and $C_{d,i}$ the discharge coefficient.
  The ratio

$$\frac{V_i}{V_o} = \frac{C_{d,i} A_{v,i}}{C_{d,o} A_{v,o}} \frac{\sqrt{h_i}}{\sqrt{h_o}}$$

Denoting $\eta_i = \dfrac{C_{d,i} A_{v,i}}{C_{d,o} A_{v,o}}$ and $\xi_i^2 = \dfrac{h_i}{h_o}$

$$V_i = V_o\, \eta_i\, \xi_i \tag{12.37}$$

and $\quad h_i = h_o\, \xi_i^2 \tag{12.38}$

Substituting into equation (12.34) yields

$$h_o\, \xi_i^2 + h_o\, \xi_{i-1}^2 - 2h_o = \frac{cV_o}{g}(\eta_{i-1}\, \xi_{i-1} - \eta_i\, \xi_i) \tag{12.39}$$

and denoting $\dfrac{cV_o}{2g\,h_o}$ by the symbol ρ (not fluid density)   (12.39a)

$$\xi_i^2 + \xi_{i-1}^2 - 2 = 2\rho\,(\eta_{i-1}\,\xi_{i-1} - \eta_i\,\xi_i)$$ (12.40)

Equation (12.40) represents a series of equations which enable the heads, at time intervals (T) $i = 1, 2, 3 \ldots$, at the valve to be calculated for a prescribed closure pattern $(A_{v,i}\ \text{v. t})$. The equations are known as the Allievi interlocking equations.

If the valve closure is instantaneous, $\eta_{1=0}$, in equation (12.40) and

$$\xi_i^2 + \xi_o^2 - 2 = 2\rho\,(\eta_o\,\xi_o - 0)$$

Now $\eta_o = 1;\ \xi_o = 1$

whence $\xi_1^2 - 1 = 2\rho = \dfrac{cV_o}{gh_o}$

$\therefore$ $h_o\left(\dfrac{h_1}{h_o} - 1\right) = \dfrac{cV_o}{g}$

or $\Delta h = h_1 - h_o = \dfrac{cV_o}{g}$

Note therefore that if the valve is closed in any time $t < \dfrac{2L}{c}$ the result is the same as that for instantaneous closure since $\eta_1 = 0$.

12.9 Alternative formulation

For a downstream valve boundary condition at discrete time intervals T, from equation (12.36)

$$V_i = \frac{C_{d,i}\,A_{v,i}}{Ap}\,\sqrt{2g\,h_i}$$

Then $V_i = B_i\,\sqrt{h_i}$; $h_i = \dfrac{V_i^2}{B_i^2}$

where $B_i = \dfrac{C_{d,i}\,A_{v,i}\,\sqrt{2g}}{Ap}$

Substituting in equation (12.32) and adding equation (12.33) yields

$$\frac{V_i^2}{B_i^2} + \frac{cV_i}{g} - h_o - 2f - \frac{cV_o}{g} = 0$$

whence $V_i = -\dfrac{B_i^2 c}{2g} + B_i\,\sqrt{\left(\dfrac{B_i c}{2g}\right)^2 + \dfrac{cV_o}{g} + h_o + 2f}$ (12.41)

Note that $f_i = -F(i-1)$ (see section 12.8)

$$F_i = \frac{c}{g}(V_o - V_i) + f_i \tag{12.42}$$

and $h_i - h_o = F_i + f_i \tag{12.43}$

12.10 The Schnyder–Bergeron graphical method

This graphical method is derived from the Riemann solution to the differential equations, (12.30) and (12.31).

It can be shown that an observer travelling along the pipeline at the wave speed, c, will experience pressure head and discharge variations given by the straight line

$$h_{x,t} - h_{x,T} = \frac{c}{g\,Ap}(Q_{x,t} - Q_{X,T}) \tag{12.44}$$

when moving in the direction of $-V$ (fig. 12.6).

Hence if $Q_{X,T}$ and $h_{X,T}$ are known for the point X at time T the values of $h_{x,t}$ and $Q_{x,t}$ at the point x, t are linearly related by equation (12.44).

Similarly, an imaginary observer travelling at speed c in the $+V$ direction will experience head and discharge variations given by

$$h_{x,t} - h_{X,T} = -\frac{c}{g\,Ap}(Q_{x,t} - Q_{X,T}) \tag{12.45}$$

Equations (12.44) and (12.45) are obtained from the Allievi equations and are called 'waterhammer lines'.

In order to locate the lines in the h, Q plane (see fig. 12.6) the boundary conditions at the ends of the pipeline are required; these must be expressible

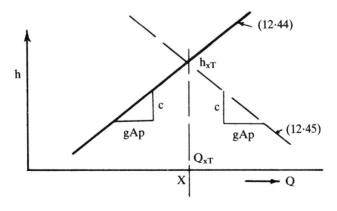

Figure 12.6 Observer waterhammer lines. Schnyder–Bergeron method

in a relationship between h and Q. For example, in the case of a downstream valve, as previously shown, the discharge through a valve of area A_v is

$$Q = C_d\, A_v\, \sqrt{2gh} \qquad\qquad (12.46)$$

where h is the pressure head behind the valve.

Since the calculations proceed at discrete time intervals, T, equation (12.46) represents a series of parabolic curves, $Q = C_d\, A_{v,i}\, \sqrt{2gh}$ each of which may be denoted by Ψ_i. The upstream boundary condition may typically be determined by reservoir. The reservoir elevation HR may be constant or a known function of time. The head is therefore independent of discharge and for a fixed level with a frictionless pipe is expressed by h_o = HR = constant and shown by a horizontal line on the h v. Q diagram.

Consider the case of a reservoir of fixed level discharging water through a frictionless pipeline terminating in a valve which is initially partially closed (fig. 12.7).

If the valve opening is reduced in such a way that it is fully closed after a number of discrete waterhammer periods the upstream and downstream boundary conditions will be as shown in fig. 12.8. The point of intersection of Ψ_o and the line HR = const. gives the initial conditions Q_o, h_o. These conditions will exist at A (the reservoir) until $t = \dfrac{L}{c}$. If at time $\dfrac{L}{c}$, when the first pressure wave arrives at A, an observer starts towards B at speed c he will experience discharge and head variations given by the waterhammer line sloping at $-\dfrac{c}{g\,Ap}$. This line is drawn through B_o, $A_o - 0{\cdot}5$ and intersects Ψ_1 to give h_1, Q_1 at the valve. The observer now returns to A, (line B_1 to $A_{1.5}$) to the reservoir. He returns to the valve, (line $A_{1.5}$ to B_2) to meet Ψ_2 the new valve boundary condition, and so on. When the valve is closed, the condition at the valve is given by B_4. The valve now represents a closed end and the pressure surges oscillate as described in fig. 12.3. Note that at B_5 the pressure, in the case shown, is subatmospheric and if this approaches -10 m vaporisation will occur. The effect of such a phenomenon on the waterhammer process is complex and is not yet fully understood. Some consider that a water column retreats leaving a vacuum behind; when the momentum of the water column is overcome it is believed to return to the closed valve resulting in a 'slamming' force. This is probably an over-simplification. Certainly low pressure water will contain a

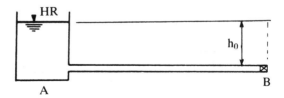

Figure 12.7 Pipeline with valve throttling

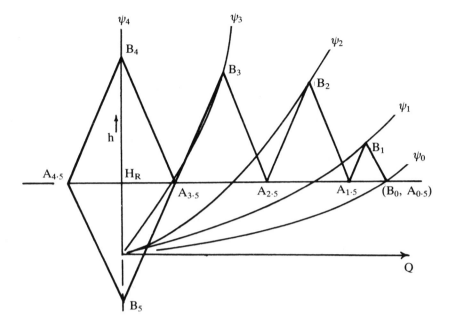

Figure 12.8 Waterhammer pressure curves

significant amount of free air and the effect of this is to reduce considerably the wave celerity.

12.11 Pipeline friction and other losses

The omission of pipeline head losses can cause serious errors in water-hammer computations. Such losses can be represented approximately in the Schnyder–Bergeron method by assuming that they are concentrated at the inlet from the reservoir to the pipeline in a device such as an orifice the head-discharge relationship of which is similar to that of the pipeline head loss discharge function i.e. $h_L = K Q^2$ (fig. 12.9).

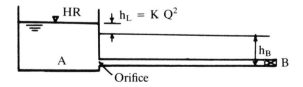

Figure 12.9   Representation of pipeline losses by orifice at inlet to pipe

The boundary condition at the reservoir is therefore $h_A = HR - K Q^2$

where $K = \left(\dfrac{\lambda L}{D} + K_m\right) \dfrac{1}{A_p^{\,2}}$ (12.47)

where K_m is the minor loss coefficient for the pipeline.

Equation (12.47) plots as shown in fig. 12.10 (curve A) on which typical valve boundary conditions and waterhammer lines have been superimposed.

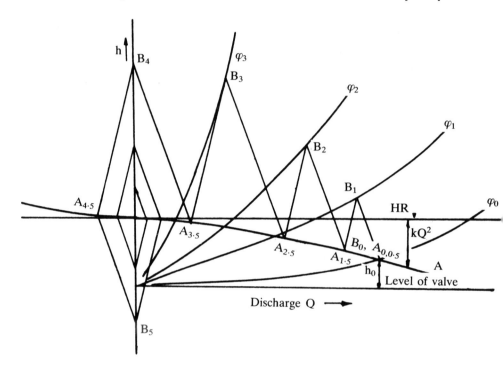

Figure 12.10 Schnyder–Bergeron diagram with pipeline losses represented

This representation gives results in close agreement with numerical solutions of the governing differential equations for the pressure head elevation at the valve. It is also possible to distribute the friction loss along the pipeline by imaginary 'orifices' placed at spatial intervals. The Schnyder–Bergeron construction then becomes more laborious and due to the superiority of numerical solutions of the governing equations in any case is rarely worth pursuing.

12.12 Pressure transients at interior points

Consider the system shown in fig. 12.11.

An observer leaving A at time, say, $T = 1$ and travelling towards B at speed c will meet a similar observer who left B at time $T = 1$, at c at time

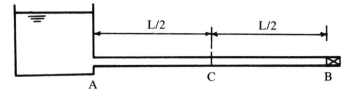

Figure 12.11   Pressure transients at intermediate points

$T + \dfrac{L}{2c}$ (i.e. 1·25 T). Using the graphical method it is necessary to obtain the location of A_1 which necessitates the construction of the $\psi_{0.5}$ valve boundary condition.

Figure 12.12 shows the construction yielding $C_{1.25}$ and this may be continued to yield $C_{2.25}$, etc.

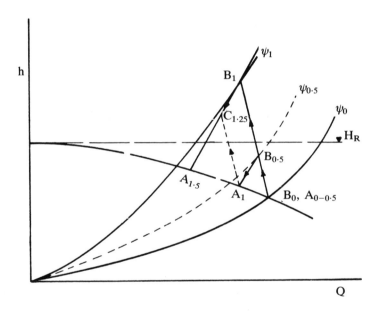

Figure 12.12   Pressure transient curves

12.13 Pressure transients caused by pump stoppage

Rapid stoppage of a pump generates a negative pressure wave to, for example, a reservoir, at which it is reflected as a positive wave. If the flow from the pump has stopped before the pressure wave has returned, the reflux valve will close and a 'dead end' condition will be created. The cycle of events occurring when stoppage occurs within a time $\dfrac{2L}{c}$ is shown by a Schnyder–Bergeron diagram (fig. 12.13) assuming that the initial negative pressure transient does not fall to vapour pressure. Note that the representation of pipeline friction simulates the decay of the waterhammer phenomenon. Long pumping mains are often provided with some form of surge suppressor such as an air vessel which consists of a closed vessel containing water maintained at a constant pressure by an air compressor. When negative pressure transients occur due to pump 'tripping' the 'air vessel' therefore introduces water into the pipeline and hence reduces the pressure drop.

If a rotodynamic pump continues to rotate after power cut due to its inertia, it may continue to provide some discharge and hence reduce the pressure transients in an analogous way to that of slow, as opposed to rapid, valve closure. It can be shown that the rate of deceleration of a pump is given by

$$\frac{dN}{dt} = \frac{3600 \, \rho g \, Q \, H_m}{4\pi^2 \, I \, N \, \eta} \tag{12.48}$$

where Q and H_m are the discharge and manometric head at rotational speed N; I is the moment of inertia of the rotating parts of the pump and motor; η is the pump efficiency at the discharge Q and speed N.

In using the graphical method and working in finite time intervals $\Delta t = T = \frac{2L}{c}$ the speed N_2 at the end of Δt is therefore given by

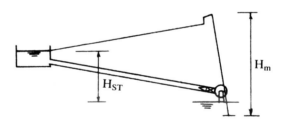

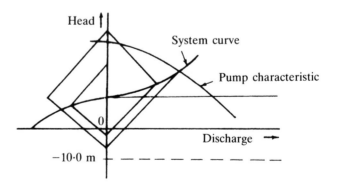

Figure 12.13 Pressure transients due to sudden pump stoppage – Schnyder–Bergeron diagram

$$N_2 = N_1 - \frac{3600 \; \rho g \; Q_1 \; H_{ml, \Delta T}}{4\pi^2 \; I \; N_1 \; \eta} \qquad (12.49)$$

The new pump characteristic curve can then be constructed using the method of Chapter 5 i.e.

$$H_2 = H_1 \left(\frac{N_2}{N_1}\right)^2; \quad Q_2 = Q_1 \left(\frac{N_2}{N_1}\right) \qquad (12.50)$$

These curves represent the upstream boundary condition. Note that, apart from the pump boundary conditions at the first and second time intervals the remainder can only be produced as the observer waterhammer lines are constructed at each time interval since they depend upon the values of Q and H_m at the end of the previous time interval (see Example 12.7).

The downstream boundary condition (for example a reservoir) is represented by

$$h = H_{ST} + \frac{Q^2}{2g \; A_p{}^2} \left(\frac{\lambda \; L}{D} + K_m\right)$$

Worked examples

Example 12.1
A surge chamber 10 m in diameter is situated at the downstream end of a low pressure tunnel 10 km long and 3 m in diameter. At a steady discharge of 36 m³/s the flow to the turbines is suddenly stopped by closure of the turbine inlet valves. Determine the maximum rise in level in the surge chamber and its time of occurrence.

Solution:

$$V_o = \frac{36}{7 \cdot 069} = 5 \cdot 093 \text{ m/s}; \quad A_T = 7 \cdot 069 \text{ m}^2; \quad A_s = 78 \cdot 54 \text{ m}^2$$

From section 12.2; $\quad T = 2\pi \sqrt{\dfrac{10\,000}{9 \cdot 81} \times \dfrac{78 \cdot 54}{7 \cdot 069}} = 668 \cdot 67 \text{ s}$

(equation (12.9)).

From equation (12.12): $C_2 = V_o \dfrac{A_T}{A_s} \dfrac{T}{2\pi} = 5 \cdot 093 \times \dfrac{7 \cdot 069}{78 \cdot 54} \times \dfrac{668 \cdot 67}{2\pi} = 48 \cdot 78$

Equation (12.8): $z = C_2 \sin \dfrac{2\pi t}{T} = 48 \cdot 78 \sin \dfrac{2\pi t}{T}$

z = maximum, when $t = \dfrac{T}{4}$, = 48·78 m occurring after 167·17 sec.

Example 12.2
A surge chamber 100 m² in area is situated at the end of a 10 000 m long, 5 m diameter tunnel; $\lambda = 0.01$. A steady discharge of 60 m³/s to the turbines is suddenly stopped by the turbine inlet valve. Neglecting surge chamber losses, determine the maximum rise in level in the surge chamber and its time of occurrence.

Solution:
Substitution of a series of values of z into equation (12.17) and using equation (12.16) enables the z v. V relationship to be plotted. When V = 0 (maximum upsurge), z is found to be about 37·1 m at a time of 123 seconds.

$$z_o = -9.42 \text{ m}, \quad V_o = 3.056 \text{ m/s}.$$

Example 12.3
A low pressure tunnel 8000 m long, 4 m diameter, $\lambda = 0.012$ delivers a steady discharge of 45 m³/s to hydraulic turbines. A surge chamber of constant area 100 m² is situated at the downstream end of the tunnel, $F_s = 1.0$. Calculate the time variation of tunnel velocity V, and level in surge chamber using the finite difference forms of the governing differential equations given by equations (12.19) and (12.20) if the flow to the turbines is suddenly stopped.

Solution:
Individual steps in the iterative procedure are not given; the solution is shown in the table on page 312. The reader should carry out a few calculations.

Note that $F_R = \dfrac{\lambda L}{2gD} + F_s \left(\dfrac{A_T}{A_s}\right)^2$

Solving Example 12.3 by the direct method: (equation (12.24)),

Take $\Delta t = 10$ s

$$A_T = 12.566 \text{ m}^2: \ A_s = 100 \text{ m}^2 \text{ (constant)}$$

$$F_R = F_s \left(\frac{A_T}{A_s}\right)^2 + \frac{\lambda L}{2g \, D_T}$$

$$= 1.0 \left(\frac{12.566}{100}\right)^2 + \frac{0.012 \times 8000}{19.62 \times 4} = 1.239$$

$$V_o = \frac{Q_o}{A_T} = \frac{45}{12.566} = 3.581 \text{ m/s}$$

$$z_o = F_T \, V_o^2 = -15.686 \text{ m}$$

In equation (12.24) a $= +\dfrac{F_R}{4} = 0.3097$

$$b = \frac{L}{g \, \Delta t} + \frac{A_T}{4A_{s.m}} \, \Delta t + F_R \, V_i$$

v_i = velocity at beginning of time step (= 3·581 m/s)

$$\therefore \qquad b = \frac{8000}{9·81 \times 10} + \frac{12·566 \times 10}{4 \times 100} + 1·239 \times 3.581$$

$$b = 86·30 \; (s^{-1})$$

$$c = z_i + \frac{A_T}{2A_{s.m}} \, V_i \, \Delta t - \frac{Q_{m.\Delta t}}{2A_{s.m}} + F_R \, V_i^2$$

$$c = -15·686 + \frac{12·566 \times 3·581 \times 10}{2 \times 100} + 1·239 \times 3·581^2$$

$$(\text{since } Q_m = 0)$$

$$c = 2·4525 \; m$$

$$\therefore \qquad \Delta V = (-86·3 + \sqrt{86·3^2 - 4 \times 0·3097 \times 2·4525})/(2 \times 0·3097)$$

$$(\text{equation (12.24)})$$

$$\Delta V = -0·0284 \; m/s$$

$$\therefore \; V_{(i+1)} = 3·552 \; m/s$$

From equation (12.21) $\Delta z = \dfrac{\Delta t}{A_{s.m}} \left(V_i \, A_T + \dfrac{A_T}{2} \Delta V - Q_m \right)$

$$\Delta z = \frac{10}{100} \left(3·581 \times 12·566 - \frac{12·566}{2} \times 0·0284 - 0 \right)$$

$$= 4·482 \; m$$

i.e. $\qquad z_{(i+1)} = (z_i + \Delta z) = -15·686 + 4·48$

$$= -11·204 \; m.$$

The values $V_{(i+1)}$; $z_{(i+1)}$ become V_i and z_i for the next time step and computations proceed in the same manner. See table on page 331.

Example 12.4
Calculate the speed of propagation of a pressure wave in a steel pipeline 200 mm in diameter with a wall thickness of 15 mm
(a) assuming the pipe to be rigid
(b) assuming the pipe to be anchored at the reservoir, with no expansion joints, and free to move longitudinally
(c) assuming the pipe to be provided with expansion joints.
    In each case determine the pressure head rise due to sudden valve closure when the initial steady velocity of flow is 1·5 m/s.

Summary of computations using iterative method

Time (s)	0	10	20	30	40	50	60	70
z (m)	−15·686	−11·204	−6·788	−2·497	1·620	5·519	9·161	12·515
V (m/s)	3·581	3·552	3·475	3·355	3·197	3·007	2·790	2·55

Time (s)	80	90	100	110	120	130	140
z (m)	15·554	18·254	20·597	22·568	24·153	25·344	26·134
V (m/s)	2·288	2·010	1·720	1·417	1·106	0·789	0·468

Time (s)	150	160	170	180
z (m)	26·52	26·49	26·06	25·23
V (m/s)	0·143	−0·182	−0·50	−0·82

Solution:

$$K = 2·1 \times 10^9 \text{ N/m}^2; \quad E = 2·1 \times 10^{11} \text{ N/m}^2; \quad \eta = 0·3$$

(a) $\quad c = \sqrt{\dfrac{K}{\rho}} = \sqrt{\dfrac{2·1 \times 10^9}{1000}} = 1450 \text{ m/s}$

$\quad \Delta h = \dfrac{cV_o}{g} = \dfrac{1450 \times 1·5}{9·81} = 221·7 \text{ m}$

(b) $\quad C_1 = \dfrac{5}{4} - \eta = 0·95$

$$c = \sqrt{\dfrac{1}{\rho\left(\dfrac{1}{K} + \dfrac{C_1 D}{TE}\right)}} = \sqrt{\dfrac{1}{1000\left(\dfrac{1}{2·1 \times 10^9} + \dfrac{0·95 \times 0·2}{0·015 \times 2·1 \times 10^{11}}\right)}}$$

$\quad c = 1365·2 \text{ m/s}; \quad \Delta h = 208·7 \text{ m}$

(c) $\quad C_1 = 1 - \dfrac{\eta}{2} = 1 - 0·15 = 0·85$

$\quad c = 1373·4 \text{ m/s}; \quad \Delta h = 210·0 \text{ m}.$

Example 12.5
A steel pipeline 1500 m long, 300 mm diameter discharges water from a reservoir to atmosphere through a valve at the downstream end. The speed of the pressure wave is 1200 m/s. The valve is closed gradually in 20 seconds and the area of gate opening varies as shown in the first table on page 317. Neglecting friction, calculate the variation of pressure head at the valve during closure if the initial head at the valve is 10 m, (a) using the Allievi method, (b) the method of section 12.9, and (c) the Schnyder−Bergeron method.

Solution:

$$\text{Waterhammer period } T = \frac{2L}{c} = \frac{3000}{1200} = 2 \cdot 5 \text{ s}$$

Working in time intervals of 2·5 s the corresponding valve areas by interpolation are shown in the second table on page 336.

(a) Allievi method
From equation (12.40)

$$\xi_i^2 + \xi_{i-1}^2 - 2 = 2\rho \left(\eta_{i-1} \xi_{i-1} - \eta_i \xi_i \right) \tag{i}$$

$$\text{where } \xi_i^2 = \frac{h_i}{h_o}; \quad \eta_i = \frac{C_{d,i} A_{v,i}}{C_{d,o} A_{v,o}}$$

$$\left(\text{where i indicates the discrete time intervals separated by } T = \frac{2L}{c} \text{ (s)} \right) \text{ and}$$

$$V_i = V_o \, \eta_i \, \xi_i \text{ (equation (12.37))} \; C_d = 0 \cdot 6 \text{ (constant)}; \; V_o = \frac{C_d \, A_{v,o}}{A_p} \sqrt{2g \, h_o} =$$

$$3 \cdot 567 \text{ m/s}, \; \rho = \frac{c V_o}{2g \, h_o} \text{ (equation (12.39a))}$$

Table of calculations

(1)	(2)	(3)	(4)	(5)	(6)
$T_{(i)}$	t(s)	η	ξ	h (m)	v (m/s)
0	0	1	1	10	3·567
1	2·5	0·780	1·2644	15·988	3·518
2	5·0	0·550	1·6908	28·588	3·317
3	7·5	0·347	2·2816	52·057	2·821
4	10·0	0·260	2·2952	52·678	2·128
5	12·5	0·193	2·1508	46·260	1·483
6	15·0	0·147	1·8753	35·166	0·981
7	17·5	0·073	2·0117	40·470	0·526
8	20·0	0	2·0952	43·897	0

Notes: At T = 1

$$\xi_1^2 + \xi_0^2 - 2 = 2\rho \left(\eta_o \xi_o - \eta_1 \xi_1 \right)$$

$$\xi_1^2 + 1 - 2 = 2\rho \left(1 - 0 \cdot 78 \, \xi_1 \right)$$

$$\xi_1^2 - 1 = 43 \cdot 632 \left(1 - 0 \cdot 78 \, \xi_1 \right)$$

whence $\xi_1 = 1 \cdot 2644$; $\quad h_1 = \xi_1^2 \times h_o = 15 \cdot 988$ m

$$V_1 = V_o \, \eta \, \xi_i = 3 \cdot 567 \times 0 \cdot 78 \times 1 \cdot 2644 = 3 \cdot 518 \text{ m/s}$$

At T = 2

$$\xi_2^2 + \xi_1^2 - 2 = 2\rho \left(\eta_1 \xi_1 - \eta_2 \xi_2 \right)$$

$$\xi_2^2 + (1 \cdot 2644)^2 - 2 = 43 \cdot 632 \left(0 \cdot 78 \times 1 \cdot 2644 - 0 \cdot 55 \times \xi_2 \right)$$

whence $\xi_2 = 1.6908$; $h_2 = 28.588$ m

$$V_2 = 3.567 \times 0.55 \times 1.6908 = 3.317 \text{ m/s}$$

(b) Alternative algebraic method

Table of calculations

(1)	(2)	(3)	(4)	(5)	(6)	(7)	(8)
$T_{(i)}$	t(s)	B	V (m/s)	F (m)	f (m)	Δh (m)	H (m)
0	0	1·128	3·567	0	0	0	10·0
1	2·5	0·880	3·518	5·988	0	5·988	15·988
2	5·0	0·620	3·317	24·577	−5·988	18·588	28·588
3	7·5	0·391	2·821	66·634	−24·577	42·057	52·057
4	10·0	0·293	2·128	109·313	−66·634	42·678	52·678
5	12·5	0·218	1·483	145·573	−109·313	36·260	46·260
6	15·0	0·165	0·991	170·739	−145·573	25·166	35·166
7	17·5	0·083	0·526	201·209	−170·739	30·470	40·470
8	20·0	0	0	235·107	−201·209	33·898	43·898

Notes: i indicates time step; ho = initial head at valve; Vo = initial velocity in pipe.

Col. (3) $B(i) = \dfrac{Cd\ A_v(i)\ \sqrt{2g}}{A_p}$

Col. (4) $V(i) = -\dfrac{B(i)^2 c}{2g} + B(i)\ \sqrt{\left(\dfrac{B(i)c}{2g}\right)^2 + \dfrac{cVo}{g} + ho + 2f(i)}$

Col. (5) $F(i) = \dfrac{c}{g}\ (Vo - V(i) + f)$

Col. (6) $f(i) = -F(i-1)$

Col. (7) $\Delta h(i) = F(i) + f(i)$

Col. (8) $h(i) = ho + \Delta h(i)$

(c) The Schnyder–Bergeron method

Generate the curves representing the h v. Q relationship for the downstream (valve) boundary condition at the discrete time intervals T.

$$Q = Cd\ A_v\ \sqrt{2g\ h}$$

See table on page 315.

Slope of waterhammer lines $= \pm \dfrac{c}{g\ Ap}$

$= \pm \dfrac{1200}{9.81 \times 0.07068} = \pm 1730.5 \text{ m/m}^3/\text{s}$

$= 17.3 \text{ m/(10 l/s)}.$

Values of h (m) calculated from the equation: $Q = Cd\ A_v\ \sqrt{2g\ h}$

Time Period	A_v (m^2)	Q l/s					
		2	5	10	20	40	60
0	0·03	112·8	178·3	252·1	356·6		
1	0·0234	86·9	139·0	196·7	278·1		
2	0·0165	62·0	98·1	138·6	196·1	277·3	
3	0·0104	39·1	61·8	87·4	123·6	174·8	214.0
4	0·0078	29·3	46·3	65·6	92·7	131·1	160·6
5	0·0058	21·8	34·5	48·7	68·9	97·4	119·4
6	0·0044	16·52	26·1	37·0	52·3	74·0	90·5
7	0·0022	8·28	13·1	18·5	26·1	37·0	45·26
8	0	—	—	—	—	·—	—

Figure 12.14 shows the graphical construction with the downstream boundary conditions at T = 0, 1 ... etc. indicated by ψ_o, ψ_1, ψ_2 ... etc. h_1, h_2 ... etc. indicate the pressure heads at the valve at the discrete water-hammer periods. (See fig. 12.14.)

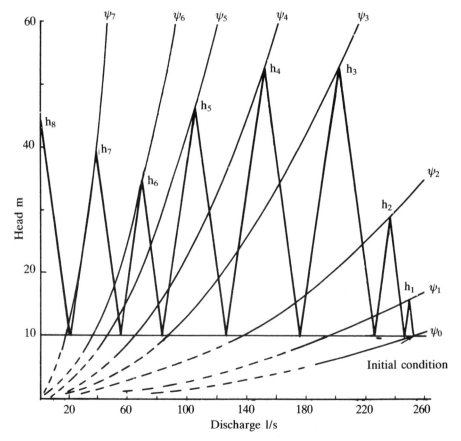

Figure 12.14 Graphical solution of waterhammer pressure transients – method of characteristics

Summary of pressure heads at the valve

Time Period	0	1	2	3	4	5	6	7	8
h (m)	10	16·0	28·5	55·5	55·0	46·5	34·5	40·0	44·0

Comment: The results of the graphical method are very close to those of the Allievi and algebraic methods. This would be expected since all three methods are based on the same basic equations.

Example 12.6
Repeat Example 12.5 incorporating pipeline friction, velocity head and entry losses. Take pipe roughness to be 0·06 mm.

Solution:
The reservoir boundary condition is computed from equation (4.10)

$$Q = -2 A_p \sqrt{2g D \frac{h_f}{L}} \log \left[\frac{k}{3 \cdot 7 D} + \frac{2 \cdot 51 \, v}{D \sqrt{2g D \frac{h_f}{L}}} \right]$$

which gives the discharge for a specified friction head loss (h_f).

The minor head loss, h_m, $\left(\text{e.g.} \ \dfrac{1 \cdot 5 \ V^2}{2g} \right)$ is added to give the head loss v. discharge relationship and hence the head v. discharge relationship for the reservoir. See last table on page 336.

The curve h_L v. Q is plotted, the values of h_L being marked off downwards from the line h = 10·0 m. The valve boundary condition lines are identical with those of Example 12.5.

Observer waterhammer lines start from $B_{0.5}$ the method of construction being identical to that of the previous Example, with the final figure having the general form of fig. 12.10.

Note that in this case the initial discharge for the given conditions is correctly represented by the point $B_{0.5}$ since this gives the solution to the pair of equations

$$Q = C_d A v_o \sqrt{2g h_o} \quad \text{and} \quad h_{B,o} = \frac{Q^2}{2g A_p^2} \left(\frac{\lambda L}{D} + K_m \right)$$

The maximum transient pressure head elevation at the valve is 51 m which is higher than that obtained using the 'frictionless pipeline' assumption.

Example 12.7
A rotodynamic pump having the tabulated characteristics and inertia 20 kg/m² delivers water to a reservoir, the level in which is 10 m above that in the pump suction well, through a 3000 m long pipeline, 300 mm in diameter and

A_v. v. t for Examples 12.5 and 12.6

Time (s)	0	2	4	6	8	10	12	14	16	18	20
A_v (m²)	0·03	0·025	0·02	0·015	0·010	0·008	0·006	0·005	0·004	0·002	0

Valve area at discrete time intervals; $\Delta t = \dfrac{2L}{c}$

Time (s)	0	2·5	5·0	7·5	10·0	12·5	15·0	17·5	20·0
A_v (m²)	0·03	0·0234	0·0165	0·0104	0·0018	0·0058	0·0044	0·0022	0·0

Head loss v. discharge relationship for pipeline (Example 12.6)

h_f (m)	1	2	4	8	10	20	40	50
Q (l/s)	32·65	47·60	68·83	99·37	111·76	160·6	230·0	258·0
h_m (m)	1·016	0·034	0·072	0·151	0·19	0·39	0·81	1·02
$h_L = (h_f + h_m)$	1·016	2·034	4·072	8·151	10·19	20·39	40·81	51·02

roughness 0·6 mm. Losses in valves etc. amount to 5 $V^2/2g$. The celerity of the pressure wave is 1200 m/s. Determine the pressure transients at the pump when the pump is 'tripped'.

Pump characteristics at 1450 rev/min (N_o)

Q (l/s)	0	20	40	60	80
H_m (m)	40·0	38·5	35·0	30·5	25·0
Efficiency (per cent)		44	64	70	60

Solution:
Calculate the head loss v. discharge relationship as in Example 12.6.

Result: (h_L includes friction and minor loss)

h_L (m)	1·02	2·04	4·08	8·16	10·20	15·31
Q (l/s)	19·54	27·95	39·92	56·84	63·66	78·10

This is plotted on the h v. Q diagram (system curve) together with the pump characteristic for 1450 rev/min (curve B (ψ_o)).

The steady state condition is given by the point of intersection of A and B, i.e.

$$Q_o = 78 \text{ l/s}; \quad H_{m,o} = 25·5 \text{ m}; \quad Q/N = 0·053; \quad \eta = 0·62$$

The slope of the observer waterhammer lines is $\pm \dfrac{c}{g \, A_p} = \pm 17·3$ m/(10 l/s).

Waterhammer period $T = \dfrac{2L}{c} = \dfrac{6000}{1200} = 5$ s

At t = 0 the pump starts to slow down, the rate of deceleration being given by

$$\frac{dN}{dt} = \frac{3600 \, \rho g \, Q \, H_m}{4\pi^2 \, I \, \eta} \quad \text{or in finite-difference form}$$

$$N_1 = N_o - \frac{3600 \, \rho g \, Q_o \, H_{m,o} \, \Delta t}{4\pi^2 \, I \, \eta_o} \quad \text{(equation (12.49))}$$

Taken over the 5 s time interval N_1 = 955 rev/min. Using equations (12.50) the new characteristic curve is drawn (shown by the broken line (X) in fig. 12.15).

A time interval of 5 s is however rather long and it may be preferable to predict the pump characteristic in smaller time intervals, say 0·25 T (1·25 s).

An observer leaving the mid point of the pipeline at t = 0, where he experiences a pressure head elevation $H_{m,o}$ and Q_o, since the pipe friction is considered concentrated at A, will reach the pump at t = 1·25 s. His water-hammer line is $+\dfrac{c}{g\,Ap}$ through B_o.

Then $N_{0\cdot25}$ = 1326·3 rev/min and the pump characteristic using equation (12.50) is obtained e.g. by the homologous points

$$H_{0\cdot25} = H_o \left(\frac{1326\cdot3}{1450}\right)^2; \quad Q_{0\cdot25} = Q_o \left(\frac{1326\cdot3}{1450}\right)$$

If we take H_o = 40; Q_o = 0; $H_{0\cdot25}$ = 33·47 m and $Q_{0\cdot25}$ = 0 and so on. The curve is shown by $\psi_{0\cdot25}$. The observer line intersects this at H_m = 20 m, Q = 75 l/s, Q/N = 0·056 whence η = 0·59.

From these values, in equation (12.49), $N_{0\cdot5}$ = 1184.19 rev/min and the corresponding pump characteristic is shown by $\psi_{0\cdot5}$. (See fig. 12.15.)

Note that an observer leaving B at t = 0 (when he experiences $H_{m,o}$ and Q_o) will reach the pump at $\dfrac{T}{2}$ where the boundary conditions are therefore obtained by the point at which the $\dfrac{+\,c}{g\,Ap}$ line through B_o meets the pump characteristic. Hence $H_{m,0\cdot5}$ = 14·5 m; $Q_{0\cdot5}$ = 72 l/s; Q/N = 0·06; η = 0·54. Proceeding in this way the pump characteristic for T = 1 is obtained (ψ_1).

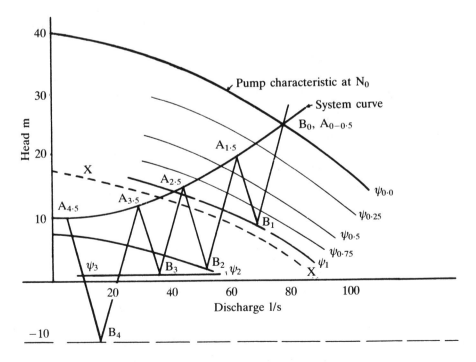

Figure 12.15 Characteristic curves

The head and discharge at the pump at this time are 9·0 m and 68·5 l/s respectively. The pump speed is 1004·4 rev/min (compared with 955 rev/min taking $\Delta t = 5$ s).

The remaining time intervals are treated in similar manner. When the pump speed becomes very low it may be realistic to consider it to have stopped since equation (12.49) neglects bearing and other forms of resistance. In this Example suppose this is assumed to occur at $T = 4$. The observer water-hammer line is proceeding towards the boundary condition $Q = 0$ (i.e. the head axis) but before reaching it experiences a head of -10 m which is approximately the vapour pressure head (depending upon ambient conditions).

The subsequent events are not well understood at the present time and it may be an oversimplification to assume, in the manner of some authors, that a distinct air gap forms. In the pumping main in question it may be desirable to fit a surge suppressor but the design of these falls outside the space limits of this book.

Recommended reading

1. Fox, J.A. (1979) *Hydraulic Analysis of Unsteady Flow in Pipe Networks.* London: Macmillan.
2. Jaeger, C. (1955) *Engineering Fluid Mechanics* (Translated from the German by P.O. Wolf). London: Blackie.
3. Pickford, J. (1969) *Analysis of Surge.* London: Macmillan.
4. Streeter, V.L. and Wylie, E.B. (1983) *Hydraulic Transients.* New York: McGraw-Hill.
5. Streeter, V.L. and Wylie, E.B. (1985) *Fluid Mechanics*, 8th edn. New York: McGraw-Hill.

Problems

1.   A low pressure tunnel 4 m in diameter, 8000 m long having a Darcy friction factor of 0·012 delivers 45 m³/s to a hydraulic turbine. A surge chamber 8 m in diameter is situated at the downstream end of the tunnel. Taking $F_s = 1·0$ and using the method of section 12·3, plot the variation of water level in the surge chamber relative to the reservoir level when the flow to the turbines is suddenly stopped.

2.   (a) Repeat problem 1 using a numerical method.
(b) If the discharge to the turbines were to be reduced linearly to zero in 90 seconds calculate the time variation of water level in the surge chamber and state the maximum upswing and time of occurrence.

3.   A steel pipeline 2000 m long, 300 mm in diameter discharges water from a reservoir to atmosphere through a control valve, the discharge coefficient of which is 0·6. The valve is closed so that its area decreases linearly from 0·065 m^2 to zero in (a) 15 s, and (b) 30 s. If the initial head at the valve is 3·0 m and the wave speed is 1333·3 m/s calculate, neglecting friction, the pressure head and velocity at the valve at the discrete water-hammer periods.

4.   A spun iron pipeline 600 mm in diameter 1500 m long, terminates in a gate valve. Water is supplied from a reservoir the level in which is 30 m above the level of the valve. The speed of propagation of a pressure wave in the pipeline is 1286 m/s. The initial area of valve opening is 0·25 m^2 and is closed completely in 14 seconds with the tabulated time area relationship. Taking λ to be constant, = 0·015, determine the pressure head elevation at the valve and at the mid-length of the pipeline at discrete waterhammer periods during a time of 14 s.

Time (s)	0	2	4	6	8	10	12	14
Valve area (m^2)	0·250	0·190	0·136	0·097	0·061	0·030	0·010	0

Chapter 13
Unsteady Flow in Channels

R.E. Featherstone

13.1 Introduction

River flood propagation, estuarial flows and surges resulting from gate operation or dam failure are practical examples of unsteady channel flows. Natural flood flows in rivers and the propagation of tides in estuaries are examples of gradually varied unsteady flow since the vertical component of acceleration is small. Surges are examples of rapidly varied unsteady flow.

Consider the two-dimensional propagation of a low wave which has a small height in relation to its wavelength. (See fig. 13.1.) The celerity, or speed of propagation relative to the water is given by $\sqrt{gy}$ where y is the water depth. Therefore the velocity of the wave relative to a stationary observer is

$$c = \sqrt{gy} \pm V \qquad\qquad\qquad (13.1)$$

Note that Froude number F_r expressed by $\dfrac{V}{\sqrt{gy}}$ is the ratio of water velocity to wave celerity. If the Froude number is greater than unity, which corresponds with supercritical flow, a small gravity wave cannot be propagated upstream. Waves of finite height are dealt with in sections 13.3 et seq.

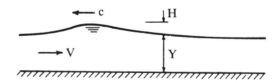

Figure 13.1 Propagation of low wave in channel

13.2 Gradually varied unsteady flow

Examples of gradually varied unsteady flow are floodwaves and estuarial flows; in such waves the rate of change of depth is gradual.

In one-dimensional form (i.e. depth and width integrated) it can be shown that the two governing continuity and dynamic partial differential equations are

$$\frac{\partial Q}{\partial x} + \frac{\partial A}{\partial t} = 0 \qquad (13.2)$$

$$\frac{\partial y}{\partial x} + \frac{V}{g}\frac{\partial V}{\partial x} + \frac{1}{g}\frac{\partial V}{\partial t} = S_o - S_f \qquad (13.3)$$

where Q is the discharge at section located at x, with cross-sectional area A at time t, y is the depth at x, t and V is Q/A. S_o is the bed slope and S_f the energy gradient.

These equations were first published by Saint-Venant. However analytical solutions of these equations are impossible unless, for example, the dynamic equation (13.3) is reduced to a 'kinematic wave' approximation by omitting the dynamic terms. In this form the equations are often applied to flood routing and overland flow computations. In general, the equations have to be evaluated at discrete space and time intervals using numerical methods such as finite-difference methods. The availability of digital (and also analogue) computers has enabled the governing equations to be applied to a wide range of practical problems. Such methods are, however, outside the scope of this text and the reader is referred to the Recommended reading for more specialist literature.

However, the case of rapidly varied unsteady flow is, with certain simplifying assumptions, amenable to direct solution.

13.3 Surges in open channels

A surge is produced by a rapid change in the rate of flow, for example, by the rapid opening or closure of a control gate in a channel.

The former causes a positive surge wave to move downstream (fig. 13.2 (a)); the latter produces a positive surge wave which moves upstream (fig. 13.2 (b)).

A stationary observer therefore sees an increase in depth as the wave front of a positive surge wave passes. A negative surge wave, on the other hand, leaves a shallower depth as the wave front passes.

Negative waves are produced by an increase in the downstream flow, for example, by the increased demand from a hydropower plant (fig. 13.3 (a)) or downstream from a gate which is being closed (fig. 13.3 (b)).

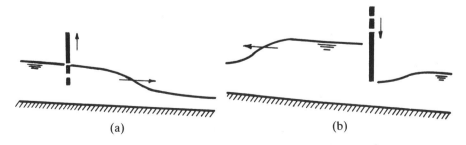

(a)   (b)

Figure 13.2 Positive surge waves

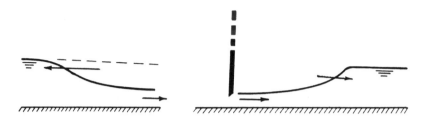

Figure 13.3 Negative surge waves

Figures 13.2 and 13.3 demonstrate that each type of surge can move either upstream or downstream.

13.4 The upstream positive surge

Consider the propagation of a positive wave upstream in a frictionless channel resulting from gate closure (fig. 13.4).

The front of the surge wave is propagated upstream at a celerity, c, relative to a stationary observer. To the observer, the flow situation is unsteady as the wave front passes; to an observer travelling at a speed, c, with the wave the flow appears steady although non-uniform. Figure 13.5 shows the surge reduced to steady state.

The continuity equation is

$$A_1 (V_1 + c) = A_2 (V_2 + c) \tag{13.4}$$

$$\text{or } V_2 = \frac{(A_1 V_1 - c (A_2 - A_1))}{A_2} \tag{13.5}$$

The momentum equation is

$$g A_1 \bar{y}_1 - g A_2 \bar{y}_2 + A_1 (V_1 + c) (V_1 - V_2) = 0 \tag{13.6}$$

where $\bar{y}_1$ and $\bar{y}_2$ are the respective depths of the centres of area. Substituting for V_2 from equation (13.5), equation (13.4) yields

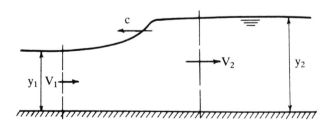

Figure 13.4 Upstream positive surge

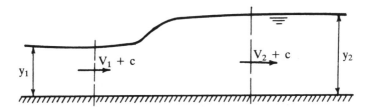

Figure 13.5 Upstream positive surge

$$g (A_2 \bar{y}_2 - A_1 \bar{y}_1) \frac{A_2}{A_1 (A_2 - A_1)} = (V_1 + c)^2$$

$$\text{whence } c = \left[g A_2 \frac{(A_2 \bar{y}_2 - A_1 \bar{y}_1)}{A_1 (A_2 - A_1)} \right]^{1/2} - V_1 \qquad (13.7)$$

In the special case of a rectangular channel,

$$A = by: \bar{y} = y/2$$

$$\text{From equation (13.7); } c = \left[\frac{g y_2}{2} \frac{(y_2^2 - y_1^2)}{y_1 (y_2 - y_1)} \right]^{1/2} - V_1$$

$$\text{whence } c = \left[\frac{g y_2}{2} \frac{(y_2 + y_1)}{y_1} \right]^{1/2} - V_1 \qquad (13.8)$$

The hydraulic jump can be shown to be a stationary surge. Putting $c = 0$ in equation (13.8),

$$V_1^2 = \frac{g y_2}{2} \frac{(y_2 + y_1)}{y_1}$$

$$\frac{2 V_1^2 y_1}{g} = y_2^2 + y_2 y_1$$

Now F_1^2 {(Froude number)^2} $= \dfrac{V_1^2}{g y_1}$

$$\therefore y_2^2 + y_2 y_1 - 2 F_1^2 y_1^2 = 0$$

$$\text{whence } y_2 = \frac{y_1}{2} (\sqrt{1 + 8 F_1^2} - 1)$$

which is identical with equation (8.17) with $\beta = 1 \cdot 0$.
In the case of a low wave where y_2 approaches y_1 equation (13.8) becomes

$$c = \sqrt{g y} - V_1 \qquad (13.9)$$

and in still water ($V_1 = 0$)

$$c = \sqrt{g y} \qquad (13.10)$$

13.5 The downstream positive surge

This type of wave may occur in the channel downstream from a sluice gate at which the opening is rapidly increased. (See fig. 13.6.)

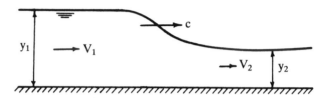

Figure 13.6   Downstream positive surge

Reducing the flow to steady state.

Continuity: $(c - V_1) A_1 = (c - V_2) A_2$ (13.11)

i.e. $V_1 = \dfrac{c A_1 - c A_2 + V_2 A_2}{A_1}$ (13.12)

Momentum: $g A_1 \bar{y}_1 - g A_2 \bar{y}_2 + (c - V_2) A_2 (V_2 - V_1) = 0$

Substituting for V_1 yields:

$$c = \left[\frac{g (A_1 \bar{y}_1 - A_2 \bar{y}_2)}{(A_1 - A_2)} \frac{A_1}{A_2} \right]^{1/2} + V_2 \tag{13.13}$$

In the case of a rectangular channel

$$c = \left[\frac{g y_1}{2 y_2} (y_1 + y_2) \right]^{1/2} + V_2 \tag{13.14}$$

13.6 Negative surge waves

The negative surge appears to a stationary observer as a lowering of the liquid surface. Such waves occur in the channel downstream from a control gate the opening of which is rapidly reduced or in the upstream channel as the gate is opened. The wave front can be considered to be composed of a series of small waves superimposed on each other. Since the uppermost wave has the greatest depth it travels faster than those beneath; the retreating wave front therefore becomes flatter (fig. 13.7).

Figure 13.8 shows a small disturbance in a rectangular channel caused by a reduction in downstream discharge; the wave propagates upstream.

Reducing the flow to steady state, the continuity equation becomes

$$(V + c) y = (V - \delta V + c) (y - \delta y)$$

Neglecting the product of small quantities

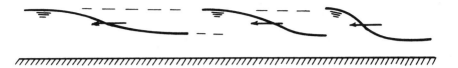

Figure 13.7 Propagation of negative surge

$$\delta y = - \frac{y \delta V}{(V + c)} \tag{13.15}$$

The momentum equation is

$$\frac{\rho g}{2} \{ y^2 - (y - \delta y)^2 \} + \rho y (V + c) \{ V + c - (V - \delta V + c) \} = 0$$

whence $\dfrac{\delta y}{\delta V} = - \dfrac{(V + c)}{g}$

or $$\delta y = - \frac{\delta V (V + c)}{g} \tag{13.16}$$

Equating (13.15) and (13.16),

$$\frac{y \delta V}{(V + c)} = \frac{(V + c) \delta V}{g}$$

whence $c = \sqrt{gy} - V$ $\tag{13.17}$

Substituting for $(V + c)$ from (13.16) into (13.17) yields

$$\delta y = - \frac{\delta V}{g} \sqrt{gy}$$

and in the limit as $\delta y \to 0$

$$\frac{dy}{\sqrt{y}} = - \frac{dV}{\sqrt{g}} \tag{13.18}$$

For a wave of finite height, integration of equation (13.18) yields

$$V = -2 \sqrt{gy} + \text{const}$$

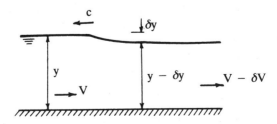

Figure 13.8 Negative surge propagation (upstream)

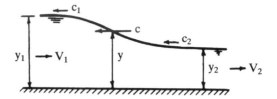

Figure 13.9 Negative surge of finite height

When $y = y_1$, $V = V_1$ whence const $= V_1 + 2\sqrt{gy_1}$

$$V = V_1 + 2\sqrt{gy_1} - 2\sqrt{gy} \tag{13.19}$$

From equation (13.17), $c = \sqrt{gy} - V$ and substituting in equation (13.19) yields

$$c = 3\sqrt{gy} - 2\sqrt{gy_1} - V_1 \tag{13.20}$$

The wave speed at the crest is therefore
$$c_1 = \sqrt{gy_1} - V_1 \tag{13.21}$$

and at the trough

$$c_2 = 3\sqrt{gy_2} - 2\sqrt{gy_1} - V_1 \tag{13.22}$$

In the case of a downstream negative surge in a frictionless channel (fig. 13.10), a similar approach yields

$$c = \sqrt{gy} + V \tag{13.23}$$

$$V = 2\sqrt{gy} - 2\sqrt{gy_2} + V_2 \tag{13.24}$$

$$c = 3\sqrt{gy} - 2\sqrt{gy_2} + V_2 \tag{13.25}$$

$$c_1 = 3\sqrt{gy_1} - 2\sqrt{gy_2} + V_2 \tag{13.26}$$

$$c_2 = \sqrt{gy_2} + V_2 \tag{13.27}$$

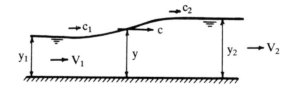

Figure 13.10 Downstream negative surge

13.7 The dam break

The dam, or gate, holding water upstream at depth y_1 and zero velocity, is suddenly removed. (See fig. 13.11.)

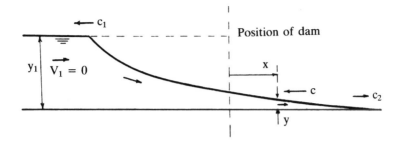

Figure 13.11   Sudden gate/dam bursting

From equation (13.20) $c = 3 \sqrt{gy} - 2 \sqrt{gy_1}$

The equation to the surface profile is therefore

$$x = (ct) = (3 \sqrt{gy} - 2 \sqrt{gy_1}) \, t$$

If $x = 0$, $y = \dfrac{4y_1}{9}$ and remains constant with time. The velocity at $x = 0$ is

$V = V_1 + 2 \sqrt{gy_1} - 2 \sqrt{gy}$ (from equation (13.19))

i.e. $V = \dfrac{2}{3} \sqrt{gy_1}$, since $V_1 = 0$

Worked examples

Example 13.1
A rectangular channel 4 m wide conveys a discharge of 25 m³/s at a depth of 3 m. The downstream discharge is suddenly reduced to 12 m³/s by partial closure of a gate. Determine the initial depth and celerity of the positive surge wave.

Solution:
Referring to fig. 13.4

$$y_1 = 3 \text{ m}; \quad V_1 = 2 \cdot 083 \text{ m/s}; \quad V_2 = \frac{12}{4 \times y_2} = \frac{3}{y_2}$$

From equation (13.4), $(V_1 + c) \, y_1 = (V_2 + c) \, y_2$ (i)

i.e. $(2 \cdot 083 + c) \times 3 = \left(\dfrac{3}{y_2} + c \right) y_2$

whence $c = \dfrac{3 \cdot 249}{(y_2 - 3)}$

Substituting in equation (13.8):

$$\frac{3 \cdot 249}{(y_2 - 3)} = \left[\frac{g y_2}{2} \frac{(y_2 + 3)}{3} \right]^{1/2} - 2 \cdot 083$$

By trial $y_2 = 3 \cdot 75$ m; $c = 4 \cdot 35$ m/s.

Example 13.2
A tidal channel, which may be assumed to be rectangular, 40 m wide, bed
slope 0·0003, Manning's roughness coefficient, n = 0·022, conveys a steady
freshwater discharge of 60 m³/s. A tidal bore is observed to propagate
upstream with a celerity of 5 m/s. Determine the depth of flow and the
discharge immediately after the bore has passed, neglecting the density
difference between the freshwater and saline water.

Solution:
The depth of uniform flow using the Manning equation = 1·52 m (= y_1)

$$V_1 = \frac{60}{40 \times 1 \cdot 52} = 0 \cdot 987 \text{ m/s.}$$

y_2 can be determined using equation (13.8)

i.e. $c = \left[\dfrac{g y_2}{2} \dfrac{(y_2 + y_1)}{y_1} \right]^{1/2} - V_1$

$5 \cdot 0 = \left[\dfrac{9 \cdot 81 \, y_2}{2} \dfrac{(y_2 + 1 \cdot 52)}{1 \cdot 52} \right]^{1/2} - 0 \cdot 987$

By trial $y_2 = 2 \cdot 66$ m.
Using the continuity equation:

$$(V_1 + c) \, y_1 = (V_2 + c) \, y_2$$

or $\qquad V_2 = \dfrac{(V_1 + c) \, y_1}{y_2} - c$

$$= \frac{(0 \cdot 987 + 5 \cdot 0) \times 1 \cdot 52}{2 \cdot 66} - 5 \cdot 0$$

$$V_2 = -1 \cdot 578 \text{ m/s}$$

and $Q_2 = V_2 \, b \, y_2 = -168 \cdot 2$ m²/s (upstream).

Example 13.3
A rectangular tailrace channel, 15 m wide, bed slope 0·0002 and Manning
roughness coefficient 0·017 conveys a steady discharge of 45 m³/s from a
hydropower installation. A power increase results in a sudden increase in
flow to the turbines to 100 m³/s. Determine the depth and celerity of the
resulting surge wave in the channel.

Solution:
Using the Manning equation the depth of uniform flow under initial conditions
at a discharge of 45 m³/s = 2·42 m.
Using equation (13.11) i.e. $(c - V_1) y_1 = (c - V_2) y_2$

$$c = \frac{V_1 y_1 - V_2 y_2}{(y_1 - y_2)}$$

$$c = \frac{\dfrac{Q_1 y_1}{by_1} - V_2 y_2}{(y_1 - y_2)}$$

$$V_2 = \frac{Q_2}{by_2} = \frac{45}{15 \times 2\cdot42} = 1\cdot24 \text{ m/s}$$

$$Q_1 = 100 \text{ m}^3/\text{s, whence } c = \frac{6\cdot67 - 3}{(y_1 - 2\cdot42)} = \frac{3\cdot67}{(y_1 - 2\cdot42)}$$

Substitution in equation (13.13),

$$\frac{3\cdot67}{(y_1 - 2\cdot42)} = \left[\frac{gy_1}{2 \times 2\cdot42} (y_1 + 2\cdot42)\right]^{1/2} + 1\cdot24$$

By trial $y_1 = 2\cdot95$ m

$$c = \frac{3\cdot67}{(2\cdot95 - 2\cdot42)} = 6\cdot92 \text{ m/s}.$$

Example 13.4
A steady discharge of 25 m³/s enters a long rectangular channel 10 m wide,
bed slope 0·0001, Manning's roughness coefficient 0·017, regulated by a
gate. The gate is rapidly partially closed resulting in a reduction of the
discharge to 12 m³/s. Determine the depth and mean velocity at the trough
of the wave, the surface profile and the time taken for the wave front to
reach a point 1 km downstream neglecting friction. (See fig. 13.12.)

Solution:
$$y_2 = 2\cdot86 \text{ m} \quad \text{(using Manning's equation)}$$

$$V_2 = \frac{25}{10 \times 2\cdot86} = 0\cdot874 \text{ m/s}$$

$$V_1 = \frac{12}{10 y_1} = \frac{1\cdot2}{y_1}$$

From equation (13.24) $V = V_2 - 2\sqrt{g}(\sqrt{y_2} - \sqrt{y})$

whence $V_1 = V_2 - 2\sqrt{g}(\sqrt{y_s} - \sqrt{y_1})$

i.e. $\dfrac{1\cdot2}{y_1} = 0\cdot874 - 6\cdot26(1\cdot691 - \sqrt{y_1})$

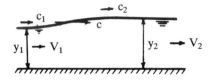

Figure 13.12 Wave front propagation

Solving by trial, $y_1 = 2.637$ m

Then $\qquad V_1 = 0.455$ m/s

$$c_2 = \sqrt{gy_2} + V_2 = 6.17 \text{ m/s}$$

Time taken to travel 1 km = 2.7 minutes.

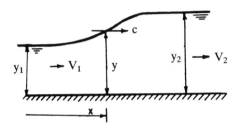

Figure 13.13 Wave front propagation

Surface profile: $x = ct = (3 \sqrt{gy} - 2 \sqrt{gy_2} + V_2) t$

$$\text{or} \quad x = (9.4 \sqrt{y} - 10.6 + 0.874) t$$

$$x = (9.4 \sqrt{y} - 9.726) t.$$

Recommended reading

1. Henderson, F.M. (1966) *Open Channel Flow*. New York: The Macmillan Company.
2. Pickford, J. (1969) *Analysis of Surge*. London: Macmillan.
3. French, R.H. (1986) *Open Channel Hydraulics*. New York: McGraw-Hill Book Co.

Problems

1. A rectangular channel 4 m wide conveys a discharge of 18 m³/s at a depth of 2.25 m. Determine the depth and celerity of the positive surge wave resulting from (a) sudden, partial gate closure which reduces the downstream discharge to 10 m³/s, and (b) sudden total gate closure.

2.   At low tide the steady freshwater flow in an estuarial channel, 20 m wide, bed slope 0·0005, Manning's roughness coefficient 0·02 is 20 m^3/s. A tidal bore forms on the flood tide and is observed to propagate upstream at a celerity of 4 m/s. Neglecting the density difference between the freshwater and the saline water determine the depth and discharge immediately after the bore has passed.

3.   A rectangular channel 10 m wide, bed slope 0·0001, Manning's roughness coefficient 0·015, receives inflow from a reservoir with a gated inlet. When a steady discharge of 30 m^3/s is being conveyed the gate is suddenly opened to release a discharge of 70 m^3/s. Calculate the initial celerity and depth of the surge wave.

4.   A steady discharge of 30 m^3/s is conveyed in a rectangular channel of bed width 9 m at a depth of 3·0 m. A control gate at the inlet is suddenly partially closed reducing the inflow to 10 m^3/s. Assuming the channel to be frictionless determine the depth behind the surge wave and the time taken for the trough of the wave to pass a point 500 m downstream.

5.   A rectangular channel 30 m wide discharges 60 m^3/s at a uniform flow depth of 2·5 m into a reservoir. The levels of water in the channel and reservoir at the reservoir inlet are initially equal. Water in the reservoir is released rapidly so that the level falls at the rate of 1 m per hour. Neglecting friction and the channel slope determine the time taken for the level in the channel to fall 0·5 m at a section 1 km upstream from the reservoir.

Chapter 14
Uniform Flow in Loose-boundary Channels

C. Nalluri

14.1 Introduction

The loose boundary (consisting of movable material) of a channel deforms under the action of flowing water and the deformed bed with its changing roughness (bed forms) interacts with the flow. A dynamic equilibrium state of the boundary may be expected when a steady and uniform flow has developed.

The resulting movement of the bed material (sediment) in the direction of flow is called sediment transport and a certain critical bed shear stress (τ_c) must be exceeded to start the particle movement. Such a critical shear stress is referred to as incipient (threshold) motion condition, below which the particles will be at rest and the flow is similar to that on a rigid boundary.

14.2 Flow regimes

Shear stresses above the threshold condition disturb the initial plane boundary of the channel and the bed and water surface assume various forms depending on the sediment and fluid flow characteristics. Two distinct regimes of flow may be identified with the increasing flows with the following bed forms:

(a) Lower regime: ripples (for smaller sediment size < 0.6 mm and low Froude no. $\ll 1$), dunes and ripples, dunes with increasing shear (τ_o) and Froude number (F_r); further increases in τ_o and F_r introduce transition to dunes/plane bed ($F_r \simeq 1$) and

(b) Upper regime: flat bed, antidunes, chutes and pools with large shear and Froude numbers (> 1).

14.3 Incipient (threshold) motion

Shields[27] introduced the concept of the dimensionless entrainment function, $F_{rd}^2 (= \tau_0/\rho g \Delta d)$ as a function of shear Reynolds number, $R_{e*} (= U_* d/\nu)$ where ρ is density of the fluid and Δ is the relative density of sediment in the fluid, d the diameter of sediment, g the acceleration due to gravity, U_*

the shear velocity $(= \sqrt{\tau_0/\rho})$ and v the kinematic viscosity of the fluid, and published a curve defining the threshold condition (see Fig. 14.1).

When the flow is fully turbulent around the bed material ($R_{e*} > 400$ and $d > \simeq 4$ mm) the Shields criterion can be written as

$$\tau_0/\rho g \Delta d = 0 \cdot 056 \qquad (14.1)$$

Combining equation (14.1) with the uniform boundary shear equation

$$\tau_0 = \rho g RS \qquad (14.2)$$

gives the limiting particle size (with $\Delta = 1 \cdot 65$) for incipient motion

$$d = 11 \, RS \qquad (14.3)$$

where R is the hydraulic radius and S is the friction gradient.

Combining equation (14.3) and Manning's equation for mean velocity

$$V = (1/n) \, R^{2/3} \, S^{1/2} \qquad (14.4)$$

with

$$n = d^{1/6}/26 \quad \text{(Strickler's equation)} \qquad (14.5)$$

gives

$$V_c/\sqrt{(gd)} \simeq 1 \cdot 9 \, \sqrt{\Delta} \, (d/R)^{-1/6} \qquad (14.6)$$

where V_c is the critical velocity for the incipient motion of sediment particles.

The recommended values of critical tractive forces and maximum permissable mean velocities for different sizes of bed material are listed in Table 8.2.

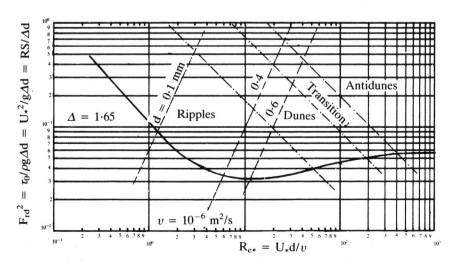

Figure 14.1 **Shields diagram**

14.4 Resistance to flow in alluvial (loose bed) channels

The resistance of an alluvial channel varies considerably with flow velocity once threshold has been passed. The bed introduces additional form drag due to bed formations and the overall friction factor λ rises rapidly to 3 to 4 times its original value. Several attempts have been made to describe a relationship between the mean velocity V, the depth y_o or hydraulic radius R, slope S and sediment size d, which can be broadly divided into two categories:

(a) Total resistance approach:
1. Regime channel equation
This was one of the earliest resistance relationship for alluvial channel flow proposed by Lacey[14] in the form

$$V = 10\cdot8\ R^{2/3}\ S^{1/3} \tag{14.7}$$

in SI units based on the regime canal data from India. Its applicability in channels or rivers with different sediment sizes and flow depths is questionable.
2. Japanese equation
Sugio[22] proposed the following equation using river data from Japan

$$V = K\ R^{0\cdot54}\ S^{0\cdot27} \tag{14.8}$$

in SI units where $K = 6\cdot51$ for ripples, $9\cdot64$ for dunes and $11\cdot28$ for transition regime.
3. Garde–Ranga Raju formula
Garde and Ranga Raju[10] analysed data from flumes, canals and natural streams and a graphical relationship (Fig. 14.2) between the parameters

$$K_1\ V/\sqrt{(\Delta gR)}\ \text{versus}\ K_2\ (R/d)^{1/3}\ S/\Delta$$

where K_1 and K_2 are functions of sediment size (see Fig. 14.3) was proposed. Figs. 14.2 and 14.3 facilitate the calculations of discharge in alluvial channels.

(b) Grain and form resistance approach
This approach either splits the overall resistance into grain resistance λ' and form resistance λ'' (Alam and Kennedy[2]) or U_* into U'_* and U''_* corresponding to grain and form resistances respectively (Einstein and Barbarossa[6]). Introducing the concept of bed hydraulic radius (R_b) charts and graphs have been produced to predict the resistance equations in alluvial channels. The proposed methods are too advanced and are out of scope of the present book.

Einstein suggested in the case of rectangular channels (bed width B) with smooth sides the following equation for the hydraulic radius of the bed

$$R_b = [1 + 2\ (y_o/B)]\ R - 2\ (y_o/B)\ R_w \tag{14.9}$$

where the hydraulic radius corresponding to the walls R_w is computed from Manning's equation assuming it is applicable to the side walls and the bed independently.

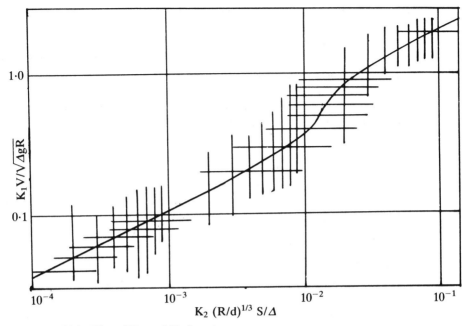

Figure 14.2 Plot of K_1 and K_2 functions

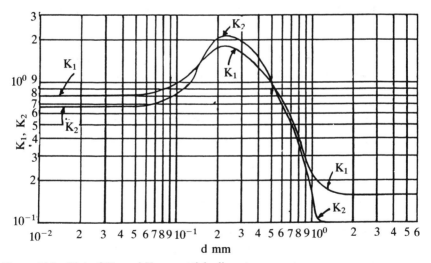

Figure 14.3 Plot of K_1 and K_2 v. particle diameter

Vanoni and Brooks[29] proposed for rough channels with smooth sides that

$$R_b = \lambda_b V^2 / 8gS \tag{14.10}$$

The bed friction factor λ_b can be found from

$$P\lambda = P_b \lambda_b + P_w \lambda_w \tag{14.11}$$

where P is the wetted perimeter with suffixes b and w for bed and walls respectively.

Resistance equations of the type Colebrook-White (or appropriate resistance plots e.g. see Fig. 14.9.) can be used to predict λ_w in equations 14.9 and 14.11, which is a function of R_e/λ where R_c is the Reynolds number and λ is the overall friction factor of the channel.

In the channels where roughness of the walls is different from the bed, the bed hydraulic radius may be used in place of the total hydraulic radius to determine the regimes, mean velocities etc.

14.5 Velocity distributions in loose-boundary channels

Einstein's equation in the form

$$u/U_*' = 5.75 \log (30.2 \, yx/k_s) \tag{14.12}$$

where u is the temporal mean velocity at a distance y from the boundary is applicable universally for smooth, transition and rough beds. The correction factor x is a function of k_s/δ' (δ' sublayer thickness given by $11.6 \, v/U_*$) given in Table 14.1

Table 14.1 Correction factor x

k_s/δ'	0.2	0.3	0.5	0.7	1.0	2.0	4.0	6.0	10.0
x	0.7	1.0	1.38	1.56	1.61	1.38	1.10	1.03	1.0

Equation 14.12 gives the mean velocity, V as

$$V = 5.75 \, U_* \log (12.27 \, Rx/k_s) \tag{14.13}$$

For $k_s/\delta' > 6.0$ the boundary is fully rough and the Manning–Strickler equation could conveniently be used to calculate the mean velocity.

14.6 Sediment transport

When flow characteristics (velocity, average shear stress etc.) in an alluvial channel exceed the threshold condition for the bed material the particles move in different modes along the flow direction. The mode of transport of the material depends on the sediment characteristics such as its size and shape, density ρ_s and movability parameter U_*/W_s where W_s is the fall velocity of the sediment particle.

Fall velocities are equally of importance in reservoir sedimentation and settling processes and may be expressed as

$$W_s = f(\text{shape and density of sediment, no. of particles falling, particle Reynolds number}).$$

The fall velocity of a single spherical particle can be written as

$$W_s = \sqrt{\{(4/3)(g \Delta d/C_D)\}} \tag{14.14}$$

where C_D is the drag coefficient. The drag coefficient is a function of the particle Reynolds number ($R_{ed} = W_s d/\nu$).

For $R_{ed} < 1$

$$C_D = 24/R_{ed} \tag{14.15}$$

For particles with a shape factor of $0 \cdot 7$ (natural sands) C_D is nearly equal to 1 when $R_{ed} > \simeq 200$ whereas it is around $0 \cdot 4$ in the case of spherical particles (shape factor = 1) for $R_{ed} > 2000$.

Figure 14.4 may be used to establish fall velocities of sediment particles of different shape factors.

Some sediment particles roll or slide along the bed intermittently and some others saltate (hopping or bouncing along the bed). The material transported in one or both of these modes is called 'bed load'. Finer particles (with low fall velocities) are entrained in suspension by the fluid turbulence and transported along the channel in suspension. This mode of transport is called 'suspended load'. The combined transport derived from the bed material is called 'total bed material load'. Sometimes finer particles from upland catchment (sizes which are not present in the bed material), called 'wash load', are also transported in suspension. The combined bed material and wash load is called 'total load'.

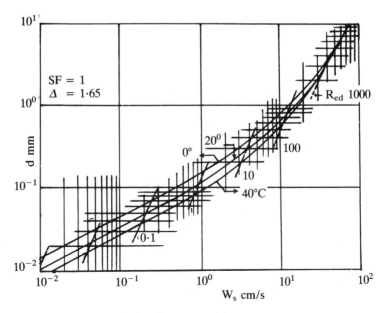

Figure 14.4 Fall velocities of sediment particles

14.7 Bed load transport

Several empirical equations from laboratory flume data have been proposed by many investigators with the basic assumptions that the sediment is homogeneous and noncohesive. The results differ appreciably and it is dangerous to transfer the information to outside the limits of the experiments. However, one can discern general trends of the transport rate by using several formulae (with some theoretical background). The following are the most commonly used equations:

1. Shields equation
Shields used the concept of excess shear responsible for the transport and presented a dimensionally homogeneous equation

$$q_b \, \Delta/qS = 10 \, (\tau_o - \tau_c)/\rho g \Delta d \tag{14.16}$$

where q_b is the bed load per unit width and q is the unit discharge in the channel. Equation 14.16 is based on the ranges of $0 \cdot 06 < \Delta < 3 \cdot 2$ and $1 \cdot 56$ mm $< d < 2 \cdot 47$ mm.

2. Schoklitsch equation
The bed load g_b in kg/m.s is given by

$$g_b = 2500 \, S^{3/2} \, (q - q_{cr}) \tag{14.17}$$

where q_{cr} is the unit discharge at threshold condition given by

$$q_{cr} = 0 \cdot 20 \, (\Delta)^{5/3} \, d^{3/2}/S^{7/6} \tag{14.18}$$

It must be noted that equation 14.17 is not dimensionally homogeneous and is valid only for q and q_{cr} in m^3/ms.

3. Kalinske equation
For $F_{rd}^2 > 0 \cdot 09$ this can be written as

$$q_b/U_* \, d = 10 \, [U_*^2/\Delta g d]^2 \tag{14.19}$$

Equation 14.19 is dimensionally homogeneous and may not be good for high transport rates.

4. Meyer–Peter and Muller formula
The energy slope, S is split into two parts and only one part (μS) is considered to be responsible for transport (grain drag; the other is expended in the form drag). The factor μ is dependent on the bed form (ripple factor) and is expressed as

$$\mu = (C_{channel}/C_{grain})^{3/2} \tag{14.20}$$

where C is the Chezy's coefficient given by

$$C = 18 \log (12R/k) \tag{14.21}$$

in which k = d for C_{grain} and k is a function of bed form ($\simeq$ dune height) for $C_{channel}$.

The ripple factor varies between 0·5 to 1·0 for dune to flat bed condition. The bed load q_b is given by

$$q_b = 8 \sqrt{(\Delta g d^3)} \, [(\mu RS/\Delta d) - 0·047]^{3/2} \qquad (14.22)$$

Equation 14.22 is dimensionally homogeneous and covers a wide range of particle sizes and is widely used.

5. Einstein's[5] equation

Introducing probability concepts of sediment movement Einstein developed an empirical relationship

$$\phi = f(\psi) \qquad (14.23)$$

where

ψ (shear intensity or flow parameter) $= \Delta d/\mu RS \; (= 1/F_{rd}{}^2) \qquad (14.24)$

ϕ (transport parameter) $\qquad\qquad = q_b/\sqrt{(g\Delta d^3)} \qquad (14.25)$

(Note: μR in equation 14.24 may be treated as grain (bed) hydraulic radius R')

Figure 14.5 shows the functional relationship (equation 14.23). For small values of ψ (< 10) (ψ is around 20 for threshold conditions) the relationship between ϕ and ψ can be expressed as

$$\phi = 40 \, (1/\psi)^3 \qquad (14.26)$$

Rearranging the Meyer−Peter and Muller equation in terms of ϕ and ψ parameters results in:

$$\phi = [(4/\psi) - 0·188]^{3/2} \qquad (14.27)$$

which agrees well with Einstein's curve in Figure 14.5.

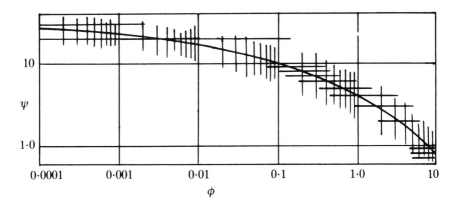

Figure 14.5 Plot of ϕ versus ψ functions

Einstein's relationship covers a very wide range of experimental data (0.785 mm $< d < 28.65$ mm; $0.052 < \Delta < 1.68$).

14.8 Suspended load transport

1. Rouse's[26] distribution equation
The vertical (suspended) mass balance equation in a two dimensional flow was first expressed by O'Brien[23] as

$$cW_s + \varepsilon_s \, dc/dy = 0 \tag{14.28}$$

where c is the volumetric concentration of the sediment and ε_s is the kinematic eddy viscosity (turbulence diffusion coefficient) in the presence of sediment, equal to $\beta\varepsilon$, ε being the eddy viscosity for clear water. β is of the order of unity in the presence of fine sediment and decreases with increasing particle size. Combining equation 14.28 with the turbulent mixing theory (log law distribution of velocity) gives the solution for sediment concentration, c at a height, y in a channel as

$$c/c_a = [a \, (y_o - y)/y \, (y_o - a)]^{W_s/\beta\chi U_*} \tag{14.29}$$

where c_a is the reference concentration at a height a from the bed and χ is Karman's constant.

The theoretical distributions of the concentration (equation 14.29) are shown in Figure 14.6 for different values of $W_s/\beta\chi U_*$ with $\beta = 1$ and $\chi = 0.4$ (i.e. clear water conditions).

Table 14.2 shows the state of suspension under different values of the movability parameter U_*/W_s.

Table 14.2 States of suspension

State of suspension	Movability parameter, U_*/W_s
Intensive saltation	0.25
Lower-half in suspension	1
Particles reach surface	3
Well developed suspension	20
Homogeneous suspension	200

The reference level a, may be assumed to be around 2d (d being the diameter of suspended particles) and c_a as bed load corresponding to this diameter. Equation 14.29 must be used with care as the parameter $W_s/\beta\chi U_*$ is not accurately computable.

The suspended load transport q_s can be obtained by summation as

$$q_s = \int_a^{y_o} c \, u \, dy \tag{14.30}$$

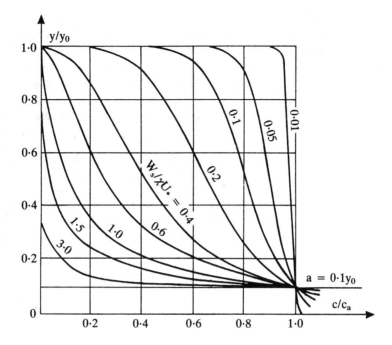

Figure 14.6 Plot of suspended load concentrations

where u is given by an appropriate velocity distribution. Equation 14.30 may be solved either numerically or graphically. Suspended load can generally be measured easily and accurately and good field measurements of both c and u predict suspended load with reasonable accuracies.

2. Lane and Kalinske's [15] approximate method
The suspended load q_s in wide channels is given by

$$q_s \simeq q\, c_a\, P\, e^{15(aW_s/y_0\, U_*)} \qquad (14.31)$$

in which P is a function of the movability parameter, U_*/W_s and $n/y_0^{1/6}$, n being Manning's coefficient. Figure 14.7 shows the plot of P in SI units.

3. Empirical equations
Several practising engineers have reported several formulae of the type

$$q_s \propto q^b \qquad (14.32)$$

the exponent b varying between 1·9 to 3.
    Engelund[8] alternatively proposed that

$$q_s = 0·5\, q\, (U_*/W_s)^4 \qquad (14.33)$$

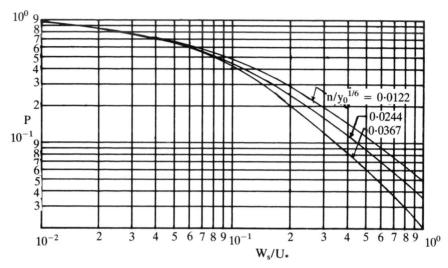

Figure 14.7 Values for the parameter P

14.9 Total load transport

Total load includes both bed material load and wash load. Wash load is usually caused by land erosion and a useful criterion for its existence may be taken as the particle Froude number (= $V/\sqrt{gd}$) around 20. Due to its small size fractions wash load moves in suspension and thus can be estimated from the total suspended load provided the suspended bed material load is known. The following approaches describe some of the available direct methods of estimating the total bed material load.

1. Laursen's[16] approach
The cross sectional mean concentration by volume C_v of the bed material load for quartz material was suggested:

$$C_v = 160\ q_b/q \qquad (14.34)$$

applicable to flume data with sand of $d < 0.2$ mm.

2. Garde's[9] equation
Field and flume data with the sediment size range of $0.011 < d$ (mm) < 0.93 gave

$$q_t \simeq 10\ (U_*/d^3)\ (RS/\Delta)^4 \qquad (14.35)$$

3. Graf's[12] approach
Similar to Einstein's bed load equation Graf et al using several flume and stream data including closed conduit data established the equation:

$$\emptyset_A = 10 \cdot 39 \, (\psi_A)^{-2 \cdot 52} \tag{14.36}$$

over a range of $10^{-2} < \emptyset < 10^3$,
where

$$\emptyset_A = C_v VR / \sqrt{(g \Delta d^3)} \tag{14.37}$$

and $\psi_A = \Delta d / RS$ (14.38)

4. Ackers–White[1] formula
The total load is predicted by a transition type equation:

$$[q_t y_0 / qd][U_* / V]^n = C [(F_{gr}/A) - 1]^m \tag{14.39}$$

where

$$F_{gr} = [U_*^n / \sqrt{(g \Delta d)}][V / (5 \cdot 75 \log (12 \cdot 2 y_0 / d))]^{1-n} \tag{14.40}$$

With the dimensionless grain diameter, d_{gr} defined by

$$d_{gr} = d \, [g \Delta / v^2]^{1/3} \tag{14.41}$$

the coefficients in equation 14.39 are shown in Table 14.3.

Table 14.3 Coefficients in equation 14.39

Coefficient	Fine	Transitional	Coarse
	$d_{gr} < 1 \cdot 0$	$1 \cdot 0 < d_{gr} < 60$	$d_{gr} > 60$
n	$1 \cdot 0$	$n = 1 \cdot 00 - 0 \cdot 56 \log d_{gr}$	$0 \cdot 00$
A	—	$A = 0 \cdot 14 + 0 \cdot 23 / \sqrt{d_{gr}}$	$0 \cdot 17$
m	—	$m = 1 \cdot 34 + 9 \cdot 66 / d_{gr}$	$1 \cdot 50$
C		$\log C = 2 \cdot 86 \log d_{gr} - (\log d_{gr})^2 - 3 \cdot 53$	$0 \cdot 025$

The above relationships have been obtained by flume data with sediments of different relative densities and size range of $0 \cdot 04 < d$ (mm) $< 4 \cdot 0$ and flow Froude numbers up to $0 \cdot 8$. This method of computing total load transport has been verified successfully with a limited amount of field data.

14.10 Regime channel design

1. Kennedy's approach
Regime equations were developed using data from stable channels in the Indian subcontinent, carrying moderate sediment loads of less than 500 ppm by weight. These equations do not consider sediment load variable and have limitations due to the fact that they are applicable to boundary characteristics similar to those found in the Indian subcontinent.

Kennedy's equation for nonsilting and nonscouring velocities, V, is given by

$$V = 0.55my_0^{0.64} \tag{14.42}$$

where y_0 is the flow depth in metres and V is in m/s; m is the critical velocity ratio (= V/V_0), a function of the sand size (m = 1 for the standard size, d ≈ 0.323 mm). Table 14.4 shows m values for other sand sizes.

Table 14.4   m values as a function of sand size

Type of sand	m	Remark
Fine silt	0.7	
Light sand silt	1.0	Standard size (data)
Coarse sand silt	1.1	
Sand loamy silt	1.2	
Coarser silt	1.3	

Equation 14.42 combined with the Manning equation (8.4b) gives two equations (Ranga Raju[24])

$$y_0 = \left(\frac{1.818Q}{(p + 0.5)m}\right)^{0.378} \tag{14.43}$$

where $p = b/y_0$, b being the bed width of a trapezoidal channel with side slopes of 0.5H:1V (the final shape of the regime channel is not truly trapezoidal, and the final side slopes are much steeper due to silt deposition on banks), and

$$\frac{SQ^{0.02}}{n^2 m^2} = 0.3 \left(\frac{(p + 2.236)^{4/3}}{(p + 0.5)^{1.31}}\right) \tag{14.44}$$

Equations 14.43 and 14.44 give any number of solutions for the three unknowns, b, y_0 and the slope S for given values of Q, m and Manning's n. Usually, the bed slope is assumed to be a reasonable value (based on past experience and surrounding terrain slope) and p and y_0, and hence b, are computed. Table 14.5 alternatively suggests the recommended values of p for stable channels as a function of Q.

Table 14.5   Recommended b/y_0 values

Q (m³/s)	5.0	10.0	15.0	50.0	100.0	200.0	300.0
p (= b/y_0)	4.5	5.0	6.5	9.0	12.0	15.0	18.0

2. Lacey's approach
Lacey proposed the following equations (Kennedy's equation does not specify the channel width, and experience suggests that this is an important parameter) for regime channel design:

$$P = 4.75 \sqrt{Q} \tag{14.45}$$

$$R = 0.47 \left(\frac{Q}{f}\right)^{1/3} \tag{14.46}$$

$$S = 3 \times 10^{-4} f^{5/3}/Q^{1/6} \tag{14.47}$$

in which the silt factor, f, is given by:

$$f = 1.76 \sqrt{d} \tag{14.48}$$

where d is in mm, P and R are in m and Q is in m^3/s. The silt factor, f, is a function of its size, as indicated in Table 14.6. Lacey, combining equations 14.45 to 14.47, suggested the resistance equation:

$$V = 10.8 R^{2/3} S^{1/3} \tag{14.49}$$

Table 14.6 Silt factor, f

Type of sand	f	Remark
Very fine silt	0·5	d ≈ 0·081 mm
Fine silt	0·6	0·12
Medium silt	0·85	0·233
Standard silt	1·0	d ≈ 0·323 mm
Medium sand	1·25	0·505
Coarse sand	1·50	0·725

Equation 14.49 is commonly used in the Indian subcontinent practice in designing stable (regime) channels. Normally an additional margin of flow depth (freeboard) is provided in the design to allow any water level fluctuations. The recommended freeboards as function of discharges are shown in Table 14.7.

Table 14.7 Recommended freeboards for canals

Q (m^3/s)	<0·75	0·75 to 1·50	1·50 to 85·0	>85·0
Freeboard (m)	0·45	0·60	0·75	0·90

3. Blench's approach

Blench developed more rational formulae (using flume and Indian subcontinent data), taking into account the effects of bank cohesiveness on channel geometry and sediment load. In a channel of mean width b and mean depth y_0, the discharge Q is written as

$$Q = Vby_0 \tag{14.50}$$

Blench introduced bed and side factors as f_b ($= V^2/y_0$) and f_s ($= V^3/b$) respectively, and wrote:

$$b = \sqrt{\frac{f_b Q}{f_s}} \tag{14.51}$$

$$y_0 = \left(\frac{f_s Q}{f_b^2}\right)^{1/3} \tag{14.52}$$

$$\frac{V^2}{g y_0 S} = 3.63 \left(\frac{Vb}{v}\right)^{1/4} \tag{14.53}$$

$$S = \frac{f_b^{5/6} f_s^{1/12} v^{1/4}}{11.91 g Q^{1/6} \left(1 + \dfrac{c}{2330}\right)} \tag{14.54}$$

where c is the sediment concentration in ppm by weight and v is the kinematic viscosity of water. He suggested $f_s = 0.1$ to 0.3 for slight to high cohesivity and $f_b = 1.9 \sqrt{d} (1 + 0.012c)$, d being in mm.

4. Simons–Albertson method

Regime channel data from the USA, Punjab and Sind (Indian subcontinent) were analysed by Simons and Albertson; their modified regime equations have a wider applicability. The channels are classified according to the nature of the bed and bank material (see Table 14.8) and following equations were suggested:

$$b = 0.92B - 0.60 \tag{14.55}$$

where b is the average width and B the water surface width (in m), and

$$P, A \text{ and } R = mQ^n \tag{14.56}$$

where the coefficients m and n are given in Table 14.8.

Table 14.8 Regime equations of Simons and Albertson (Garde and Ranga Raju[11])

Type of channel		Sand bed and banks	Sand bed and cohesive banks	Cohesive bed and banks	Coarse noncohesive boundary
P	m	6.33	4.74	4.63	3.44
	n	0.512	0.512	0.512	0.512
A	m	2.57	2.25	2.25	0.939
	n	0.873	0.873	0.873	0.873
R	m	0.403	0.475	0.557	0.273
	n	0.361	0.361	0.361	0.361

The following resistance equations were also proposed by Simons and Albertson:

Sand bed and banks: $\qquad\qquad V = 9.33 (R^2 S)^{1/3} \tag{14.57}$

$$V^2/g y_0 S = 0.885 (Vb/v)^{0.37} \tag{14.58}$$

Sand bed and cohesive banks: $\quad V = 10\cdot8\ (R^2S)^{1/3}\quad$ (14.59)

$$V^2/gy_0S = 0\cdot525\ (Vb/v)^{0\cdot37}\quad (14.60)$$

Coarse noncohesive material: $\quad V = 4\cdot75\ (R^2S)^{0\cdot286}\quad$ (14.61)

$$V^2/gy_0S = 0\cdot324\ (Vb/v)^{0\cdot37}\quad (14.62)$$

The slope equations (14.58, 14.60 and 14.62) are recommended for $Vb/v < 2 \times 10^7$ and equations 14.57, 14.59 and 14.61 may be preferred to equation 14.54 for the determination of the slope when $Vb/v > 2 \times 10^7$. It has been suggested that the flow Froude number be kept less than 0·30 for stability considerations.

5. Nonscouring erodible boundary channel design

This method approaches the criterion that the bed material (coarse) does not move when the channel carries either clear water or water with fine silt in suspension (not depositing). The principle of design is to achieve a cross section in which the boundary material is on the verge of motion (initiation criterion). The method utilises the information on boundary shear distribution and the Shields initiation criterion (both on bed and banks) and establishes either permissable depth or slope (given one or the other). The Manning resistance equation with the appropriate n value ($= d^{1/6}/26$) further establishes the bed width required to transport the design discharge. (See worked example 14.6 for the detailed design procedures.)

In the above approach it is to be noted that not all the boundary particles are on the verge of motion (side slopes are less sustainable) and such a section is not economical/efficient. The most desirable section (bed and bank material at the incipient motion) is of the following profile (Glover and Florey[13]):

$$y = y_0 \cos \left(\frac{0\cdot8x}{y_0} \tan \phi \right) \tag{14.63}$$

The design procedures using equation 14.63 are illustrated by worked example 14.7.

6. Design of stable erodible boundary channel

The most important physical processes in the formation of stable channels are now well documented and White et al.[32] have proposed a solution using the Ackers–White sediment transport model coupled with the White et al.[31] resistance equation for alluvial channels.

For a defined channel boundary material (i.e. known d and ρ_s) and water viscosity there are six channel/sediment parameters: Q, Q_s, V, B, y_0 and S. Three equations (continuity, the Ackers–White transport equation and the White et al. resistance equation) and a fourth one based on the variational principle – minimum stream power, i.e. maximised transport with least energy expenditure – would then facilitate the design procedures if two of the six parameters are stipulated. For further detailed information and design tables based on this more rational method the reader is referred to White *et al.*[32]

14.11 Rigid bed channels with sediment transport

Rigid bed channels are the conveyances with no boundary erosion and the sediment is fed from external sources, e.g. lined irrigation canals carrying silt, sewers and outfalls, etc. The channel is designed for no deposition criteria. The mode of transport and design criteria largely depend on the sediment and channel characteristics. A great deal of research into the areas of sediment initiation, transport, cohesivity aspects of sediments, etc. has taken place and is still currently in progress.[3,18,21,22]

Studies of noncohesive suspended silt reveal (Garde and Ranga Raju[11]) that the limiting (for no deposition) concentration (C_v) is a function of sediment size (d), density (ρ_s), fall velocity (W_s), water discharge (Q), flow depth (y_0), water surface width (B), channel slope (S) and bed friction factor (λ_b). Figure 14.8 shows the proposed relationship valid for circular, rectangular and trapezoidal channels.

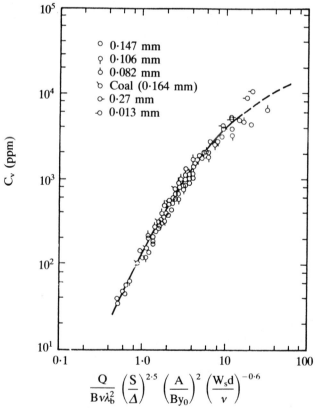

$$\frac{Q}{Bv\lambda_b^2} \left(\frac{S}{\Delta}\right)^{2.5} \left(\frac{A}{By_0}\right)^2 \left(\frac{W_s d}{v}\right)^{-0.6}$$

Figure 14.8 Limit-deposition concentration of suspended silt in circular, rectangular and trapezoidal channels

Novak and Nalluri[22] suggested for initiation of noncohesive coarser sediment (bedload):

$$\frac{V_c}{\sqrt{gd_{50}\Delta}} = 0.50 \left(\frac{d_{50}}{R}\right)^{-0.4} \tag{14.64}$$

The limit-deposition criterion in the case of rectangular channels is given by (Mayerle *et al.*[17]):

$$\frac{V_s}{\sqrt{g\Delta d_{50}}} = 11.59 \ D_{gr}^{-0.14} \ C_v^{0.15} \left(\frac{d_{50}}{R_b}\right)^{-0.43} \lambda_s^{-0.18} \tag{14.65}$$

where C_v is the limiting sediment concentration by volume that can be transported with a velocity V_s (self-cleansing). The overall friction factor λ_s (Nalluri and Kithsiri[20]) is given by:

$$\lambda_s = 0.851 \ \lambda_c^{0.86} \ C_v^{0.04} \ D_{gr}^{0.03} \tag{14.66}$$

where λ_c is the channel's clear water friction factor given by Colebrook-White's equation (equation 4.15). Equations 14.65, 14.66 and 4.15 will give (by iterative solution) the design velocity for the self-cleansing criterion in rigid boundary rectangular channels with bedload transportation.

In the case of clean pipe channels, the limit-deposition criterion may be written as (Nalluri *et al.*[19]):

$$\frac{V_s}{\sqrt{g\Delta d_{50}}} = 3.08 \ D_{gr}^{-0.09} \ C_v^{0.21} \left(\frac{d_{50}}{R}\right)^{-0.53} \lambda_s^{-0.21} \tag{14.67}$$

with the friction factor λ_s given by

$$\lambda_s = 1.13 \ \lambda_c^{0.98} \ C_v^{0.02} \ D_{gr}^{0.01} \tag{14.68}$$

where λ_c is the clear water friction factor given by equation 4.15.

The limit-deposition criterion in the case of pipe channels with deposited flat beds of width b is given by

$$\frac{V_s}{\sqrt{g\Delta d_{50}}} = 1.94 \ C_v^{0.165} \left(\frac{b}{y_0}\right)^{-0.4} \left(\frac{d_{50}}{D}\right)^{-0.57} \lambda_{sb}^{0.10} \tag{14.69}$$

The bed friction factor with transport, λ_{sb}, is given by

$$\lambda_{sb} = 6.6\lambda_s^{1.45} \tag{14.70}$$

where λ_s is given by

$$\lambda_s = 0.88 \ C_v^{0.01} \left(\frac{b}{y_0}\right)^{0.03} \lambda_c^{0.94} \tag{14.71}$$

where λ_c is the clear water friction factor given by equation 4.15.

It must be stressed that the proposed equations 14.64 to 14.71 are developed by analysing experimental data and are valid within the experimental ranges of the data used.

Worked examples

Example 14.1

(a) Starting from first principles show that the fall (sedimentation) velocity W_s of a particle size d in a fluid is given by

$$W_s = A \sqrt{(g\Delta d)}$$

where A is a function of the drag coefficient C_D and Δ is the relative density of the particle in water $\{= (\rho_s - \rho)/\rho\}$.

Assuming that for particles of shape factor (SF) = 1

$$C_D = \text{Constant} (= 2) \text{ for large diameters}$$

and $\quad C_D = 24/R_{ed}$ for very fine particles

give the full equation for W_s in each of these cases.

(b) Examine the stability of the bed material ($\rho_s = 2650$ kg/m³), mean diameter = 1 mm) of a wide stream having a slope of 10^{-3} and carrying a flow at a depth of 0·3 m.

(c) What type(s) of transport and bed form, if any, do you expect in this stream?

Solution:

(a)   Equating gravity force (weight of the particle) to drag force

$$(\rho_s - \rho) g \, \pi d^3/6 = \frac{1}{2} C_D \rho (\pi d^2/4) W_s^2$$

or $\qquad\qquad W_s^2 = 4\Delta gd/3C_D$

or $\qquad\qquad W_s = \sqrt{(4/3C_D)(g\Delta d)}$

Comparing this with the given equation

$$A = \sqrt{(4/3C_D)} = f(R_{ed})$$

since the drag coefficient C_D is a function of R_{ed}

Coarse sediment: $C_D = 2$ (given)

$\therefore \qquad\qquad W_s = \sqrt{\{(2/3)g\Delta d\}}$

Fine sediment: $\quad C_D = 24/R_{ed} = 24v/W_s d$

$\therefore \qquad\qquad A = \sqrt{(W_s d/18v)}$

Hence $\qquad\qquad W_s = g\Delta d^2/18v$

(b) Threshold condition: Shields criterion

Wide channel $\Rightarrow R \simeq y_o = 0\cdot 3$ m

By constructing a d (= 1 mm) line on Fig. 14.1 (Shields diagram) we obtain

$\tau_c/\rho g \Delta d = 0.035$

or $\tau_c = \rho U_*^2 = 0.035 \times \rho g \times 1.65 \times d$

Available boundary shear stress

$\tau = \rho g RS = 2.943 \text{ N/m}^2$

Thus minimum d for stability $= 5.2 \text{ mm} > 1 \text{ mm}$

Hence the bed material is not stable.

(c) Available $\tau/\rho g \Delta d = U_*^2/g \Delta d = RS/1.65 \times 0.001 = 0.181$ and from Fig. 14.1 this relates to bed dunes to high regime plane bed transition.

Example 14.2
It is intended to stabilise a river bed section with the following data by depositing a layer of gravel or stone pitching:

Channel width $= 20 \text{ m}$

Bed slope $= 0.0045$

Max. discharge $= 500 \text{ m}^3/\text{s}$

Chezy's C $= 18 \log (12R/d)$; (d is the mean diameter of the material)

Relative density of the bed material $\Delta = 1.65$

Determine the depth of flow assuming the section to be rectangular and the minimum size of stone required for stability. Use Shields criterion for stability: $\tau/\rho g \Delta d = 0.05$

Solution:
From Shields criterion: $d = 12RS$

$\therefore$ Chezy's $C = 18 \log (12R/12RS) = 18 \log (1/S) = 42.24 \text{ m}^{1/2}/\text{s}$

Mean velocity, $V = Q/A = 500/20 \times y_o = 42.24 \sqrt{(RS)}$ (i)

Hydraulic radius, $R = 20y_o/(20 + 2y_o)$ (ii)

Equations (i) and (ii) give $y_o \simeq 4.9 \text{ m}$ (by iteration)

$\therefore R = 3.29 \text{ m}$ giving $d (= 12 RS) = 147 \text{ mm}$

$\therefore$ Provide an armour layer with 150 mm size stones.

Example 14.3
The following data relates to a wide stream:

Slope $= 0.0001$

Bed material: Size, $d = 0.4 \text{ mm}$

Density, $\rho_s = 2650 \text{ kg/m}^3$

(a) Find the limiting depth of flow at which the bed material just begins to move.

(b) Find the corresponding mean velocity in the stream.

Solution:

(a) Problems of this kind may explicitly be solved by using the modified Shields diagram (Fig. 14.9):

Modified R_{e*} can be written as

$$R'_{e*} = (R_{e*})^{2/3}/(\tau/\rho g \Delta d)^{1/3}$$

$$= (\Delta g)^{1/3} d/v^{2/3} \ (= d_{gr})$$

$$R'_{e*} = (1·65 \times 9·81)^{1/3} \times 0·0004/(10^{-6})^{2/3}$$

$$= 10·12$$

$\therefore$ From Fig. 14.9 $\tau_c/\rho g \Delta d = 0.035$

giving $\tau_c = 0·227 \ N/m^2$

Boundary shear in wide channel, $\tau_c = \rho g y_o S$

Hence for critical condition, $\rho g y_o S = 0·227$

$\therefore$ The limiting flow depth, $y_o \qquad = 0·231 \ m$

(b) At the threshold condition the bed is plane with roughness, $k = d$

Chezy's C (in transition) $= 18 \log \{12R/(k + 2\delta'/7)\}$ (8.41)

where δ' is sublayer thickness given by $\delta' = 11·6v/U_*$

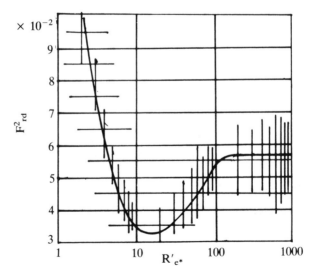

Figure 14.9 **Modified Shields curve**

$$k = d = 0.0004 \text{ m}; \quad \delta' = 11.6 \times 10^{-6}/\sqrt{(9.81 \times 0.231 \times 0.0001)}$$

$$= 7.7 \times 10^{-5} \text{ m}$$

$$\therefore k + 2\delta'/7 = 0.0004 + 0.000022 \quad = 0.000422 \text{ m}$$

$$\therefore C = 18 \log (12 \times 0.231/0.000422) = 54.7 \text{ m}^{1/2}/s$$

Hence mean velocity, $V = C \sqrt{(RS)}$

$$= 54.7 \times \sqrt{(0.231 \times 0.0001)}$$

$$= 0.263 \text{ m/s}$$

Example 14.4
A wide alluvial stream carries water with a mean depth of 1 m and slope of
0.0005. The mean diameter of the bed material is 0.5 mm with a relative
density of 2.65. Examine the bed stability and bed form, if any. Also
calculate the sediment transport rate that may exist in the channel.

Solution:
Wide channel $\Rightarrow R \simeq y_o = 1.0$ m; $\quad S = 0.0005; \quad d = 0.0005$ m

$\quad \therefore d_{gr} = 12.5$

From Fig. 14.9 $\tau/\rho g \Delta d = 0.032$

$\quad \therefore \tau_c = 0.032 \times 1000 \times 9.81 \times 1.65 \times 0.0005$

$\quad = 0.234 \text{ N/m}^2$

Channel boundary shear stress, $\tau_o = \rho gRS = 4.9 \text{ N/m}^2$

Since $\tau_o > \tau_c$ sediment transport exists.

Type of transport:

Fall velocity of sediment particle from Figure 14.4

$$W_s = 0.075 \text{ m/s}$$

$\quad \therefore$ Movability parameter, $U_*/W_s = 0.93$

Referring to Table 14.2, most part of the transport may be treated as bed
load which may be computed using Shields equation (Equation 14.16)

Discharge computations:

Referring to Garde and Ranga Raju's plots (Figures 14.2 and 14.3)
mean velocity in the channel may be determined.

For $d = 0.5$ mm; $\quad K_1 = K_2 = 0.95$ (Fig. 14.3)

$\quad \therefore K_2(R/d)^{1/3}S/\Delta = 3.63 \times 10^{-3}$

From Figure 14.2 $K_1 V/\sqrt{(g\Delta R)} = 0.20$

Hence $V = 0.847$ m/s giving $q = Vy_o = 0.847$ m³/ms

From Shields equation 14.19, the bed load transport

$q_b = 10 (\tau_o - \tau_c) qS/\rho g\Delta^2 d$

$= 10 \times (4.9 - 0.292) \times 0.847 \times 0.0005/1000 \times 9.81 \times 1.65^2 \times 0.0005$

$= 1.46 \times 10^{-3}$ m³/ms

$\simeq 40$ N/ms or 4 kg/ms.

Bed form:

$\tau_o/\rho g\Delta d = 4.9/1000 \times 9.81 \times 1.65 \times 0.0005 = 0.61$

From Shields curve (Fig. 14.1) the bed form may be in high regime transition to antidunes; however, the flow Froude no. ($= V/\sqrt{(gy_o)}$) is less than 1 and the bed may be in dune to high regime transition form.

Equivalent bed roughness (k) or dune height:

Using Chezy's equation, $V = C \sqrt{(RS)}$

Chezy's $C = 0.847/\sqrt{(1 \times 0.0005)} = 37.88$ m^{1/2}/s

Hence from $C = 18 \log (12R/k)$ (assuming fully rough bed)

$k = 9.5 \times 10^{-2}$ m or 95 mm

Example 14.5
A laboratory rectangular flume with smooth sides and rough alluvial bed (d = 6.5 mm) of the following data carries 0.1 m³/s of water:

Bed width $= 0.5$ m

Depth of flow $= 0.25$ m

Slope $= 3 \times 10^{-3}$

Find the bed hydraulic radius using the resistance curve for smooth sides (Fig. 14.9.), bed shear stress and Manning's coefficient. Also examine the stability of the bed.

Solution:

Overall hydraulic radius, $R = A/P = 0.5 \times 0.25/(0.5 + 2 \times 0.25)$

$= 0.125$ m

Velocity, $V = Q/A = 0.1/(0.5 \times 0.25) = 0.8$ m/s

∴ Reynolds no., $R_e = 4VR/\nu = 4 \times 10^5$

Overall friction factor, $\lambda = 8gRS/V^2 = 0.046$

∴ $R_e/\lambda = 8.7 \times 10^6$

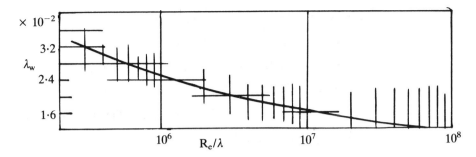

Figure 14.10 Resistance curve for smooth walls

and from Figure 14.10 wall friction factor, $\lambda_w = 0.017$

From equation 14.11 $\Rightarrow P\lambda = P_b\lambda_b + P_w\lambda_w$

bed friction factor, $\lambda_b = 0.075$

∴ Bed hydraulic radius, $R_b = \lambda_b V^2/8gS$ (equation 14.10)

$$= 0.204 \text{ m}$$

Bed shear stress, $\tau_b = \rho g R_b S = 6 \text{ N/m}^2$

From Manning's equation $V = (1/n_b)(R_b)^{2/3}S^{1/2}$

$n_b = 0.0237$

Bed stability:

$d_{gr} = 167.5$

Hence $\tau_c/\rho g\Delta d = 0.056$ (Shields criterion) (see Fig. 14.9)

or critical shear stress for stability, $\tau_c = 5.9 \text{ N/m}^2$

As $\tau_b \simeq \tau_c$ the bed is just stable.

(Note: Overall Manning's $n = R^{2/3}S^{1/2}/V = 0.017$ which may be compared with Strickler's equation giving $n = d^{1/6}/26 = 0.0166$ with $k = d$)

Example 14.6
Design a stable alluvial channel of trapezoidal cross section with the following data:

Discharge	$= 50 \text{ m}^3/\text{s}$
Bed material size	$= 4 \text{ mm}$
Angle of repose ϕ	$= 30°$
Bed slope	$= 10^{-4}$
Channel side slopes	$= 2H:1V$ ($\tan\theta = 1/2$)

Solution

$$R'_{e*} = (\Delta g)^{1/3} d / v^{2/3} = 100 \cdot 8$$

From Figure 14.9 $F^2_{rd} = \tau/\rho g \Delta d = 0 \cdot 054$

$\therefore$ Critical bed shear stress, $\tau_{bc} = 0 \cdot 054 \times 1000 \times 9 \cdot 81 \times 1 \cdot 65 \times 0 \cdot 004$

$$= 3 \cdot 496 \text{ N/m}^2$$

Hence critical shear on slopes, $\tau_{sc} = K \tau_{bc}$

where $K = \{1 - (\sin^2 \theta / \sin^2 \phi)\}^{1/2}$ (8.9)

$$= \{1 - (0 \cdot 2 / 0 \cdot 25)\}^{1/2}$$

$$= 0 \cdot 447$$

$\therefore \tau_{sc} = 0 \cdot 447 \times 3 \cdot 496 = 1 \cdot 563 \text{ N/m}^2$

Mean boundary shear distribution in a 2H : 1V trapezoidal channel (see Figure 8.5):

For $B/y_o > 10$ $\tau_{bm}/\rho g y_o S \simeq 0 \cdot 985$

$$\tau_{sm}/\rho g y_o S \simeq 0 \cdot 78$$

Using critical shear on bed as the criterion for stability

$\tau_{bc} = \tau_{bm} = 0 \cdot 985 \times \rho g y_o S$

Limiting flow depth, $y_o = 3 \cdot 62$ m

Using critical shear on sides as the criterion

$\tau_{sc} = \tau_{sm} = 0 \cdot 78 \times \rho g y_o S$

Limiting flow depth, $y_o = 2 \cdot 04$ m

$\therefore$ Choose $y_o = 2$ m (smaller of the two criteria)

The channel boundary is in threshold condition thus the bed is plane, with roughness $k = d$.

$\therefore$ Manning's $n = d^{1/6}/26$ (Strickler's formula)

$$= 0 \cdot 0153$$

$\therefore$ Mean velocity, $V = (1/n)R^{2/3}S^{1/2}$ (Manning's formula)

Select various B/y_o values and calculate Q (see Table 14.9).

Table 14.9 Design of stable channel

B/y_o	y_o m	B m	A m²	P m	R m	V m/s	Discharge, Q m³/s
10	2	20	48	28·94	1·659	0·916	43·97
11·5	2	23	54	31·04	1·691	0·924	50·09
12	2	24	56	32·94	1·700	0·931	52·14

Adopt a trapezoidal section of bed width = 23 m

flow depth = 2 m

Example 14.7

The most economical section (i.e. in which the particles are at threshold all over its perimeter) of an alluvial channel is given by (see Fig. 14.11):

$$y = y_o \cos (0{\cdot}8 \tan \phi \, x/y_o) \tag{14.63}$$

where ϕ is the angle of repose of the bed material.

Calculate the maximum possible discharge that such a channel in an alluvium ($d = 4$ mm, $\phi = 30°$) with a bed slope of 10^{-3} can carry.

Design the most economical section to carry a discharge of 1 m³/s.

Solution:
The parameters like area (A), perimeter (P), hydraulic radius (R), water surface width (T) are all functions of flow depth y_o and ϕ (see Table 14.10).

Table 14.10   Economic alluvial section as a function of ϕ

ϕ	15°	20°	25°	30°	35°	40°	45°
A/y_o^2	7·5	5·4	4·21	3·46	2·79	2·31	2·00
P/y_o	12	8·8	7·0	5·8	4·9	4·2	3·85
R/y_o	0·625	0·615	0·602	0·588	0·570	0·550	0·520
T/y_o	11·72	8·63	6·74	5·44	4·49	3·74	3·14

(Note: In order to accommodate lift forces on particles the working value of ϕ may be taken as $\tan^{-1} (0{\cdot}8 \tan \phi)$)

Limiting bed shear for $d = 4$ mm

$$\tau_{bc} = 3{\cdot}496 \text{ N/m}^2 \text{ (Shields)}$$

As the shear distribution is uniform and equal to $\rho g y_o S$

$$\rho g y_o S = 3{\cdot}496$$

$\therefore$ Flow depth, $y_o = 0{\cdot}356$ m

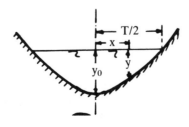

Figure 14.11   Cosine profile – most economical section

∴ The boundary cross section of the channel is given by equation 14.63:

y = 0·356 cos (1·3x)

From Table 14.10 we can get for $\phi = \tan^{-1}$ (0·8 tan 30°) = 24·8°

A = 0·54 m², R= 0·214 m and T = 2·43 m

Manning's n = $d^{1/6}/26$ = 0·0153 and

Mean velocity, V = $(0·214)^{2/3}(0·001)^{1/2}/0·0153$

$= 0·739$ m/s

∴ Maximum safe discharge, Q_o = AV = 0·4 m³/s

Design discharge Q = 1 m³/s
As Q > Q_o provide an additional central rectangular section (Fig. 14.12) to accomodate the excess flow.

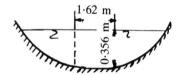

Figure 14.12 Cosine profile with flat bed for large discharges

Velocity in the central section using Manning's equation

V = $(0·356)^{2/3}(0·001)^{1/2}/0·0153$ = 1·038 m/s

Additional bed width, b = (1·0 − 0·4)/(1·038 × 0·356) = 1·62 m

(Note: For discharges Q < Q_o the section may be shortened by reducing T to (T − 2x) as per Table 14.11).

Table 14.11 Design table for Q < Q_o

2x/T	0	0·2	0·4	0·6	0·8	1·0
Q/Q_o	1	0·615	0·31	0·11	0·015	0

Example 14.8
The following data refer to an alluvial canal:

Average water surface slope = 5×10^{-4}

Average water depth = 4·82 m

Width = 52·5 m

Average velocity = 2·43 m/s

Grain size distribution: d_{90} = 50 mm

d_{50} = 20 mm

Estimate the bed load transport in the canal.

Solution:

Hydraulic radius, $R = 52 \cdot 5 \times 4 \cdot 82/(52 \cdot 5 + 2 \times 4 \cdot 82) = 4 \cdot 07$ m

$C_{channel} = V/\sqrt{(RS)} = 53 \cdot 87$ m$^{1/2}$/s

$C_{grain} = 18 \log (12R/d_{90})$

(Note: Grain roughness exposed to flow is equivalent to d_{90} as d_{50} is eroded and kept in motion as sediment load)

$\therefore C_{grain} = 53 \cdot 8$ m$^{1/2}$/s

As $C_{channel} = C_{grain}$, ripple factor, $\mu = 1 \Rightarrow$ flat bed.

Discharge in canal $= 4 \cdot 82 \times 2 \cdot 43 = 11 \cdot 7$ m^3/ms

Bed shear, $\tau_o = \rho gRS = 19 \cdot 96$ N/m^2

Critical shear stress for d_{50}

$\tau_c = 0 \cdot 056 \, \rho g \Delta d = 18 \cdot 1$ N/m^2

Bed load transport:

(i) Meyer–Peter and Muller equation (equation 14.22):

$q_b = 8 \, (1 \cdot 65 \times 9 \cdot 81)^{1/2}(0 \cdot 02)^{3/2}(\mu RS/\Delta d - 0 \cdot 047)^{3/2}$

$\mu RS/\Delta d = 0 \cdot 062$

$\therefore q_b = 1 \cdot 67 \times 10^{-4}$ m^3/ms

or total bed load $\simeq 23$ kg/s

(ii) Einstein's curve (Fig. 14.5)

$\psi = \Delta d/\mu RS = 16 \cdot 12$ giving $\emptyset = 0 \cdot 015$

$\therefore q_b = 0 \cdot 015 \, (g\Delta d^3)^{1/2} = 1 \cdot 71 \times 10^{-4}$ m^3/ms

or total bed load $\simeq 24$ kg/s

(iii) Schoklitsch equation (equation 14.17):

Bed load in kg/ms, $g_s = 2500 \, S^{3/2} \, (q - q_{cr})$

q_{cr} from equation 14.18 $= 9 \cdot 27$ m^3/ms

$\therefore g_s = 6 \cdot 79 \times 10^{-2}$ kg/ms

or total load $= 3 \cdot 6$ kg/s

(iv) Shields equation (equation 14.16)

$q_b = 10qS \, ((\tau_o - \tau_c)/\rho g \Delta^2 d)$

$= 5 \cdot 2 \times 10^{-4}$ m^3/ms

or total load $= 72$ kg/s

Shields equation is applicable for the sediment range of $1 \cdot 56 < d$ (mm) $<$

2·47 and is not reliable in this case where d = 20 mm. Also Schoklitsch's equation is not dimensionally homogeneous and does not include coarse material. Hence the probable total bed load in the canal is around 30 kg/s.

Example 14.9
The following data refer to a wide river:

Depth   = 3 m

Mean velocity   = 1 m/s

Chezy coefficient   = 50 m$^{1/2}$/s

Density of sediment = 2650 kg/m^3

A sample of wash load (mean diameter of 0·02 mm) taken at half depth of flow showed a concentration of 200 mg/l (200 ppm).
   (i) Establish the wash load concentration distribution as a function of depth and calculate the rate of suspended load assuming a homogeneous distribution.
   (ii) Determine the transport rate using Lane and Kalinske's approximate method and compare with the above result.

Solution:
   From Chezy's formula, $(RS)^{1/2}$ = V/C = 0·02

   $\therefore$ Shear velocity, U_* = $(gRS)^{1/2}$ = 0·063 m/s

   Fall velocity, W_s = 0·00035 m/s (Fig. 14.4)

   $\therefore$ Movability parameter, U_*/W_s = 0·063/0·00035 = 180

Referring to Table 14.2. the sediment is almost in homogeneous suspension.

   $\therefore$ Total suspended load = 200 × 3 × 1 × 1000/10^6 = 0·6 kg/ms

Distribution of sediment concentration:

   From equation 14.29 $\Rightarrow$ a = 1·5 m; c_a = 200 mg/l; y_o = 3m

   W_s = 0·00035 m/s (Fig. 14.4.); β= 1; χ = 0·4

   $\therefore$ c (y) = 200 $[1·5 (3 - y)/1·5 \ y]^{0·00035/0·4 \times 0·063}$

              = 200 $[(3 - y)/y]^{0·0139}$ mg/l

Equation 14.31 (Lane and Kalinske) $\Rightarrow$ q_s/q = 200 P e$^{15(1/2)0·00555}$

   P from Fig. 14.7: Manning's n = $R^{2/3}S^{1/2}$/V

   $S^{1/2}$ = 0·02/3$^{1/2}$ = 0·01155

   $\therefore$ n = 3$^{2/3}$ × 0·01155/1 = 0·024

   $\therefore$ n/$y_o^{1/6}$ = 0·02

   and hence P = 1 at W_s/U_* = 0·00555

$$\therefore q_s/q = 200 \times 1 \times e^{0.0416} = 200 \times 1.04$$

Hence $q_s = 208 \times 1 \times 3 \times 1000/10^6 = 0.624$ kg/ms

Example 14.10

An alluvial river with the following data discharges water into a downstream reservoir of 10×10^6 m^3 capacity.

Width	= 12 m
Depth	= 4 m
Slope	= 3 × 10^{-4}
Discharge	= 75 m^3/s
Bed material size, d_{50}	= 0.5 mm
Density of bed material	= 2600 kg/m^3

Determine the total sediment transport rate by Ackers–White formula in the river and compare the result with that of Graf's formula. What is the life expectancy of the reservoir fed by this river?

Solution:

$$R = A/P = 48/20 = 2.4 \text{ m}$$

Graf's formula: Equation 14.36 $\Rightarrow \psi_A = \Delta d/RS = 1.146$

$$\therefore \emptyset_A = 10.39 \times (1.146)^{-2.52} = 7.37$$

Hence $C_v VR/\sqrt{(g\Delta d^3)} = 7.37$

or $1.526 \times 2.4\, C_v/(4.5) \times 10^{-5} = 7.37$

æ $C_v = 9.05 \times 10^{-5}$

Total transport rate $= 9.05 \times 10^{-5} \times 75 = 6.79 \times 10^{-3}$ m^3/s

Ackers–White's approach:

$$d_{gr} = 0.0005 (9.81 \times 1.6/10^{-12})^{1/3} = 12.4 \text{ (Equation 14.41)}$$

For d_{gr} in transition $(1 < d_{gr} < 60)$:

$$n = 1 - 0.56 \log d_{gr} = 0.388$$

$$A = 0.14 + 0.23/\sqrt{d_{gr}} = 0.205$$

$$m = 1.34 + 9.66/d_{gr} = 2.12$$

$$\log C = 2.66 \log d_{gr} - (\log d_{gr})^2 - 3.53 = -1.865$$

$$\therefore C = 0.0136$$

$$U_* = (9.81 \times 2.4 \times 0.0003)^{1/2} = 0.084 \text{ m/s}$$

$$U_*^n = 0.382; \quad (g\Delta d)^{1/2} = 0.0886$$

$\log (12 \cdot 2 \ y_o/d) = 4 \cdot 99$

$\therefore F_{gr} = [0 \cdot 382/0 \cdot 0886] [1 \cdot 562/5 \cdot 75 \times 4 \cdot 99]^{1 - 0 \cdot 388} = 0 \cdot 726$ (Equation 14.40)

$F_{gr}/A = 0 \cdot 726/0 \cdot 205 = 3 \cdot 54$

$\therefore C[F_{gr}/A - 1]^{2 \cdot 12} = 0 \cdot 098$

$q_t/q = [0 \cdot 098 \times 0 \cdot 0005/4] [1 \cdot 562/0 \cdot 084]^{0 \cdot 388} = 3 \cdot 82 \times 10^{-5}$ (Equation 14.39)

$\therefore Q_t = 3 \cdot 82 \times 10^{-5} \times 75 = 2 \cdot 86 \times 10^{-3} \ m^3/s$

Reservoir life expectancy:

Ackers–White $\Rightarrow$ Annual deposit $= 2 \cdot 86 \times 10^{-3} \times 365 \times 24 \times 60 \times 60$

$$= 9 \times 10^4 \ m^3/year$$

$\therefore$ Reservoir life $= 10 \times 10^6/9 \times 10^4 \simeq 100$ years

Graf $\Rightarrow$ Annual deposit $= 6 \cdot 79 \times 10^{-3} \times 365 \times 24 \times 60 \times 60$

$$= 2 \cdot 14 \times 10^5 \ m^3/year$$

Reservoir life $= 10 \times 10^6/2 \cdot 14 \times 10^5 \simeq 50$ years.

(Porosity and consolidation of sediments will affect the filling rate)

Recommended reading

1. Ackers, P. and White, W.R. (Nov. 1973) *Sediment Transport: New Approach and Analysis.* Journal of Hydraulic Engineering, Proc. ASCE Vol. 99, HY−11.
2. Alam, Z.U. and Kennedy, J.F. (Nov. 1969) *Friction Factors for Flow in Sand-bed Channels.* Journal of Hydraulic Engineering, Proc. ASCE Vol. 95, HY−6.
3. Arora, A.K. et al. (April 1984) *Criterion for Deposition of Sediment Transported in Rigid Boundary Channels.* Proc. Int. Conf. on Hydraulic Design in Water Resourcs Engineering: Channels and Channel Control Structures (Ed. Smith K.V.H.) Univ. of Southampton, UK.
4. Bettess, R. and White, W.R. (1979) *A One-Dimensional Morphological River Model.* Hydraulics Research Ltd., U.K., IT 194.
5. Einstein, H.A. (1950) *The Bed Load Function for Sediment Transportation in Open Channel flows.* U.S. Dept. Agric., Soil Conserv. Serv., T.B. no. 1026.
6. Einstein, H.A. and Barbarossa, N.L. (1952) *River Channel Roughness.* Trans. ASCE Vol. 117.
7. Einstein, H.A. and Chien, Ning (1955) *Effects of Heavy Sediment Concentration near the Bed on Velocity and Sediment distribution*, Uni. Cali. Inst. Eng. Res. no. 8.
8. Engelund, F. (1970) *Instability of Erodible Beds.* Journal of Fluid Mechanics. Vol. 42, Part 2.

9. Garde, R.J. (Oct. 1968) *Analysis of Distorted River Models with Movable Beds.* Journal of Irrigation and Power, India, Vol. 25 No. 4.

10. Garde, R.J. and Ranga Raju, K.G. (July 1966) *Resistance Relationships for Alluvial Channel Flow.* Journal of Hydraulic Division, Proc. ASCE. Vol. 92, HY-4.

11. Garde, R.J. and Ranga Raju, K.G. (1991) *Mechanics of Sediment Transportation and Alluvial Stream Problems*, (2nd ed.) Wiley Eastern Ltd, New Delhi.

12. Graf, W.H. (1984) *Hydraulics of Sediment Transport.* Water Resources Publications.

13. Glover, R.E. and Florey, Q.L. (1951) *Stable Channel Profiles.* USBR Hyd. Lab. Report No. 325.

14. Lacey, G. (1930) *Stable Channels in Alluvium.* Proc. ICE. Vol. 229.

15. Lane, E.W. and Kalinske, A.A. (1939) *The Relation of Suspended Load to Bed Material in Rivers.* Trans. Am. Geo. Union, Vol. 20.

16. Laursen, E.M. (1958) *Total Sediment Load of Streams.* Journal of Hydraulic Division, Proc. ASCE. Vol. 84. HY-1.

17. Mayerle, R., Nalluri, C. and Novak, P. (1991) *Sediment Transport in Rigid Bed Conveyances.* Journal of Hydraulic Research, Vol. 29, Part 4, pp. 5-495.

18. Nalluri, C. (1986) *Sediment Transport in Rigid Boundary Channels.* Proc. Euromech 192: Transport of Suspended Solids in Open Channels. Balkema.

19. Nalluri, C., Ab Ghani, A. and El-Zaemey, A.K.S. (1994) *Sediment Transport Over Deposited Beds in Sewers.* Water Science Technology, Vol. 29, No. 1, pp. 125-33.

20. Nalluri, C. and Kithsiri, M.M.A.U. (1992) *Extended Data on Sediment Transport in Rigid Bed Rectangular Channels.* Journal of Hydraulic Research, Vol. 30, Part 6, pp. 851-6.

21. Novak, P. and Nalluri, C. (Sept. 1975) *Sediment Transport in Smooth Fixed Bed Channels.* Journal of Hydraulic Division, Proc. ASCE. Vol. 101, HY-9.

22. Novak, P. and Nalluri, C. (1984) *Incipient Motion of Sediment Particles over Fixed Beds.* Journal of Hydraulic Research, Vol. 22, No. 3.

23. O'Brien, M.P. (1933) *Review of the Theory of Turbulent Flow and its Relation to Sediment Transportation.* Trans. Am. Geo. Union.

24. Ota, J.J. (1999) Effect of particle size and gradation on sediment transport in storm sewers, PhD thesis, University of Newcastle, UK.

25. Ranga Raju, K.G. (1993) *Flow through Open Channels*, (2nd ed.). Tata-McGraw Hill, New Delhi.

26. Raudkivi, A.J. (1990) *Loose Boundary Hydraulics*, (3rd ed.) Pergamon.

27. Rouse, H. (1937) *Modern Conceptions of Mechanics of Fluid Turbulence.* Trans. ASCE. Vol. 102.

28. Shields, A. (1936) *Unwendung der Aehnlichkeitsmechanik und der Turbulenzforschung auf die Geschiebebewegung Mitteilungen der Pruessischen Versuchsanstalt für Wasserbau und Schiffbau.* Berlin.

29. Task Committee on the Bed Configuration and Hydraulic Resistance of

Alluvial Streams (Nov. 1974) *The Bed Configuration and Roughness of Alluvial Streams* ASCE.

30. Vanoni, V.A. and Brooks, N.H. (Dec. 1957) *Laboratory Studies of the Roughness and Suspended load of Alluvial Streams*. Caltech Report No. E–68.

31. White, W.R., Paris, E. and Bettess, R. (Sept. 1980a) *The Frictional Characteristics of Aluvial Streams*, Proc. Institution of Civil Engineers, Part 2, Vol. 69.

32. White, W.R., Paris, E. and Bettess, R. (1980b) *The Frictional Characteristics of Alluvial Streams: A New Approach*. Proc. ICE, Part 2, Vol. 69, pp. 737–50.

33. White, W.R., Paris, E. and Bettess, R. (1981) *Tables for the Design of Stable Alluvial Channels*. Hydraulics Research Ltd, Wallingford, Report No. IT 208.

34. White, W.R., Bettess, R. and Shiqiang, Wang. (Dec. 1987) *Frictional Characteristics of Alluvial Streams in the Lower and Upper Regimes*. Proc. Institution of Civil Engineers, Part 2, Vol. 83.

Problems

1. (a) Using Shields threshold criterion $\tau/\rho g \Delta d = 0.056$ and Strickler's equation for Manning's n $= d^{1/6}/26$ show, for wide channels

   (i) $d = 11RS$ (Equation 14.3)

   (ii) $V_c/\sqrt{(gd)} = 1.9 \sqrt{\Delta} (d/R)^{-1/6}$ (Equation 14.6)

   (iii) $q_{cr} = 0.2 (\Delta)^{5/3} d^{3/2}/S^{7/6}$ (Equation 14.18)

   (b) A flood plain river bank in fine silty sand is experiencing extensive erosion. The bank full discharge of the river is 60 m³/s and the section is approximately 10 m wide and 2 m deep at this discharge. The flood plain is protected by a cover of 30 mm size stone rip-rap whose friction coefficient, $\lambda = 0.022$. Examine the stability of the rip-rap cover using Shields criterion.

2.   A river bed of the following data is stabilised by the deposition of a gravel layer:

   Channel width $= 12$ m

   Bed slope $= 5 \times 10^{-3}$

   Max. discharge $= 15$ m³/s

   Determine the limiting depth of flow assuming the section to be rectangular if the gravel size is 30 mm.

3.   A straight canal with side slopes of $2\frac{1}{2}$H : 1V is carrying water with a

mean depth of 1 m and mean velocity of 0·87 m/s. The Chezy's coefficient is around 25 $m^{1/2}$/s. Determine the minimum size of broken gravel that can be used as a protective layer around the periphery of the canal. The angle of internal friction for the stone whose density is 2600 kg/m³ may be assumed as 35°.

4. A channel of trapezoidal cross section (side slopes of 2H : 1V, bed width of 8 m and flow depth of 2 m) is excavated in gravel (mean diameter of 4 mm and ø of 30°). Determine the limiting bed slope of the channel.

5. A channel bed is protected with 40 Kg stones (ρ_s = 2800 kg/m³). If the flow depth in the channel is 4 m calculate the critical velocity at which the stability of the protective layer is in danger. If the velocity in the channel exceeds this critical velocity by 20% determine the size of the stones that will be needed for its protection.

6. A long and wide laboratory flume is to be prepared to carry out experiments to check Shields diagram. It is proposed to cover the bed of the flume with a layer of homogeneous noncohesive material.
  Determine the unit discharge rate and slope of the bed required for a flow depth of 2 m to investigate the studies using (i) sand with d = 0·125 mm and (ii) gravel with d = 4 mm. The density of both the materials may be assumed as 2650 kg/m³ with water temperature of 12°C.

7. The following data refer to a wide river:

Flow depth	= 2 m
Mean velocity	= 0·71 m/s
Slope	= 1/12 000
Grain size	= 1 mm
Density of grains	= 2000 kg/m³
Settling velocity	= 0·10 m/s
Kinematic viscosity of water	= 10^{-6} m²/s

(a) (i) Calculate the rate of sediment transport in N/day using the Meyer−Peter and Muller formula: $ø = (4/\psi - 0·188)^{3/2}$
  (ii) Determine the k value of the bed and identify the possible bed formation and explain whether the bed is hydraulically smooth or rough.
  (iii) Check whether there will be suspended load or not.
(b) If this river is discharging into a lake of constant water level and the sediment transport is interrupted at 10 km upstream of the lake, discuss, with the help of neat sketches, the consquences of the river regime along this 10 km stretch of the river.

8.   In a wide stream of 3 m depth, the shear stress on the bed is estimated to be 2·4 N/m². The concentration of suspended sediment is found to be 25·8 kg/m³ at a point 0·03 m from its bed. The settling velocity of the sediment is 9·14 mm/s in still water.

   (i) Plot the profile of the sediment concentration through the depth with Karman's constant, $\chi = 0.4$.
   (ii) Assuming the velocity profile as

$$u/u_* = 5.8 + 2.5 \ln (u_* y/v)$$

estimate the suspended load transport per unit width of the channel.

9.   Suspended particles 50% by weight with $W_s = 9.1$ mm/s and 50% with $W_s = 15.2$ mm/s are admitted to a sedimentation tank with a mean velocity of 0·152 m/s and a depth of 1·52 m. Find the fraction of removal (F) of the total load if a tank length of 30 m with a bed slope of 10^{-4} is used.

   [Hint: Use Sumer's equation: $x/y_o = 12V \log (1 - F)/(U_* - 10W_s)$]

10.   A mountainous creek is monitored to measure the suspended load by sampling the concentration of silt at its mid-depth.
   For a flow depth of 1 m the mean concentration is found to be 21 N/m³ (dry weight) with $d = 30\mu$ and $\rho_s = 2650$ kg/m³.
   The average width of the creek at the sampling section is 22 m and from the topographic map of the area its slope was determined to be 3·6 m/km. Stage-discharge measurements of the creek indicated that its bed roughness (k-value) could be taken as 0·12 m. The water temperature during sampling was 20°C.
   Calculate the suspended load transport for this flow in the creek.

Chapter 15
Hydraulic Structures

J. Saldarriaga

15.1 Introduction

A variety of hydraulic structures are available to control water levels and regulate discharges for purposes of water supply, water storage, flood alleviation, irrigation, etc. These range from the weirs/sluices of small channels to the overflow spillways of large dams. Many of these structures may also be used as discharge measuring devices (BS 3680, 1981). The present chapter mainly deals with the spillways and their associated energy dissipators, as examples of common hydraulic structures.

15.2 Spillways

A spillway is the overflow device of a dam project which is essential to evacuate excess of water, which otherwise may cause upstream flooding, dam overtopping and eventually dam failure. Basically there are three types of spillways – overfall (ogee or side channel), shaft (morning glory), and siphon. The overfall types are the most frequently encountered and which are described here in detail. The reader may refer to Novak et al. (1997)[4] for other types and further information.

(a) Overfall spillways

The basic shape of the overfall spillway (ogee spillway) is derived from the lower envelope of the overfall nappe (see fig. 15.1) flowing over a high vertical rectangular notch with an approaching velocity V_a close to zero and a fully aerated space beneath the nappe (see also Chapter 3, Section 3.17 (e)). For a notch of width b, head h, and discharge coefficient C'_d the discharge equation (see also equation 3.34) is:

$$Q = \frac{2}{3}\sqrt{2g}bC'_d\left[\left(h + \frac{\alpha V_a^2}{2g}\right)^{3/2} - \left(\frac{\alpha V_a^2}{2g}\right)^{3/2}\right] \qquad (15.1)$$

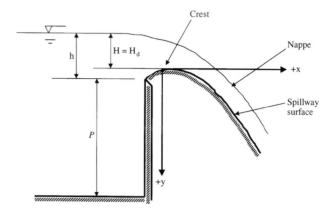

Figure 15.1   Crest of an ogee overfall spillway

which, for $V_a = 0$, reduces to:

$$Q = \frac{2}{3}\sqrt{2gb}C_d'h^{3/2} \tag{15.2}$$

This coefficient C_d' ($= 0.62$) is valid for a rectangular sharp crested weir.

Scimeni (see USBR, 1987[5]) expressed the shape of the nappe in coordinates x and y, measured from the highest point (at a distance $0.282\ H_d$ away from the sharp crested sill) of the water jet, for $H/H_d = 1.0$ as:

$$y = Kx^n \tag{15.3}$$

with

$$K = 0.5;\ n = 1.85$$

For other values of H the nappes are similar and the equation can be rewritten as:

$$y/H = K(x/H)^n$$

or

$$y = Kx^n H^{1-n} = 0.5x^{1.85}\ H^{-0.85}$$

Towards upstream of the summit point the nappe has the shape of two circular arcs, one with a radius $R = 0.5\ H_d$ up to a distance of $0.175\ H_d$ and the other with a radius of $R = 0.2\ H_d$ up to the sharp sill (Novak et al., 1997[4]).

In fig. 15.1 it is clear that the head H above the new crest is smaller than the head h above the crest of the sharp-edge notch from which the shape of the overfall spillway was derived. For an overfall spillway one can rewrite the discharge equation as:

$$Q = \frac{2}{3}\sqrt{2gb}C_{d_0}\ H_{d_e}^{3/2} \tag{15.4}$$

or, if $V_a = 0$:

$$Q = \frac{2}{3}\sqrt{2g}bC_{d_0}H_d^{3/2}$$

In equation (15.4) H_{d_e} is the design energy head given by:

$$H_{d_e} = H_d + \alpha\frac{V_a^2}{2g} \tag{15.5}$$

H_d being the design head and C_{d_0} being the design discharge coefficient equal to 0·745 for spillways of $P/H_{d_e} > 3\cdot0$.

For any other head (H) the discharge coefficient varies suggesting for $H/H_d < 1\cdot0, 0\cdot58 < C_d < 0\cdot745$, and for $H/H_d > 1\cdot0, C_d > 0\cdot745$ (see figs 3.19 and 3.20).

Negative pressures and cavitation

When $H/H_d > 1\cdot0$ negative pressure exists underside of the nappe which may lead to cavitation problems. In order to avoid this problem it is suggested that $H < 1\cdot65\ H_d$. To find the pressure under the downstream nappe one can use the Cassidy's (see USBR, 1987[5]) relation:

$$\frac{p_m}{\rho g} = -1\cdot17H\left(\frac{H}{H_d} - 1\right)$$

in which p_m is the gauge pressure under the nappe.

Gated spillways

Gates are used in spillways to increase the reservoir capacity. For gated spillways, the placing of the sill by $0\cdot2H_d$ (see Novak *et al.*, 1997[4]) downstream of the crest substantially reduces the tendency towards negative pressures for outflow under partially raised gates. The discharge through partially raised gates can be calculated from:

$$Q = \frac{2}{3}\sqrt{2g}bC_{d_1}\left(H^{3/2} - H_1^{3/2}\right) \tag{15.6}$$

with $C_{d_1} = 0\cdot6$, in which H is the distance from the spillway crest until the upstream (reservoir) water level and H_1 is the distance from the lower gate lip until the same water level. Alternatively an equation (orifice) of the type,

$$Q = C_{d_2}ba(2gH_e)^{1/2} \tag{15.7}$$

where a is the distance of the gate lip from the spillway surface, and H_e is the effective head on the gated spillway, which is very similar to H could also be used.

Offset spillways

For slender dam sections as in the case of arch dams, it may be necessary to offset the upstream spillway face into the reservoir in order to gain enough space to develop its shape. This upstream offset has no effects on the discharge coefficient.

Effective spillway length

In all the previous equations, b refers to the spillway length. If the crest has piers (to support gates) this length must be reduced to (see worked example 3.18):

$$b_e = b - 2(nk_p + k_a)H_e \qquad (15.8)$$

where n is the number of piers, k_p is the pier contraction coefficient, and k_a is the abutment contraction coefficient. The pier contraction coefficient is a function of the pier shape; for example

Point nose pier: $k_p = 0$
Square nose pier: $k_p = 0.02$
Round nose pier: $k_p = 0.01$

The abutment contraction coefficient is: $0 < k_a < 0.2$.

Self aeration

An important design feature is the point at which self-aeration of the overfall nappe (in contact with the spillway) starts. Steep slope implies high velocities which in turn entrain air from the atmosphere into the flow. The appearance of the water is white with a violently agitated free surface. The air entrapment occurs at a location (say C, fig. 15.2) where:

$$\frac{\rho h u^2}{\sigma}\left(\frac{\nu}{u_* h}\right)^{1/2} \geq 56 \qquad (15.9)$$

The location of point C (incipient point) is a function of the discharge per unit width (q), the surface roughness (k_s), and the slope (S) of the downstream channel:

$$L_C = f(q, k_s, S) \cong 15\sqrt{q} \qquad \text{(as an approximation)}$$

The entrainment of air has an effect of increasing (bulking effect) the flow depth downstream of the incipient point. To calculate the new depth one can use the following process:
1. The theoretical velocity at C:

$$u_t = \sqrt{2gH_1}$$

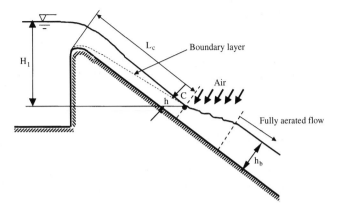

Figure 15.2   Air entrainment on overfall spillway surface

and the actual velocity is:

$$u = 0.875u_t$$

2.   With this velocity, the depth at point C is:

$$h = q/u$$

3.   The air concentration in the fully aerated flow is given by:

$$\bar{C} = \frac{1.35nFr_c^{3/2}}{1 + 1.35nFr_c^{3/2}}$$

where n is the Manning's coefficient for the channel (spillway surface) and Fr_c is the Froude number at C,

$$Fr_c = \frac{u}{\sqrt{gh}}$$

4.   Now, the bulked depth (fully aerated) is:

$$h_b = \frac{h_{n_a}}{1 - \bar{C}}$$

where h_{n_a} is the depth without air entrainment given by Manning's equation:

$$h_{n_a} = \left(\frac{nq}{\sqrt{S}}\right)^{3/5}$$

(b) Side channel spillways

In the case of a side channel spillway (in case part of the main body of the dam can not be used as a spillway), the proper spillway is designed as a normal overfall spillway. The side channel itself must be designed in such a way that the maximum flood discharge passes without affecting the free flow operation of the spillway crest.

The flow in the side channel is an example of a spatially varied flow that is best solved by the application of the momentum principle, assuming that the lateral inflow into the channel (see fig. 15.3) has no momentum in the direction of the main flow towards the chute or tunnel (see also Chapter 8, Section 8.16).

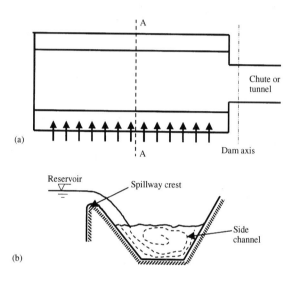

Figure 15.3   Side channel spillway; (a) plan view, and (b) section A–A

Equation (8.23) suggests the location of the control point (critical depth) in the side channel and in a given length of the channel, the control section, exists only if its slope satisfies equation (8.24). Equations (8.23) and (8.24) may be modified for the use of the Manning's n instead of Chezy's C by the relation:

$$C = \frac{R^{1/6}}{n} \tag{15.10}$$

To determine the spatially varied flow profiles, one can use a numerical integration technique with a trial-and-error process (see Henderson, 1966[3]; Chow, 1959[2]). The change in water surface elevation can be written as:

$$\Delta h' = -\Delta h + S_0 \Delta x \tag{15.11}$$

We can also write (using finite difference approach) from the momentum equation

$$\Delta h' = \frac{Q_1}{g} \frac{V_1 + V_2}{Q_1 + Q_2} \left[\Delta V + \frac{V_2}{Q_1} \Delta Q \right] + S_f \Delta x \tag{15.12}$$

Equations (15.11) and (15.12) can be used to determine the spatially varied water surface profile for the case of increasing (longitudinal direction) discharge that occurs in the side channel of the spillway (see worked example 15.3).

15.3 Energy dissipators and downstream scour protection

Creation of a hydraulic jump downstream of the spillway toe is the most effective way of dissipating the high energy (supercritical flow) of the incoming water (see Chapter 8, Section 8.8 and worked example 8.16). In addition, numerous devices such as stilling basins, aprons and vortex shafts are known in general as energy dissipators. The most common type of energy dissipator is through the formation of a hydraulic jump (free or forced) in a stilling basin (horizontal apron downstream of the toe of spillway), which at times may consist of impact blocks, ramps, steps, baffles, etc. (see USBR, 1987[5]; Vischer and Hager, 1999[6]).

Equation (8.17) suggests a sequential (conjugate) downstream subcritical depth (y_2) to incoming supercritical depth (y_1). The necessary measures to control erosion and dissipate energy (equation (8.18)) of the incoming flow depend on the available tail water depth (y_T) and the conjugate depth (y_2) (see figs 15.4 to 15.6 illustrating various scenarios of downstream conditions of energy dissipation). Figure 15.4(a) shows the classic formation of a hydraulic jump downstream of the toe when $y_2 = y_T$ (a condition rarely encountered) and in this case a simple 'horizontal apron' is adequate for downstream scour protection. Figure 15.4(b) suggests a remedial method to achieve this condition, if necessary.

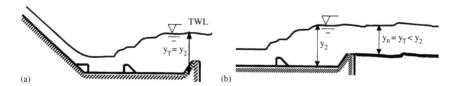

Figure 15.4 Hydraulic jump with conjugate depth equal to tail water level; (a) simple horizon (stilling basin) apron, and (b) stilling basin with lowered bed to achieve $y_2 = y_T$

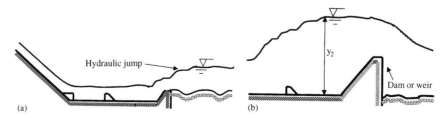

Figure 15.5 Hydraulic jump with conjugate depth greater than tail water level; (a) hydraulic jump outside of the stilling basin when $y_2 > y_T$, and (b) downstream prominent weir pushing hydraulic jump into the stilling basin

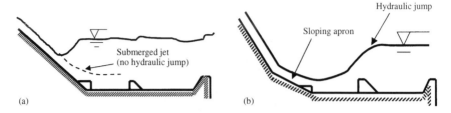

Figure 15.6 Hydraulic jump with conjugate depth smaller than tail water level; (a) submerged jump at the dam toe when $y_2 < y_T$, and (b) extension of the toe by a sloping apron (free jump in stilling basin)

Figure 15.5 shows the scenario when the tail water level is less than the conjugate depth. In this case the hydraulic jump would occur downstream of the stilling basin (see fig. 15.5(a)) without achieving the energy dissipation and scour protection. To avoid this, one can use a stilling basin with lowered level, as before, or a secondary dam or weir to increase the tail water depth causing the jump to form at the toe of the main dam.

A third possible scenario is shown in fig. 15.6, the case when the tail water level is greater than the conjugate hydraulic jump depth. This case is not desirable as a submerged jet (highly inefficient for energy dissipation) is likely to form downstream of the spillway face (see fig. 15.6(a)). Figure 15.6(b) suggests an alternative solution when $y_2 < y_T$, thus creating adequate energy dissipation.

The USBR (*Design of Small Dams*, 1987[5]) suggests elaborate layouts of these dissipating structures and the worked example 15.4 illustrates various design elements of such structures (stilling basin).

In order to chose the appropriate lowered bed level of a stilling basin, to achieve $y_2 > y_T$, it is necessary to develop the following design procedure, which uses the hydraulic jump and energy conservation equations and the geometry shown in fig. 15.7. The hydraulic jump equation is, y_1 being the incoming supercritical flow depth for a given (design) discharge:

$$y_2 = \frac{y_1}{2}\left(\sqrt{1 + 8Fr_1^2} - 1\right) \tag{8.17}$$

where $\quad Fr_1^2 = \dfrac{V_1^2}{gy_1}$

Making use of fig. 15.7, the following equations can be established:

$$y_2' > y_2 \Rightarrow y_2' = \sigma y_2 \Leftarrow \sigma > 1\cdot 0$$
$$y^+ = y_2' - y_n = \sigma y_2 - y_n \tag{15.13}$$

The energy conservation equation between the reservoir upstream and the point just downstream of the basin entrance, including all the losses, is given by (see fig. 15.7):

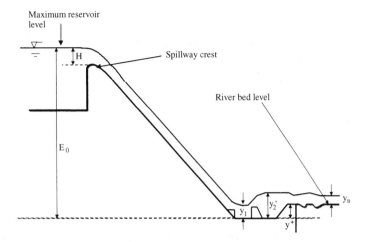

Figure 15.7 Design of a stilling basin bed level

$$E_0 = y_1 + \frac{V_1^2}{2g} + \xi \frac{V_1^2}{2g}$$

ξ being the spillway (control and chute) loss coefficient. By writing:

$$\zeta^2 = \frac{1}{1 + \xi}$$

we can have:

$$E_0 = y_1 + \frac{q^2}{2gy_1^2 \zeta^2} \tag{15.14}$$

The value of σ, which represents the safety factor for the conjugate jump depth, must be between the limits: $1 \cdot 05 < \sigma < 1 \cdot 20$, $1 \cdot 10$ being the best value. The value of ξ increases with the crest height and roughness of the spillway, which implies more losses. The range of ζ, $1 > \zeta > 0 \cdot 5$, is suggested for the design considerations. Example 15.4 illustrates the design process of a stilling basin.

Worked examples

Example 15.1
An overflow spillway is to be designed to pass a discharge of 2000 m³/s of flood flow at an upstream water surface elevation of 200·0 m. The crest length (spillway width) is 75·0 m and the elevation of the average bed is 165·0 m. Determine:

(a)   The design head
(b)   The spillway profiles, upstream and downstream
(c)   The discharge through the spillway if the water surface elevation reaches
      202·0 m

Solution:

(a): The discharge over the spillway per unit width is:

$$q_d = \frac{Q}{b} = \frac{2000 \text{ m}^3/\text{s}}{75 \text{ m}} = 26\cdot67 \text{ m}^2/\text{s}$$

For the first calculation, one can assume a discharge coefficient corresponding
to a deep upstream approaching channel. For such a case:

$$\frac{P}{H_{d_e}} \geq 3\cdot0 \Rightarrow C_{d_0} = 0\cdot74$$

Now, the design head can be computed using the discharge equation:

$$q_d = \frac{2}{3} C_{d_0} \sqrt{2g} H_{d_e}^{3/2}$$

$$26\cdot67 = \frac{2}{3} 0\cdot74\sqrt{2 \times 9\cdot81} H_{d_e}^{3/2}$$

$$\Rightarrow H_{d_e} = 5\cdot30 \text{ m}$$

The approach velocity (channel upstream of the spillway crest) is:

$$V_a = \frac{q}{P + H_d} = \frac{26\cdot67}{200 - 165} \text{ m/s} = 0\cdot762 \text{ m/s}$$

and the velocity head is:

$$\frac{V_a^2}{2g} = \frac{0\cdot762^2}{2 \times 9\cdot81} \text{ m} \approx 0\cdot03 \text{ m}$$

Therefore, the spillway crest elevation is:

$$200 \text{ m} + \frac{V_a^2}{2g} - H_{d_e} = 194\cdot73 \text{ m}$$

Now, the previous assumption of deep approaching channel can be checked:

$$P = 194\cdot73 - 165 = 29\cdot73 \text{ m}$$

$$\Rightarrow \frac{P}{H_{d_e}} = \frac{29\cdot73}{5\cdot30} = 5\cdot6 > 3 \Rightarrow \text{assumption} \rightarrow \text{ok}$$

The design head is:

$$H_d = H_{d_e} - \frac{V_a^2}{2g} = 5\cdot30 \text{ m} - 0\cdot03 \text{ m} = 5\cdot27 \text{ m}$$

(b): The upstream spillway profile is given by equation (i) that can replace the two circular curves mentioned earlier. The result of this profile must be extended until it reaches the same slope of the upstream face of the dam.

$$\frac{y}{H_d} = 0.724\left(\frac{x}{H_d} + 0.27\right)^{1.85} - 0.432\left(\frac{x}{H_d} + 0.27\right)^{0.625} + 0.126 \tag{i}$$

i.e. $\frac{y}{5.27} = 0.724\left(\frac{x}{5.27} + 0.27\right)^{1.85} - 0.432\left(\frac{x}{5.27} + 0.27\right)^{0.625} + 0.126$

and the downstream spillway profile is given by the Scimeni's equation (ii):

$$\frac{y}{H_d} = 0.50\left(\frac{x}{H_d}\right)^{1.85} \tag{ii}$$

i.e. $\frac{y}{5.27} = 0.50\left(\frac{x}{5.27}\right)^{1.85}$

The resulting upstream and downstream profiles are shown in fig. 15.8.

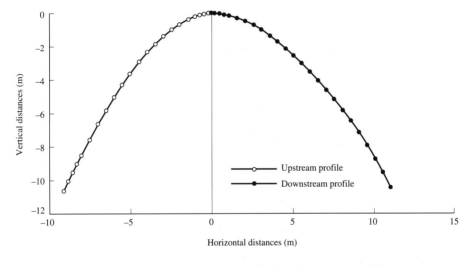

Figure 15.8 Upstream and downstream profiles for the proposed spillway

(c): The new head causing the flow is:

H = 202·0 m − 194·73 m = 7·27 m

As one does not know the approach velocity, it can be assumed, for first iteration, say:

$$\frac{V_a^2}{2g} = 0.05 \text{ m}$$

$\Rightarrow H_e = 7.27 \text{ m} + 0.05 \text{ m} = 7.32 \text{ m}$

The discharge coefficient and the new discharge per unit width can now be found (see fig. 3.20):

$$\frac{H_e}{H_{d_e}} = \frac{7 \cdot 32}{5 \cdot 30} = 1 \cdot 38 \Rightarrow \frac{C_d}{C_{d_0}} = 1 \cdot 04$$

$$\Rightarrow C_d = 0 \cdot 768$$

$$\Rightarrow q = \frac{2}{3} C_d \sqrt{2g} H_e^{3/2}$$

$$q = \frac{2}{3} 0 \cdot 768 \sqrt{2 \times 9 \cdot 81} \times 7 \cdot 32^{1 \cdot 5}$$

$$q = 45 \ \mathrm{m^2/s}$$

Using this new discharge per unit width, the approach velocity and the corresponding velocity head are:

$$V_a = \frac{q}{P + H} = \frac{45}{29 \cdot 73 + 7 \cdot 27} = 1 \cdot 216 \ \mathrm{m/s}$$

$$\Rightarrow \frac{V_a^2}{2g} = \frac{1 \cdot 216^2}{2 \times 9 \cdot 81} \mathrm{m} = 0 \cdot 08 \ \mathrm{m}$$

Second iteration: Using the new velocity head it can be found that:

$$H_e = 7 \cdot 27 \ \mathrm{m} + 0 \cdot 08 \ \mathrm{m} = 7 \cdot 35 \ \mathrm{m}$$

$$\Rightarrow \frac{H_e}{H_{d_e}} = 1 \cdot 386 \Rightarrow C_d = 0 \cdot 7683$$

therefore:

$$q = 45 \cdot 28 \ \mathrm{m^2/s}$$

and finally:

$$Q = qb = 45 \cdot 28 \ \mathrm{m^2/s} \times 75 \ \mathrm{m} = 3396 \ \mathrm{m^3/s}$$

Example 15.2
A vertical faced overfall spillway of large height (deep approaching channel) is to be spanned by a bridge deck. The design discharge is 450 m³/s under a maximum design head of 3·45 m. The pier contraction coefficient = 0·012 and the abutment contraction coefficient = 0·1.

The following table suggests the variation of C_d with H for this spillway:

H/H_d	C_d/C_{d_0}
1·000	1·000
0·908	0·988
0·682	0·952
0·455	0·920
0·227	0·855

(a) Find the number of spans required if the maximum span length is 7 m.
(b) Establish the stage–discharge relationship for the proposed design.

Solution:

(a): To find the number of spans required, the spillway discharge equation, with the design discharge coefficient, can be used:

$$Q = \frac{2}{3} C_{d_0} b_e \sqrt{2g} H_d^{3/2}$$

$$450 = \frac{2}{3} 0 \cdot 74 b_e \sqrt{2 \times 9 \cdot 81} \times 3 \cdot 45^{3/2}$$

$$\Rightarrow b_e = 32 \cdot 136 \text{ m}$$

As the maximum span between piers is 7 m, 4 piers (5 spans of 7 m) are needed. The real effective width can now be found:

$$b_e = b - 2(nK_p + K_a)H$$

$$b_{e_R} = 35 \text{ m} - 2(4 \times 0 \cdot 012 + 0 \cdot 1) \times 3 \cdot 45 \text{ m}$$

$$b_{e_R} = 33 \cdot 98 \text{ m} > b_e \Rightarrow \text{therefore safe.}$$

(b): Stage–discharge relationship for the proposed design.

Using the data given for the variation of the discharge coefficient it is possible to construct the following graph (see fig. 15.9), using an Excel spreadsheet.

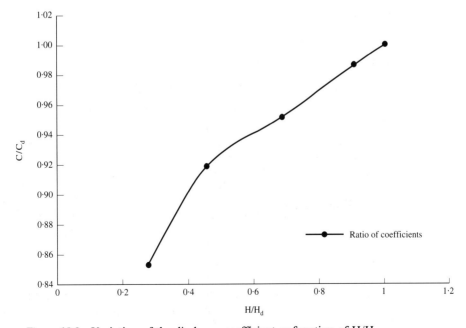

Figure 15.9 Variation of the discharge coefficient as function of H/H_d

Using fig. 15.9 and the discharge equation it is possible to establish the stage–discharge relationship for the proposed design (again an Excel spreadsheet was used to construct the graph – see table 15.1 and fig. 15.10).

Table 15.1   Stage–discharge data of the proposed spillway

H (m)	H/H_d (–)	C_d/C_{d_0} (–)	C_d (–)	b_e (m)	Q (m^3/s)
0·00	0·00000	0·7903050	0·58483	35·0000	0·0000
0·30	0·08696	0·8150876	0·60316	34·9112	10·2174
0·60	0·17391	0·8398702	0·62150	34·8224	29·7021
0·90	0·26087	0·8913478	0·65960	34·7336	57·7631
1·20	0·34783	0·9041304	0·66906	34·6448	89·9769
1·50	0·43478	0·9169130	0·67852	34·5560	127·1970
1·80	0·52174	0·9296957	0·68797	34·4672	169·1010
2·10	0·60870	0·9424783	0·69743	34·3784	215·4650
2·40	0·69565	0·9552609	0·70689	34·2896	266·1280
2.70	0·78261	0·9680435	0·71635	34·2008	320·9720
3·00	0·86957	0·9808261	0·72581	34·1120	379·9020
3.45	1·00000	1·0000000	0·74000	33·9788	475·8020
3·80	1·10145	1·0149130	0·75104	33·8752	556·5140
4·10	1·18841	1·0276957	0·76049	33·7864	629·9010
4.40	1·27536	1·0404783	0·76995	33·6976	707·1330
4·70	1·36232	1·0532609	0·77941	33·6088	788·1810
5·00	1·44928	1·0660435	0·78887	33·5200	873·0190
5·30	1·53623	1·0788261	0·79833	33·4312	961·6280
5·70	1·65217	1·0958696	0·81094	33·3128	1085·6100

The proposed design, finally, has the following results:

$$H = 3\text{·}45 \text{ m} = H_d \Rightarrow C_d = 0\text{·}74 \Rightarrow Q = 475\text{·}8 \text{ m}^3/\text{s} > Q_d$$

$$H = 5\text{·}7 \text{ m} = 1\text{·}65 \ H_d \Rightarrow C_d = 0\text{·}81 \Rightarrow Q_{Max} = 1085\text{·}6 \text{ m}^3/\text{s}$$
(to avoid cavitation problems)

Example 15.3
A rectangular lateral spillway channel 77 m long is designed to carry a discharge which increases at a rate of 6 $m^3/s/m$. The cross-section has a width of 20 m. The longitudinal slope of the channel is 0·12 and begins at an upstream elevation of 100 m. If Manning's n is equal to 0·013, determine the water surface profile.

Solution:
The first step is to determine if a critical cross-section exists, and if so, the second step is to determine its longitudinal position. This requires a trial-and-

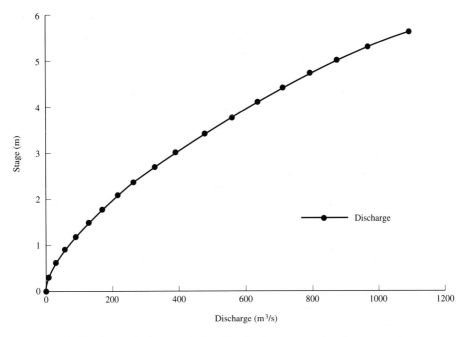

Figure 15.10 Stage–discharge relationship for the proposed spillway

error method to solve equation (8.23). Table 15.2 (Excel spreadsheet) suggests the final results with the following procedure:

Column 1: An assumed longitudinal position

Column 2: The total discharge at the assumed longitudinal position. In this case is equal to column 1 times 6 m³/s/m

Column 3: Critical depth of flow corresponding to the discharge of column 2,

$$y_c = \sqrt[3]{\frac{Q^2}{gb^2}}$$

Column 4: Wetted area $= 20y_c$

Column 5: Wetted perimeter $= 20 + 2y_c$

Column 6: Hydraulic radius $=$ column 4 divided by column 5

Column 7: The calculated distance to the critical section (equation (8.23), with $\beta \approx 1{\cdot}0$),

$$x_c = \frac{8q^2}{gb^2\left(S_0 - \frac{gP}{C^2b}\right)^3} \tag{8.23}$$

C being Chezy's coefficient given by equation (15.10).

Table 15.2 Location of the control section

1 Trial X_c (m)	2 Q (m³/s)	3 y_c (m)	4 A (m²)	5 P (m)	6 R (m)	7 X_c (m)
60·0000	360·000	3·2084	64·1685	26·417	2·4291	44·2515
50·0000	300·000	2·8412	56·8244	25·682	2·2126	44·2568
40·0000	240·000	2·4485	48·9697	24·897	1·9669	44·2719
44·0000	264·000	2·6091	52·1823	25·218	2·0692	44·2643
44·2600	265·560	2·6194	52·3876	25·239	2·0757	44·2639
44·2640	265·584	2·6195	52·3908	25·239	2·0758	44·2639
44·2639	265·583	2·6195	52·3907	25·239	2·0758	44·2639

With the critical flow section located 44·264 m downstream of the beginning of the channel, the water surface profile both upstream (subcritical) and downstream (supercritical) of this point (control point) can be estimated by equation (15.11). The calculations for the water surface profile upstream of the critical flow section are contained in table 15.3 while those for the profile downstream are contained in table 15.4 (see also fig. 15.11). In these tables:

Column 1: Distance between the point of computation and the beginning of the channel
Column 2: Incremental distance between adjacent points of calculation
Column 3: Elevation of the channel bottom
Column 4: Assumed depth of flow
Column 5: Elevation of water surface (column 3 + column 4)
Column 6: Change in water surface elevation,

$$\Delta h' = -\Delta h + S_0 \Delta x \qquad (15.11)$$

Column 7: Wetted area
Column 8: Discharge = column 1 times the discharge per unit length of the channel
Column 9: Velocity = column 8 divided by column 7
Column 10: Addition of discharges
Column 11: Addition of velocities
Column 12: Change in discharge
Column 13: Change in velocity
Column 14: Drop in the water surface due to impact loss,

$$\Delta h'_m = \frac{Q_1(V_1 + V_2)}{g(Q_1 + Q_2)} \left(\Delta V + \frac{V_2}{Q_1} \Delta Q \right) \qquad (15.12a)$$

Column 15: Hydraulic radius associated with the assumed depth of flow
Column 16: Head loss due to friction computed from,

Table 15.3 Water surface profile calculations upstream of the control section

1 x (m)	2 Δx (m)	3 Z_o (m)	4 h (m)	5 Z (m)	6 Δh' (m)	7 A (m²)	8 Q (m³/s)	9 V (m/s)	10 Q_1+Q_2 (m³/s)	11 V_1+V_2 (m/s)	12 ΔQ (m³/s)	13 ΔV (m/s)	14 $Δh'_m$ (m)	15 R (m)	16 S_r (−)	17 Δh' (m)
44.26	—	94.688	2.620	97.3079	—	52.39	265.58	5.069	—	—	—	—	—	—	—	—
40.00	4.26	95.200	2.550	97.7500	0.442	51.00	240.00	4.706	505.583	9.7752	25.583	0.3634	0.4275	2.0319	0.006201	0.4539
40.00	4.26	95.200	2.570	97.7700	0.462	51.40	240.00	4.669	505.583	9.7385	25.583	0.4000	0.4432	2.0446	0.006054	0.4690
40.00	4.26	95.200	2.580	97.7800	0.472	51.60	240.00	4.651	505.583	9.7204	25.583	0.4181	0.4508	2.0509	0.005983	0.4764
40.00	4.26	95.200	2.590	97.7900	0.482	51.80	240.00	4.633	505.583	9.7025	25.583	0.4361	0.4584	2.0572	0.005912	0.4837
40.00	4.26	95.200	2.595	97.7950	0.487	51.90	240.00	4.624	505.583	9.6936	25.583	0.4450	0.4622	2.0603	0.005878	0.4873
30.00	10.00	96.400	2.550	98.9500	1.155	51.00	180.00	3.529	420.000	8.1537	60.000	1.0949	0.9391	2.0319	0.008180	1.0209
30.00	10.00	96.400	2.500	98.9000	1.105	50.00	180.00	3.600	420.000	8.2243	60.000	1.0243	0.9218	2.0000	0.008692	1.0088
30.00	10.00	96.400	2.400	98.8000	1.005	48.00	180.00	3.750	420.000	8.3743	60.000	0.8743	0.8838	1.9355	0.009853	0.9823
30.00	10.00	96.400	2.370	98.7700	0.975	47.40	180.00	3.797	420.000	8.4217	60.000	0.8268	0.8713	1.9159	0.010242	0.9737
30.00	10.00	96.400	2.367	98.7670	0.972	47.34	180.00	3.802	420.000	8.4266	60.000	0.8220	0.8701	1.9140	0.010282	0.9729
20.00	10.00	97.600	2.200	99.8000	1.033	44.00	120.00	2.727	300.000	6.5296	60.000	1.0750	0.7924	1.8033	0.005727	0.8496
20.00	10.00	97.600	2.000	99.6000	0.833	40.00	120.00	3.000	300.000	6.8023	60.000	0.8023	0.7498	1.6667	0.007697	0.8268
20.00	10.00	97.600	1.980	99.5800	0.813	39.60	120.00	3.030	300.000	6.8326	60.000	0.7720	0.7447	1.6528	0.007942	0.8241
20.00	10.00	97.600	1.990	99.5900	0.823	39.80	120.00	3.015	300.000	6.8174	60.000	0.7872	0.7473	1.6597	0.007818	0.8255
20.00	10.00	97.600	1.993	99.5930	0.826	39.86	120.00	3.011	300.000	6.8128	60.000	0.7917	0.7481	1.6618	0.007782	0.8259
10.00	10.00	98.800	1.800	100.6000	1.007	36.00	60.00	1.667	180.000	4.6772	60.000	1.3439	0.6920	1.5254	0.002673	0.7188
10.00	10.00	98.800	1.700	100.5000	0.907	34.00	60.00	1.765	180.000	4.7752	60.000	1.2458	0.6906	1.4530	0.003198	0.7226
10.00	10.00	98.800	1.600	100.4000	0.807	32.00	60.00	1.875	180.000	4.8855	60.000	1.1355	0.6883	1.3793	0.003870	0.7270
10.00	10.00	98.800	1.530	100.3300	0.737	30.60	60.00	1.961	180.000	4.9713	60.000	1.0498	0.6859	1.3270	0.004456	0.7304
10.00	10.00	98.800	1.524	100.3240	0.731	30.48	60.00	1.969	180.000	4.9790	60.000	1.0420	0.6856	1.3225	0.004511	0.7307
0.00	10.00	100.000	1.300	101.3000	0.976	26.00	00.00	0.000	60.000	1.9685	60.000	1.9685	0.3950	1.1504	0.000000	0.3950
0.00	10.00	100.000	1.100	101.1000	0.776	22.00	00.00	0.000	60.000	1.9685	60.000	1.9685	0.3950	0.9910	0.000000	0.3950
0.00	10.00	100.000	0.800	100.8000	0.476	16.00	00.00	0.000	60.000	1.9685	60.000	1.9685	0.3950	0.7407	0.000000	0.3950
0.00	10.00	100.000	0.720	100.7200	0.396	14.40	00.00	0.000	60.000	1.9685	60.000	1.9685	0.3950	0.6716	0.000000	0.3950
0.00	10.00	100.000	0.719	100.7190	0.395	14.38	00.00	0.000	60.000	1.9685	60.000	1.9685	0.3950	0.6708	0.000000	0.3950

Table 15.4 Water surface profile calculations downstream of the control section

1 x (m)	2 Δx (m)	3 Z_o (m)	4 h (m)	5 Z (m)	6 $\Delta h'$ (m)	7 A (m²)	8 Q (m³/s)	9 V (m/s)	10 Q_1+Q_2 (m³/s)	11 V_1+V_2 (m/s)	12 ΔQ (m³/s)	13 ΔV (m/s)	14 $\Delta h'_m$ (m)	15 R (m)	16 S_r (-)	17 $\Delta h'$ (m)
44·26	—	94·688	2·620	97·308	—	52·391	265·58	5·069	—	—	—	—	—	—	—	—
50·00	5·74	94·000	2·650	96·650	0·7188	53·000	300·00	5·660	565·58	10·730	34·42	0·59109	0·6803	2·09	0·01159	0·74678
50·00	5·74	94·000	2·660	96·660	0·7288	53·200	300·00	5·639	565·58	10·708	34·42	0·56981	0·6666	2·10	0·01145	0·73235
50·00	5·74	94·000	2·661	96·661	0·7298	53·220	300·00	5·637	565·58	10·706	34·42	0·56769	0·6653	2·10	0·01144	0·73092
50·00	5·74	94·000	2·662	96·662	0·7308	53·240	300·00	5·635	565·58	10·704	34·42	0·56558	0·6639	2·10	0·01143	0·72949
50·00	5·74	94·000	2·662	96·662	0·7303	53·230	300·00	5·636	565·58	10·705	34·42	0·56663	0·6646	2·10	0·01144	0·73020
60·00	10·00	92·800	2·700	95·500	1·2385	54·000	360·00	6·667	660·00	12·303	60·00	1·03075	1·3476	2·13	0·02748	1·62238
60·00	10·00	92·800	2·800	95·600	1·3385	56·000	360·00	6·429	660·00	12·064	60·00	0·79265	1·1618	2·19	0·02460	1·40777
60·00	10·00	92·800	2·810	95·610	1·3485	56·200	360·00	6·406	660·00	12·042	60·00	0·76977	1·1443	2·19	0·02433	1·38760
60·00	10·00	92·800	2·820	95·620	1·3585	56·400	360·00	6·383	660·00	12·019	60·00	0·74706	1·1270	2·20	0·02407	1·36765
60·00	10·00	92·800	2·823	95·623	1·3615	56·460	360·00	6·376	660·00	12·012	60·00	0·74028	1·1218	2·20	0·02399	1·36171
70·00	10·00	91·600	3·000	94·600	1·3770	60·000	420·00	7·000	780·00	13·376	60·00	0·62380	1·1268	2·31	0·02715	1·39833
70·00	10·00	91·600	3·100	94·700	1·4770	62·000	420·00	6·774	780·00	13·150	60·00	0·39800	0·9448	2·37	0·02459	1·19070
70·00	10·00	91·600	3·010	94·610	1·3870	60·200	420·00	6·977	780·00	13·353	60·00	0·60055	1·1078	2·31	0·02688	1·37660
70·00	10·00	91·600	3·005	94·605	1·3820	60·100	420·00	6·988	780·00	13·365	60·00	0·61216	1·1173	2·31	0·02702	1·38744
70·00	10·00	91·600	3·007	94·607	1·3840	60·140	420·00	6·984	780·00	13·300	60·00	0·60751	1·1135	2·31	0·02696	1·38310
77·00	7·00	90·760	3·200	93·960	1·0330	64·000	462·00	7·219	882·00	14·202	42·00	0·23505	0·6597	2·42	0·01893	0·79221
77·00	7·00	90·760	3·100	93·860	0·9330	62·000	462·00	7·452	882·00	14·435	42·00	0·46791	0·8500	2·37	0·02083	0·99582
77·00	7·00	90·760	3·120	93·880	0·9530	62·400	462·00	7·404	882·00	14·388	42·00	0·42014	0·8105	2·38	0·02043	0·95351

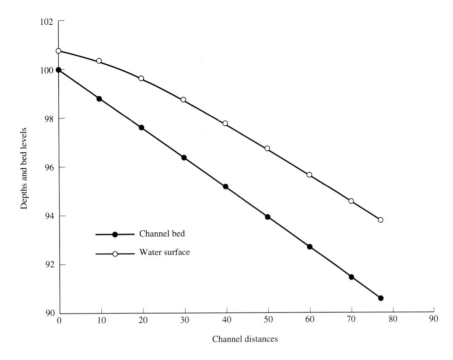

Figure 15.11 Water surface profile of spatially varied flow along the side channel

$$h_f = S_f \Delta x = \left(\frac{nQ}{AR^{2/3}} \right)^2 \Delta x$$

Column 17: Drop in the water surface between two adjacent sections calculated by

$$\Delta h' = \frac{Q_1}{g} \frac{V_1 + V_2}{Q_1 + Q_2} \left[\Delta V + \frac{V_2}{Q_1} \Delta Q \right] + S_f \Delta x \qquad (15.12b)$$

At each station a trial-and-error process is used until the values of columns 6 and 17 agree.

Example 15.4
In a small dam for water supply purposes it is necessary to design a baffle stilling basin. The level of the reservoir is located 30 m above the original river bed. The chute has a width of 15 m, a rectangular section, and moves a maximum discharge of 170 m³/s. Select the basin floor level and carry out its design. At this maximum flow rate the normal depth in the river is 4·5 m and for this chute assume $\sigma = 1 \cdot 10$ and $\zeta = 0 \cdot 7$.

Solution:

For the first iteration y^+ (see fig. 15.12) may be assumed as 8 m. Thus, the energy head and the discharge per unit width are:

$$E_0 = 30 \text{ m} + 8 \text{ m} = 38 \text{ m}$$

$$q = \frac{Q}{b} = \frac{170}{15} \text{ m}^2/\text{s} = 11 \cdot 33 \text{ m}^2/\text{s}$$

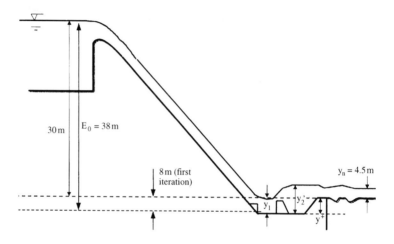

Figure 15.12 Stilling basin design example

Using equation (15.14) we get:

$$E_0 = y_1 + \frac{11 \cdot 33^2}{2 g y_1^2 \, 0 \cdot 7^2} = y_1 + \frac{13 \cdot 36}{y_1^2} \qquad (i)$$

Also equation (8.17) results in:

$$y_2 = \frac{y_1}{2} \left(\sqrt{1 + \frac{104 \cdot 74}{y_1^3}} - 1 \right) \qquad (ii)$$

and equation (15.13) gives:

$$y^+ = y_2 - 4 \cdot 5 \qquad (iii)$$

where $y_2 = \sigma y_2$

Using equations (i), (ii) and (iii) it is possible to follow the design procedure, which is iterative. The following Excel spreadsheet (table 15.5) shows the result for the final basin floor level; note that the method converges very fast.

Thus, the stilling basin floor level is $2 \cdot 12$ m below the river bed. Using USBR, 1987,[5] it is possible to chose one of the standard stilling basin designs. For this case, the Froude number and the velocity of the incoming supercritical flow are:

Table 15.5 Stilling basin floor level design

Trial y^+ (m)	E_0 (m)	y_1 [eq (i)] (m)	V_1 (m/s)	Fr_1 (–)	y_2 [eq (ii)] (m)	y^+ [eq (iii)] (m)
8·000	38·000	0·5976	18·96475	7·832624	6·327557	2·460313
2·460	32·460	0·6480	17·48971	6·936824	6·041229	2·145351
2·145	32·145	0·6513	17·40110	6·884169	6·023559	2·125915
2·126	32·126	0·6515	17·39575	6·881000	6·022492	2·124741

$$Fr_1 = 6\cdot88 > 4\cdot5$$

$$V_1 = 17\cdot39 \text{ m/s} < 20 \text{ m/s}$$

Using these results, it is clear that USBR type III stilling basin must be used (see USBR, 1987[5]). The details of the stilling basin are shown in fig. 15.13:

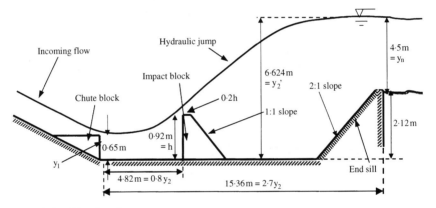

Chute block: width and spacing = y_1
Impact block: width and spacing = 0.75 h, h being its height given by,

$$h = \frac{y_1(4 + Fr_1)}{6}$$

Figure 15.13 Scheme of the stilling basin final design including its floor level and general dimensions (USBR, 1987[5]; Novak *et al.*, 1997[4])

Recommended reading

1. BS 3680: Part 4A: (1981) *Methods of measurement of liquids in open channels*. London: British Standards Institution.
2. Chow, V.T. (1959) *Open Channel Hydraulics*. New York: McGraw-Hill.
3. Henderson, F.M. (1966) *Open Channel Flow*. New York: McMillan.
4. Novak, P., Moffat, A.I.B., Nalluri, C. and Narayanan, R. (1997) *Hydraulic Structures*. 2nd edn. London: E & FN Spon.

5. United States Bureau of Reclamation – USBR (1987) *Design of Small Dams,*
   3rd edn. US Department of the Interior, Washington D.C.
6. Vischer, D.L. and Hager, W.H. (1999) *Dam Hydraulics.* Chichester, UK:
   John Wiley & Sons.

Problems

1.   An overflow spillway is to be designed to pass a discharge of 1700 m^3/s of
flood flow with an upstream water surface elevation of 230·0 m. The crest length
is 22·0 m and the elevation of the average river bed is 183·0 m.

   (a) Determine the design head.
   (b) Determine the spillway profiles, upstream and downstream.
   (c) Calculate the discharge through the spillway if the water surface
        elevation reaches the point of maximum allowable head.
   (d) What would be the minimum pressure downstream (underside of the
        nappe) of the spillway crest under part (c) discharge condition? Use the
        Cassidy's relation given in the text.

For the discharge coefficients:

C_d/C_{d_0}	H/H_d
1·06	1·50
1·07	1·60

2.   If the spillway of problem 1 is required to have piers to support a bridge
deck over it and if the maximum free span between them is 6·3 m, what would be
the discharge corresponding to the design head? What is the new allowable
maximum discharge? The pier coefficient = 0·011 and the abutment contraction coefficient = 0·12.

3.   A trapezoidal lateral spillway channel 127 m long is designed to carry a
discharge which increases at a rate of 11 m^3/s/m. The cross-section has a bottom
width of 18 m and lateral slopes of 2V:1H. The longitudinal slope of the channel
is 0·08 and starts at an upstream elevation of 520 m. If Manning's n is equal to
0·014 (concrete), determine if there is a critical section and the corresponding
water surface profile.

4.   A hydraulic jump stilling basin is to be designed to dissipate the energy of
a maximum discharge of 750 m^3/s. The water surface in the upstream reservoir
is located at a level of 350 m. The chute channel has a slope of 45°, rectangular in
cross-section with a width of 22·5 m. At this discharge the river downstream of
the basin has a normal depth of 4·4 m, and the bed is located at an elevation of
307 m. For this chute σ = 1·05 and ζ = 0·75.

   (a) Determine the sequent depth of the hydraulic jump and the bed level of
        stilling basin.

(b) Choose the stilling basin design according to the USBR (1987) standards. Make a sketch of the basin.

(c) Determine the percentage of energy dissipation in the basin.

Answers

1. Properties of Fluids

(1(b)) 5 Ns/m^2; **(2)** 2 m/s, 0·4 Ns/m^2; **(3)** 1·7 kW; **(4)** 21·3 m/s; **(5)** 4·9 mm, $-1·9$ mm; **(6)** 80 N/m^2.

2. Fluid Statics

(1(a)) 57·63 kN/m^2; **(1(b))** 31·83 m; **(2)** 36·9 kN/m^2, 10·2 kN/m^2; **(3)** 32 mm, 31·6; **(4)** 2·94 kN/m^2; **(5)** 0·25 N/mm^2; **(6)** 0·67 m, 2·0 m, 3·53 m, 26·2 kN/m; **(7)** 0·190 m; **(8)** 100·7 kN, 3·49 m below water level, 53·3 kN; **(9)** 171·2 kN, 39·23° to horizontal, 1·9 m below water surface; **(10)** 4·83 MN, 66°, 26·21 m from heel; **(11)** 0·85; **(13)** 12·82 kN; **(14)** $\dfrac{L^2}{4h} - \dfrac{h}{2}$ above water level; **(15)** 0, 11·92 sin θ kNm; **(16)** 1·41 m, 2·54°; **(17)** 244·6 mm; **(18)** 9·10°, 11·12°; **(19)** 17·17 kN.

3. Fluid Flow Concepts and Measurements

(1) 15 m/s^2, 150 m/s^2; **(2)** 21 m/s^2; **(4)** 3·62 kW, towards 450 mm diam. section; **(5)** $-50·6$ kN/m^2, 36·7 kN/m^2, 1·34 kW; **(6)** 3·98 m, 76·8°; **(7)** 23·7 kN, 45°, 16·75 kN; **(8)** 12·91 kN, 9·4° to the horizontal; **(9)** 811 kN; **(10)** 1·35 kN, 67·5° to horizontal; **(11)** 856 kN, 12·54° to the vertical; **(12)** 130 mm; **(13(a))** 46·36 mm; **(13(b))** No change; **(14)** 25 km/h; **(15)** $C_v = 0·96$, $C_c = 0·62$, $C_d = 0·596$; **(16)** 7 min 51·7 s; **(17)** 25·5 mm, 39·5 mm; **(18)** m$^{1/2}$/s, 62·3 mm, 0·6%; **(19(b))** 1·6%; **(20)** 1·48, 2·5, 0·626; **(21)** 3 h 10 min; **(22)** 25·3 m^3/w, **(23(a))** 5·27 m; 29·72 m; **(23(b))** 3396 m^3/s; $-32·57$ kN/m^2.

4. Flow of Incompressible Fluids in Pipelines

(1(a)) 179·3 l/s; **(1(b))** 20·0 m; **(2)** 8700 m, 22·78 kW; **(3)** 158 l/s; **(4(a))** 350 mm, 214·7 l/s; **(4(b))** 12·47 m; **(5)** 58·8 l/s, 0·28 mm; **(6(a))** 215·5 l/s; **(6(b))** 9350 m; **(7)** 114·7 mm, 13·4 kW; **(8(a))** 9·5 l/s, laminar; **(8(b))** 24·77 l/s, turbulent; **(9(a)(i))** 62·25 l/s, **(ii)** 64·82 l/s; **(9(b)(i))** 50·13 l/s, **(ii)** 49·10 l/s; **(10)** 0·0144.

5. Pipe Network Analysis

(1) $Z_B = 91\cdot48$ m, $Q_{AB} = 107\cdot5$ l/s, $Q_{BC.1} = 52\cdot1$ l/s, $Q_{BC.2} = 55\cdot$ l/s; **(2)** $Z_B = 91\cdot0$ m, $Q_{AB} = 110\cdot8$ l/s, $Q_{BC.1} = 55\cdot8$ l/s, $Q_{BC.2} = 55\cdot0$ l/s; **(3)** $Z_B = 75\cdot31$, $Q_{AB} = 156\cdot4$ l/s, $Q_{BC} = 56\cdot3$ l/s, $Q_{BD.1} = 59\cdot2$ l/s, $Q_{BD.2} = 40\cdot9$ l/s; **(4)** $Z_B = 132\cdot3$ m, $Q_{AB} = 118\cdot3$ l/s, $Q_{BC} = 40$ l/s, $Q_{BD} = 78\cdot3$ l/s. Pump total head $= 21\cdot75$ m, power consumption $= 14\cdot22$ kW; **(5)** $Z_B = 126\cdot7$ m, $Z_D = 109\cdot4$ m, $Q_{AB} = 505\cdot2$ l/s, $Q_{BC} = 133\cdot2$ l/s, $Q_{BD} = 372\cdot0$ l/s, $Q_{DE} = 221\cdot2$ l/s, $Q_{DF} = 150\cdot8$ l/s; **(6)** $Z_A = 200$ m, $Z_C = 100$ m, $Z_E = 60$ m, $Z_F = 50$ m, $Z_B = 100\cdot19$ m, $Z_D = 71\cdot71$ m, $Q_{AB} = 279\cdot1$ l/s, $Q_{BC} = 126\cdot7$ l/s, $Q_{BD} = 152\cdot4$ l/s, $Q_{DE} = 52\cdot1$ l/s, $Q_{DF} = 100\cdot3$ l/s; **(7)** Q_{AB} $104\cdot8$ l/s, $Q_{BC} = 45\cdot4$ l/s, $Q_{CD} = 4\cdot6$ l/s, $Q_{DE} = 44\cdot6$ l/s, $Q_{EA} = -95\cdot2$ l/s, $Q_{BE} = -0\cdot6$ l/s, $Z_A = 60$ m, $Z_B = 39\cdot03$, $Z_C = 17\cdot37$, $Z_D = 18\cdot14$ m, $Z_E = 39\cdot06$; **(8)** $Q_{BCE} = 62$ l/s, $Q_{BE} = 45$ l/s, $Q_{BDE} = 93$ l/s, head loss in AF $12\cdot5$ m; **(9)** $Q_{AB} = 106\cdot4$ l/s, $Q_{BC} = 52\cdot5$ l/s, $Q_{CD} = 2\cdot5$ l/s, $Q_{DE} = -37\cdot5$ l/s, $Q_{EA} = -93\cdot6$ l/s, $Q_{BE} = -6\cdot1$ l/s, $Z_A = 60$ m, $Z_B = 38\cdot50$ m, $Z_C = 25\cdot03$, $Z_D = 24\cdot77$ m, $Z_E = 39\cdot78$ m; **(10)** Discharges to nearest $0\cdot5$ l/s

Pipe	AB	BH	HF	FG	GA
Discharge (l/s)	136·5	56.5	2·5	53·5	93·5

Pipe	BC	CD	DH	DE	EF
Discharge (l/s)	30·0	10·0	−24·0	14·0	−26·0

Junction	A	B	C	D	E	F	G	H
Head elevation (m)	100·00	69·44	64·18	63·85	63·22	67·31	88·94	67·34

(11) Flows given to nearest $0\cdot1$ l/s

Pipe	AB	BC	CD	DE	EF	BE
Flow (l/s)	95·3	5·2	44·7	−49·9	−44·7	−5·2

Junction	A	B	C	D	E	F
Head elevation (m)	100	87·70	71·67	65·48	87·66	90·0

(12) See **(10)**; **(13)** $Q_{AB} = 131\cdot55$ l/s, $Q_{BE} = 25\cdot02$ l/s, $Q_{FE} = 48\cdot45$ l/s, $Q_{AF} = 88\cdot45$ l/s, $Q_{BC} = 46\cdot53$ l/s, $Q_{CD} = 6\cdot55$ l/s, $Q_{ED} = 23\cdot47$ l/s; $H_A = 40\cdot00$ m, $H_B = 31\cdot29$ m, $H_C = 11\cdot65$ m, $H_D = 10\cdot12$ m, $H_E = 14\cdot79$ m, $H_F = 38\cdot41$ m.

6. Pump-pipeline System Analysis and Design

(1) 136 l/s, 88·9 kW; **(2(a)(i))** 182 l/s, **(ii)** 192 l/s; **(2(b)(i))** 138·4 kW, **(ii)** 186·3 kW; **(3)** Mixed flow; **(4)** 31 l/s, 20·8 l/s, 10·0 l/s; **(5)** $N_s = 5120$, Axial flow, 125 l/s, 6·44 kW; **(6(a))** 4·81 m, 0·12; **(6(b))** 4·75 m; **(7)** 1390 rev/min; **(8)** 27·5 l/s;

(9(a)) 137·6 l/s; **(9(b))** 166·0 l/s; **(10(a))** Z_B = 90·46 m, Q_{AB} = 57·1 l/s, Q_{BC} = 38·5 l/s, Q_{BD} = 18·6 l/s; **(10(b))** Z_B = 89·02 m, Q_{AB} = 61·5 l/s, Q_{BC} = 37·5 l/s, Q_{BD} = 24·0 l/s. Head delivered by pump = 26·5 m, Power consumption = 12·4 kW.

7. Boundary Layers on Flat Plates and in Ducts

(1) 381·6 N, 0·186 m, 4·56 N/m^2; **(2(a))** 770·8 N; **(2(b))** 748·9 N; **(3)** 22·23 m/s; **(4)** 1·95 mm, 3·7 m/s; **(5)** 16·38 N/m^2, 0·067, 23·65 l/s, 0·091 mm; **(6)** 3·0 mm, 0·0114, 93·9 l/s, $\dfrac{k}{\delta}$ = 20·8.

8. Steady Flow in Open Channels

(1) 4·9 N/m^2; **(2(a))** k = 2·027 mm, n = 0·0149; **(2(b))** Q (Darcy) = 56·21 m^3/s, Q (Manning) = 56·57 m^3/s; **(3)** 2·74 m; **(4)** 3·776 m^3/s, 1·83 m/s, 5·48 N/m^2; **(5)** 3·6 m^3/s, 0·00324; **(7)** 17·6 m^3/s, 1·67 m; **(8(a))** 30·11 m^3/s; **(8(b))** 29·29 m^3/s; **(8(c))** 33·95 m^3/s; **(9)** 2·5 m; **(10)** Width = 12·25 m, depth = 6·13 m; **(11)** Bed width = 3·33 m, depth = 4·02 m; **(12)** 1·45 : 1; **(13)** Bed width = 20·5 m, depth = 2·297 m; **(14)** 2·09 m;

(15(a))

z (m)	0·1	0·2	0·3	0·4	0·5
y_1 (m)	2·50	2·50	2·50	2·50	2·50
y_2 (m)	2·37	2·24	2·10	1·94	1·74

z (m)	0·6	0·7	0·8	0·9	1·0
y_1 (m)	2·54	2·66	2·78	2·89	3·01
y_2 (m)	1·45	1·45	1·45	1·45	1·45

(15(b)) y_c = 1 : 451 m; **(15(c))** z_c = 0·568 m; **(16)** y_1 = 1·54 m, y_2 = 1·047 m, Q = 0·877 m^3/s; **(17)** Initial depth = 0·639 m, Upstream depth = 7·42 m, Force = 786 kN; **(18)** Submerged flow at gate, 5·157 m, 2·176 m, 256·93 kN; **(19)** y_n = 3·5 m, 13 000 m, 4·45 m; **(20)** See table below; **(21)** y_n = 1·48 m, y_c = 0·714 m, y = 1·32 at x = 350 m; **(22)** 10 410 m; **(23(a))** 103·20 m AOD; **(23(b))** submerged inlet and reduced flowrate. **(24(a))** 2·25 m; **(24(b))** 2·00 m; **(25)** 46 m for s = 2 m.

(20)

Depth (m)	4·0	3·9	3·8	3·7	3·6	3·5	3·4
Distance (m)	0·00	151·4	308·3	471·8	643·3	824·9	1019

Depth (m)	3·3	3·2	3·1	3·0	2·9	2·8
Distance (m)	1230	1464	1733	2057	2468	3110

9. Dimensional Analysis, Similitude and Hydraulic Models

(1) 85 l/s, 0·01445; **(2)** 35·2 kN/m^2; **(3)** 2·42 l/s, 1 : 3–4; **(4)** (Length scale)$^{3/2}$; **(6)** 0·4645 m, 118·11 l/s; **(7)** 93·75 m/s, 11·2 kN;

(8)

Q_p (l/s)	0	103·1	206·2	309·3	412·4
H_p (m)	83·0	78·6	66·4	43·6	10·3

(9) $N_s = \dfrac{N\sqrt{P}}{H^{5/4}}$; **(9(b))** 1549·2 rev/min, 27·89 MW, 21·069 m³/s;

(10) 2·85 s, 7·2 m, 0·5 m, 800 kN/m.

10. Ideal Fluid Flow and Curvilinear Flow

(1) $V(y\cos\alpha - x\sin\alpha)$, $V_r(\sin\theta\cos\alpha - \cos\theta\sin\alpha)$; **(2)** $\phi = \dfrac{1}{2}(x^2 - y^2)$;

(9) 74·6 m³/s; **(10)** 2·2 m, 2·66 m; **(11)** 105·8 m³/s, −5·36 m, −4·40 m;
(12) 0·4 l/s; **(13)** 200 litres.

11. Gradually Varied Unsteady Flow from Reservoirs

(1) 175·91 h; **(2)** 36·69 h, 131·39 h; **(3)** 602·78 h; **(4)** Peak outflow = 45 m³/s
at t = 11 h.

12. Mass Oscillations and Pressure Transients in Pipelines

(1) z max = 40·5 m at t = 95 s; **(2(a))** z max = 40·8 m after 100 s;
(2(b)) z max = 37·2 m after 150 s;

(3(a))

time (s)	0	3	6	9	12	15
V (m/s)	4·23	4·22	4·17	4·031	3·56	0
h (m)	3·0	4·66	8·09	17·00	53·03	436·74

(3(b))

time (s)	0	3	6	9	12	15	18	21	24	27	30
V (m/s)	4·23	4·23	4·21	4·17	4·12	4·02	3·87	3·62	3·17	2·27	0
h (m)	3·0	3·68	4·64	5·96	7·89	10·85	15·71	24·42	42·24	86·54	228·46

(4) Waterhammer period = 2·33 s (T), Z = head elevation (above datum
through valve) (m), Q = discharge at valve.

T	0	1	2	3	4	5	6
Z (valve)	2·6	4·8	10·7	23·5	62·5	190·2	37·6
Z (mid-length)	16·3	17·4	20·3	25·5	42·9	95·2	66·5
Q (l/s)	1071	1066	1046	991	840	367	0

13. Unsteady Flow in Channels

(1(a)) $y_2 = 2\cdot815$ m, $c = 3\cdot54$ m/s; **(1(b))** $y_2 = 3\cdot295$ m, $c = 4\cdot311$ m/s; **(2)** $y_1 = 0\cdot98$ m, $y_2 = 1\cdot807$ m, $Q_2 = 46\cdot15$ m³/s (upstream); **(3)** $y_2 = 2\cdot98$ m, $y_1 = 3\cdot54$ m, $c = 7\cdot14$ m/s; **(4)** $y_1 = 2\cdot631$ m, $1\cdot515$ min; **(5)** $36\cdot45$ min.

14. Uniform Flow in Loose-boundary Channels

(1) Safe; **(2)** $0\cdot57$ m; **(3)** 15 mm; **(4)** $9\cdot2 \times 10^{-5}$; **(5)** $6\cdot95$ m/s, $0\cdot6$ m; **(6(i))** $0\cdot686$ m³/s.m, $8\cdot2 \times 10^{-6}$; **(6(ii))** $2\cdot52$ m³/s.m, $1\cdot73 \times 10^{-4}$; **(7(a))** 15 kN/day/m, 21 mm, dunes, rough; **(8)** 21 kg/s.m; **(9)** $83\cdot4\%$; **(10)** 1060 N/s.

15. Hydraulic Structures

(1(a)) $10\cdot60$ m; **(1(c))** 3787 m³/s; **(1(d))** $-11\cdot90$ m of water; **(2)** 1424 m³/s; 2764 m³/s; **(3)** $X_c = 105\cdot38$ m; $y_c = 7\cdot01$ m; **(4(a))** $y_1 = 1\cdot423$ m; $y_2 = 11\cdot924$ m; $y^+ = 8\cdot120$ m; **(4(c))** 61%.

Index

absolute pressure, 10
acceleration, 23
  centrifugal, 24
  convective, 50
  horizontal, 23
  local, 50
  normal, 51
  radial, 23
  tangential, 50
  uniform linear, 23
  vertical, 23
Ackers-White, 373
alluvial, 362
angular velocity, 23
antidune, 362
Archimedes, principle of, 18

back water curves in channels, 195, 245
bed friction, 364
bed hydraulic radius, 364
bed load, 367, 368
bed material load, 367
bed shear, 362
Bernoulli's equation, 54
Blasius, 177
  smooth friction factor, 93
boundary layers, in turbulent pipe flow,
    180
  boundary shear stress, 178
  combined drag due to laminar and
    turbulent boundary layers, 179
  displacement thickness, 179
  drag, 178
  effect of plate roughness, 177
  Karman-Prandtl equations, 182
  laminar boundary layers, 177
  laminar sub-layer, 182
  mixing length, 180
  Nikuradse, 180, 182
  on flat plates, 178, 179
  Prandtl mixing theory, 180
  shear velocity, 180, 362
  thickness, 177
  turbulent boundary layers, 178
  velocity distribution, 177

Boussinesq, coefficient of, 58, 85, 206
bulk modulus, 2
  effect on wave speed, 325
bouyancy, 18
  centre of, 18

canal delivery, 210, 247, 248
capillarity, 3
cavitation, 57, 61, 158
  number, 160, 168
celerity, 323, 350, 352
centrifugal pumps, 154
channels (*see* open channel flow)
characteristic curves, pumps, 156
Chezy equation, 196
Colebrook-White equation, 93, 182, 195
column separation, 333, 339, 348
compressibility, 2
concentration, 370
continuity, equation of, 49, 121, 319
  two dimensional flow, 277
contraction, coefficient of, 59
Coriolis, coefficient of, 57, 85, 194
corresponding speed, 267
critical depth, 203
critical shear, 362
critical velocity, 203
Culvert, 212, 249
curvilinear flow
  in duct bend, 290, 293
  in channel bend, 298
  in rotating cylinder, 294
  in siphon spillway, 295
curved surface, force exerted due to
    hydrostatic pressure, 15

density, mass, 1
  specific volume, 4
  specific weight, 1
dimensional and model analysis, 250, 258,
    259
  Buckingham π theorem, 260
  coresponding speed, 267
  dimensional forms, 259
  for pipelines, 261

for rectangular weir, 265
for rotodynamic pumps, 263
for V-notch, 264
Froude number, 260, 261, 269, 270
model studies, 260, 268, 270, 271, 274
non-dimensional groups, 259
Reynolds number, 260, 264, 267, 271, 274
similitude, 260
Weber number, 260, 265, 268, 269
discharge, 50
coefficient of, 63, 398, 399
under varying head, 64, 302
displacement thickness, 179
drag, flow over flat plate, 177
dunes, 362
dynamic similarity, 260

eddy viscosity, 52
Einstein's equation, 369
energy
equation for ideal fluid flow, 53
equation for real, incompressible flow, 55
kinetic, 55
potential, 54
pressure, 55
total, 54
energy dissipators, 403,
chute blocks, 417
end sill, 417
impact block, 417
energy equation, 54, 55, 91, 194, 203
energy losses, 55, 96
losses in pipes, 50, 91, 96, 97
of flowing fluid, 91, 194, 202
Engeleund, 371
equilibrium
neutral, 19
relative, 22
stable, 18
unstable, 18
equipotential lines, 280, 282
estuary models, 271
Euler's equation of motion, 54

fall velocity, 366
floating bodies
equilibrium of, 18
periodic time of oscillation, 21
stability of, 18
stability of vessel carrying liquid, 21
flow
curvilinear, 277

gradually varied, steady, 206, 207
gradually varied, unsteady, (*see* quasi-steady flow)
ideal, 277
in channels, 194, 350
in pipe networks, 91, 121
of compressible fluids in pipelines, 91–120
rapidly varied, steady, 202
rapidly varied, unsteady, (*see* pressure transients, waterhammer, surges, in channels)
flow nets, 282
flow regimes, 362
fluid
definition of, 1
ideal, definition, 53
Newtonian, 1, 2
fluid flow, 47
dynamics of, 53
Eulerian description of, 47
ideal, 53
irrotational, 49
kinematics of, 47
Lagrangian description of, 47
measurement of, 59–70
non-uniform, 48
one dimensional, 49
real, 55
rotational, 49
steady, 48
three dimensional, 49
two dimensional, 49
uniform, 48
unsteady, 48
fluids in relative equilibrium, 22
effect of acceleration, 22, 33
forced vortex, 46
fluids, real and ideal, 53, 55, 277
fluid statics, 7
force exerted by a jet on a flat plate, 79
force on pipe bend and conduits, 77
forced vortex, 287, 294
free surface flow models, 268, 269
free vortex, 287
friction factor, 91
dependence on Reynolds number, 93, 95
in laminar flow, 92
in quasi-steady flow, 302
in smooth pipes, 93
friction losses in pipes, 91, 92, 93
Froude number, 203, 205, 260, 269, 271, 273, 362

Garde-Ranga Raju formula, 364
gas, definition of, 1
geometric similarity, 260

Hagen-Poiseuille equation, 92
head, 9
    potential or elevation, 54
    pressure, 9
    variable, 65
    velocity or kinetic, 55
hydraulic gradient, 91
hydraulic jump, 204, 205, 403
    location of, 243, 404,
hydraulic radius, 92, 195, 196
hydraulic structures, 397
hydrostatic thrust (force), on a plane
        surface, 11
    on a curved surface, 15

ideal fluid flow, 277
    boundary conditions, 287
    circulation, 279
    combination of basic flow patterns, 281
    curvilinear flow, 274, 286, 290, 295, 298
    equipotential lines, 280, 282
    flow nets, 282
    forced vortex, 287, 294
    free vortex, 287
    graphical methods, 288
    Laplace equation, 283
    line sink, 281
    line source, 280
    numerical methods, 283, 290
    pathline, 48, 277
    pressure distribution, 282
    radial velocity component, 279
    siphon spillway, 294
    streamlines, 48, 277
    stream function, 277, 281
    streamtube, 47, 277
    tangential velocity component, 279
    uniform flow pattern, 280
impeller, 154
    efficiency of, 154
incipient, 362
irrotational flow, 49, 272

Kalinske equation, 368
kinetic energy correction factor, 56, 57

Lacey, 364
lamimar boundary layer, 177
laminar flow, 52

laminar sub-layer, 182
Lane, 371
Laplace equation, 283
Laursen, 372
lined canal, 373
liquid, definition of, 1
logarithmic velocity distribution, 181, 217
loose-boundary, 362
losses, 58, 60
    in pipe fittings, 96, 97
    sudden contraction, 59
    sudden enlargement, 58

Manning formula, 196, 363
    coefficient of, 196, 198
manometer
    differential, 10
    inclined, 10
    U-tube, 10
mass oscillations in pipelines, 318
    finite difference methods, 321
    sudden discharge stoppage, 320, 321
    surge chamber operation, 318
metacentre, 19, 20
metacentric height, determination of, 19,
        22
Meyer-Peter Muller equation, 368
model studies, (*see* dimensional analysis)
model testing (*see* similitude)
modulus, bulk, 2
momentum, 58
    correction factor, 58
    equation, 58, 204
Moody's chart, 94
mouthpieces, 63
    hydraulic coefficients of, 63
movability parameter, 366

negative surge, 354
net positive suction head, 159
Newtonian fluid, 1, 2
Newton's law of motion, 56
Newton's law of viscosity, 2
Nikuradse, 93, 180, 181, 182
notches, 66
    Bazin formula, 67
    end contractions, 67
    Francis formula, 67
    rectangular, 66
    Rehbock formula, 67
    suppressed, 67
    velocity of approach, 64
    V or triangular, 67

ogee spillway, 68
open channel flow (steady)
    channel design, 198
    Chezy equation, 196, 219
    Colebrook-White equation, 195
    composite roughness, 194
    compound section, 194
    critical depth flume, 240
    critical tractive force, 199, 200, 201
    Darcy-Weisbach equation, 195
    De Marchi coefficient, 214
    economic section, 198
    energy components, 203
    energy principles, 203
    gradually varied flow, 206, 207, 208
    hydraulic jump, 204
    Manning formula, 196
    mobile boundary, 199
    momentum equation, 204
    part-full circular pipes, 200
    rapidly varied flow, 202
    rigid boundary, 198
    sewers, 200
    spatially varied flow, 250, 251, 252
    specific energy, 203
    storm sewer, 200, 227
    uniform flow, 195
    uniform flow resistance, 196
    velocity distribution, 217
    venturi flume, 232
    wastewater sewer, 226
    water surface profiles, 246
    wetted perimeter, 197
open channel flow (unsteady)
    celerity (wave), 350, 352
    dam break, 356
    downstream positive surge, 354
    gradually varied, 350
    negative surge waves, 354
    upstream positive surge, 352
orifice
    constant head, 62
    hydraulic coefficients of, 63
    large, 63
    small, 62
    submerged, 65
    varying head, 65
orifice meter, 60

Pascal's law, 7
pathline, 48, 277
piezometer, 10
piezometric pressure head, 9

pipelines
    Colebrook-White equation, 93, 195
    Darcy-Weisbach equation, 92
    design, 108
    effective roughness size, 92, 182
    friction factor, 93
    Hagen-Poiseuille equation, 92
    incompressible steady flow resistance, 91
    Karman-Prandtl equations, 93, 182
    local losses, 96
    Moody diagram, 93, 94
    Moody formula, 93
    networks (*see* pipe networks)
    pipes in series, 101
    resistance in non-circular sections, 95
    with laterally distributed outflow, 102
pipe networks, 121
    effect of booster pump, 131
    gradient method, 126
    head balance method, 122
    quantity balance method, 123
pitot tube, 61
plane bed, 362
pressure
    absolute, 10
    at a point, 7
    atmospheric, 10
    centre of, 11, 12
    diagram, 14
    distribution, 14
    gauge, 10
    hydrostatic pressure distribution, 9
    measurement of, 9
    piezometric pressure head, 9
    saturated vapour pressure, 3
    stagnation pressure, 61
    vacuum, 10
    vapour pressure, 3
    variation with depth, 8
    within a droplet, 4
pressure transients in pipelines, 323
    Allievi equations, 327
    at interior points, 334
    basic differential equations, 324
    due to pump stoppage, 335
    due to valve closure, 323, 336
    pipeline friction and other losses, 333
    Schnyder-Bergeron graphical method,
      331
    waterhammer lines, 331
    wave celerity, 323
pumps
    cavitation in, 158, 160

characteristic curves, 158
efficiency, 156
impeller, 154
in pipe network, 131, 171
manometric head, 155
manometric suction head, 159
matching, 162
net positive suction head, 159
operation point, 163
parallel operation, 156
pipeline selection in pumping system, 163
power input, 156
pump-pipeline system, 155
rotodynamic types, 154
selection, 154
series operation, 157
specific speed, 154
static lift, 155
system curve, 163
Thoma cavitation number, 160, 168
variable speed, 157

quasi-steady flow, 302
between reservoir, 302
establishment in pipeline, 305
over a spillway, 304

reference concentration, 343
regime channel design, 373
Kennedy's approach, 373
Lacey's approach, 374
Blench's approach, 375
Simons—Albertson method, 376
nonscouring boundary, 377
stable erodible boundary, 377
relative density, 2
relative roughness, 92
Reynolds number, 52, 92, 93, 98, 260, 261, 264, 265, 268
Riemann, 327
rigid-boundary, 373
rigid boundary (fixed bed) channels, 378
suspended load, 378
bed load, 379
limit deposition, 379
friction factors, 379
ripples, 362
river models, 268, 269, 270, 271
rotational flow, 277
Rouse's distribution, 370
Runge-Kutta method, 251

Schnyder-Bergeron graphical method, 331

Schoklitsch equation, 368
sea outfall, 373
separation, 59
sewers, 373
shear stress, 1, 7, 91, 178, 199, 200, 219, 343
shear velocity, 180, 362
Shields criterion, 363
Shields transport equation, 368
side weir, 252
similitude, similarity, 260
dynamic, 260
geometric, 260
laws for river models, 268, 270, 271
laws for rotodynamic machines, 272
laws for weirs and spillways, 267, 274
sink, 281
sluice gate, 237
source, 280
spatially varied flow, 213, 402, 410–415
specific energy, in channels, 204
specific speed, rotodynamic machines, 154
specific volume, 5
specific weight, 2
spillway, 68, 397
cavitation, 399
effective spillway length, 400
gated spillway, 399
negative pressures, 399
offset spillway, 400
ogee spillway, 69, 397, 398
profile (ogee spillway), 398
self-aeration, 400
shaft (morning glory) spillway, 397
side channel spillway, 401, 402
siphon spillway, 295, 397
static lift, 155
steady flow energy equation, 54, 91
stilling basin, 403
stream function, 279, 280
streamline, 48, 277
patterns of, 51
streamtube, 47, 277
Strickler's equation, 363
surface tension, 3, 259, 260, 264, 268, 269, 270, 274
surge chambers, 318, 319, 321
suspended load, 367, 370

threshold, 363
tidal flow, 271
tidal period, 271
total energy line, 92, 204

total load, 367, 372
turbulent flow, in pipes, 52, 93

units
 force, 1
 power, or energy, 1
 S I system, 1
 work, 1
unsteady flow
 negative surge wave, 354
 open channels, 350
 pipe flow, 302, 318, 323
 positive surge wave, 352, 354

vapour pressure, 3
velocity,
 approach, 64
 average, 50
 coefficient of, 62
 components of, 48
 distribution, in laminar flow, 177; in
  open channel, 217; in pipe, 179–182;
  on plates, 177, 178
 fluctuations of, 48
 temporal mean, 48
vena-contracta, 64
venturi flume, 232

venturi meter, 60
 coefficient of discharge, 63
viscosity, 2
 dynamic, 2
 kinematic, 2
 Newton's law of, 2
volume, specific, 5
vortex
 forced, 287, 294
 free, 287

wash load, 367
wave propagation speed, 328
 effect of free air, 333
wave reflections, dead end, 328
Weber number, 260, 266, 268, 269
weirs, 68
 Cipolletti, 67
 proportional or sutro, 68
 submergence of measuring structures,
  70
 trapezoidal, 68
wetted perimeter, 91, 96, 197

Young's modulus, effect on wave speed,
 325

Conversion Table (Metric to Imperial units)

To convert:	To:	Multiply by:
millimetres (mm)	inches (in)	0·03937
metres (m)	inches (in)	39·37
metres (m)	feet (ft)	3·28
kilometres (km)	miles (miles)	0·621
metres (m)	yards (yd)	1·094
square millimetres (mm^2)	square inches (sq in)	0·00155
square metres (m^2)	square feet (sq ft)	10·765
square metres (m^2)	square yards (sq yd)	1·196
square kilometres (km^2)	square miles (sq miles)	0·386
cubic metres (m^3)	cubic feet (cu ft)	35·32
cubic metres (m^3)	cubic yards (cu yd)	1·307
litres (l)	US gallons (US gal)	0·264
litres (l)	Imperial gallons (Imp gal)	0·220
kilograms (kg)	pounds (lb)	2·207
kilograms (kg)	tons (ton)	0·0011
newtons (N)	pound force (lbf)	0·2247
newtons (N)	kilogram force (kgf)	0·102
newtons per square metre (N/m^2)	pounds per square foot (psf)	0·0209
kilonewtons per square metre (kN/m^2)	pounds per square inch (psi)	0·145
cubic metres (m^3)	gallons (US gal)	263·16
cubic metres (m^3)	gallons (Imp gal)	220·0
cubic metres (m^3)	acre-feet (acre-ft)	0·000811
cubic metres per minute (m^3/min)	gallons per minute (US gal/min)	250·0
litres per second (l/s)	gallons per minute (US gal/min)	15·584
litres per second (l/s)	gallons per minute (Imp gal/min)	13·20
viscosity (Ns/m^2)	viscosity (lbfs/ft^2)	0·0209
kilowatts (kW)	horse power (hp)	1·3404